Gerhard Beutler

Methods of Celestial Mechanics

Volume II:
Application to Planetary System,
Geodynamics and Satellite Geodesy

In Cooperation with
Prof. Leos Mervart and Dr. Andreas Verdun

With 266 Figures Including 14 Color Figures,
28 Tables and a CD-ROM

 Springer

Professor Dr. Gerhard Beutler

Universität Bern
Astronomisches Institut
Sidlerstrasse 5
3012 Bern, Switzerland
e-mail: gerhard.beutler@aiub.unibe.ch

Cover picture:
Oppolzer motion of the Earth's rotation pole, generated by program system "CelestialMechanics"

Library of Congress Control Number: 2004105868

ISSN 0941-7834
ISBN 3-540-40750-2 Springer Berlin Heidelberg New York

Springer is a part of Springer Science+Business Media

springeronline.com

© Springer-Verlag Berlin Heidelberg 2005
Printed in Germany

The use of general descriptive names, registered names, trademarks, etc. in this publication does not imply, even in the absence of a specific statement, that such names are exempt from the relevant protective laws and regulations and therefore free for general use.

Typesetting by the author
Final layout: Frank Herweg, Leutershausen
Cover design: *design & production* GmbH, Heidelberg

Printed on acid-free paper   55/3141/ba - 5 4 3 2 1 0

To my family
and my friends and co-workers
at the Astronomical Institute of the University of Bern

Preface of Volume II

This is the second of the two volumes entitled Methods of Celestial Mechanics: Application to Planetary System, Geodynamics and Satellite Geodesy. It consists of Part II: Applications and Part III: Program System.

Part II focuses on applications of astrodynamics. The developments are based on lecture notes about Celestial Mechanics of the planetary system and of artificial satellites, and about the rotation of Earth and Moon. The lectures were intended for diploma students of astronomy, physics, mathematics, and geography at the University of Bern in their first three academic years. In view of the broad and inhomogeneous audience, the lectures had to be self-consistent and based on simple, generally known physical and mathematical facts and concepts.

Earth rotation (Chapter 2), satellite motion (Chapter 3) and the development of the planetary system (Chapter 4) are the constituents of Part II. The three chapters should be viewed as applications of the theoretical developments in Chapter I- 3. It is recommended to briefly review the corresponding sections of this introductory chapter before studying one of the three application chapters. Volume I also should be consulted whenever subtleties of the equations of motion are in the focus of interest. The mathematical foundations of the developments in this second Volume are (of course) those of Volume I. It should be nevertheless possible to get a good overview of the three applications of astrodynamics without consulting Volume I too often – provided one is ready to accept the underlying equations of motion and the mathematical tools used as known ("black boxes").

Part III gives an overview of the program system accompanying and illustrating this work. The algorithms are those developed in Chapter I- 7. The program system is meant to illustrate the mathematical concepts, but also to introduce the three central topics *rotation of Earth and Moon, satellite motion*, and *evolution of the planetary system*. The three programs ERDROT, SATORB and PLASYS are, from the design point of view, key elements for a thorough understanding of the three topics. They are in particular helpful to develop a proper understanding of the order of magnitude of the effects considered.

This Volume is accompanied by a Compact Disk (CD) containing the computer programs as executable modules for Personal Computers (PC). The program system is easy to install and to use on PCs with a Windows operating system.

Prof. Leoš Mervart of the Technical University of Prague designed and wrote the menu system accompanying the computer programs. It is in essence his merit that the computer-programs are easy to understand and to use.

My colleague and co-worker Dr. Andreas Verdun was the design expert concerning the structure and the formal appearance of this work. In addition, his collaboration was paramount in all aspects related to his specialization, the history of astronomy, in particular of Celestial Mechanics. He screened and proof-read the entire manuscript. His expertise and never ending encouragement was of greatest importance for the realization and completion of this work.

This work never could have been completed without the assistance of the two young colleagues. Their contribution is acknowledged with deep gratitude.

Prof. Paul Wild, my predecessor as director of the Astronomical Institute of the University of Bern (AIUB), contributed in many respects to this book. Paul Wild adapted his fabulous skill to screen Schmidt-plates for new objects (minor planets, comets, supernovae, etc.) to the manuscript of this book by performing an amazingly thorough proof-reading of major parts of the manuscript. The final result is undoubtedly very much improved thanks to his effort.

Chapter 2 (rotation of Earth and Moon) was proof-read by Claudia Urschl, Chapter 3 (satellite motion) by Michael Meindl. The two young colleagues are Ph.D.-candidates at our institute. Dr. Thomas Schildknecht reviewed the chapter 4 (planetary system). Dr. Urs Hugentobler received his diploma in theoretical physics, then joined the CCD group and wrote a Ph.D. thesis in the field of astrometry and Celestial Mechanics. After a longer research stay at ESOC in Darmstadt, he joined the AIUB team as head of AIUB GPS research group. With his broad background and his sharp mind he was perfectly suited to proof-read the entire Part II of this work.

Profs. Robert Weber from the Technical University of Vienna, Markus Rothacher from the Technical University of Munich, and Prof. Werner Gurtner, director of the Zimmerwald Observatory, also read and commented major parts of the manuscript. Dr. Jan Kouba from the National Geodetic Survey of Canada thoroughly read the major part of Part II. The comments by the four distinguished colleagues are very much appreciated. A final proof-reading of the entire manuscript was performed by Ms Edith Stöveken and Ms Claudia Urschl.

The editing and reviewing process of a treatise of this extent is a crucial aspect, at times even a nightmare. The reviewing work was a considerable addition to the normal professional duties of the colleagues mentioned above and to those of the author. It is my sincere desire to thank my friends and colleagues for their assistance. I can only promise to assist them in a similar way, should they decide to achieve something similar. I cannot recommend this to anybody, on the other hand: My sabbatical leave from the University of Bern in spring and summer 2001 and the following two years were in essence sacrificed to the purpose of writing and completing this two volume work.

The author hopes that the two volumes will be helpful to and stimulating for students and researchers – which in turn would help him to forget the "(blood), sweat and tears" accompanying the creative act.

Bern, February 2004   *Gerhard Beutler*

Contents

Contents of Volume I

Part II

Applications

1. Volume II in Overview

The work *Celestial Mechanics: Theory and Applications* consists of two volumes and three parts. This is Volume II, containing Parts II *Applications* and III *Program System*. Three key applications are discussed in the applications part, eight programs are described in Part III, *Program System*, which is used throughout the two volumes of this work.

1.1 Review of Volume I

Chapter I-2 of Volume I briefly reviews the development of classical Celestial Mechanics, but also the developments related to the motion of artificial satellites.

In Chapter I-3 the equations of motion were derived for three types of problems, namely

- the classical N-body problem with point masses in general and our planetary system in particular,
- the N-body problem with extended mass distributions in general and the three body problem Earth-Moon-Sun in particular, and
- the motion of an artificial Earth satellite,

which are considered in detail in this Volume.

In Chapter I-4 the classical two- and three-body problems were developed and the extensions required for the relativistic treatment of these problems were specified. The definition of the classical orbital elements and the concepts of osculating and mean orbital elements will be assumed known in this Volume.

The variational equations, i.e., the differential equations for the partial derivatives of a particular solution of the equations of motion w.r.t. one of the parameters defining the initial (or boundary) values or the force field acting on the celestial body considered, were derived in Chapter I-5. The solutions of the variational equations have to be known when determining the orbit of a celestial body or when studying the stability of a trajectory. Variational equations will be needed in particular in Chapter 4 of this Volume.

In Chapter I-6 we derived differential equations for the osculating orbital elements. The set of the six first order differential equations for the six osculating elements (per point mass considered) is mathematically equivalent to the set of three second order differential equations for the Cartesian coordinates of the same body. The advantage of using the equations for the elements resides in the fact that they may be solved approximately. In this Volume the technique is used to interpret the osculating (and mean) elements emerging from the numerical solution of satellite orbits.

Numerical analysis, in particular the numerical solution of ordinary differential equations, was reviewed in Chapter I-7. Starting from the most general problem, that of numerically solving a non-linear system of ordinary differential equations of order $n \geq 1$, algorithms for solving linear equations and for evaluating definite integrals (numerical quadrature) were developed. The so-called collocation methods were found to be very fruitful from the theoretical and from the practitioner's point of view. A collocation method provides an approximating function of the true solution, allowing it to compute (approximations of) the solution vector (and its derivatives) for any epoch contained within the integration interval. Multistep methods, but also the famous Gaussian methods for numerical quadrature, were recognized as special cases of collocation methods. Numerical integration techniques are the basis of the computer programs ERDROT, SATORB, and PLASYS accompanying the three main applications to be dealt with below. Not reviewing Chapter I-7 before studying one of the chapters of Part II just implies that numerical integration is used as a "black box".

Orbit determination and parameter estimation is the concluding chapter of Volume I. As a matter of fact, this topic contains aspects of both, theory and application. Chapter I-8 makes the distinction between first orbit determination (a non-linear parameter estimation problem) and orbit (parameter) improvement, which may be dealt with by linearizing and iteratively improving the orbits. In satellite geodesy orbit determination (improvement) often cannot be separated from the determination of other parameters. Many of the results discussed in Chapter 2 originally stem from such general parameter estimation processes.

1.2 Part II: Applications

Rotation of Earth and Moon. Chapter 2 deals with all aspects of the N-body problem Earth-Moon-Sun-planets. All developments and analyses are based on the corresponding equations of motion developed in Chapter I-3; the illustrations, on the other hand, are based almost exclusively on the computer program ERDROT (see section 1.3).

In order to fully appreciate the general characteristics of Earth (and lunar) rotation, it is necessary to understand the orbital motion of the Moon in the first place. This is why the orbital motion of the Moon is analyzed before discussing the rotation of Earth and Moon.

The main properties of the rotation of Earth and Moon are reviewed afterwards under the assumption that both celestial bodies are rigid. Whereas the characteristics of Earth rotation are well known, the rotational properties of the Moon are usually only vaguely known outside a very limited group of specialists. Despite the fact that the structure of the equations is the same in both cases, there are noteworthy differences, some of which are discussed in this chapter. The analysis pattern is the same for the two bodies: The motions of the rotation axis in the body-fixed system and in the inertial system are established by computer simulations (where it is possible to selectively "turn off" the torques exerted by the respective perturbing bodies); the simulation results are then explained by approximate analytical solutions of the equations of motion. The simulations and the approximate analytic solutions are compared to the real motion of the Earth's and Moon's rotation poles. Many, but not all aspects are explained by the rigid-body approximation.

This insight logically leads to the discussion of the rotation of a non-rigid Earth. This discussion immediately leads in turn to very recent, current and possible future research topics. Initially, the "proofs" for the non-rigidity of the Earth are provided. This summary is based mainly on the Earth rotation series available from the IERS (International Earth Rotation and Reference Systems Service) and from space geodetic analysis centers. Many aspects of Earth rotation may be explained by assuming the Earth to consist of one solid elastic body, which is slightly deformed by "external" forces. Only three of these forces need to be considered: (1) the centrifugal force due to the rotation of the Earth about its figure axis, (2) the differential centrifugal force due to the rotation of the Earth about an axis slightly differing from this figure axis, and (3) the tidal forces exerted by Sun and Moon (and planets). The resulting, time-dependent deformations of the Earth are small, which is why in a good approximation they may be derived from Hooke's law of elasticity.

The elastic Earth model brings us one step closer to the actual rotation of the Earth: The difference between the Chandler and the Euler period as well as the observed bi-monthly and monthly LOD (Length of Day) variations can be explained now.

The elastic Earth model does not yet explain all features of the observed Earth rotation series. There are, e.g., strong annual and semi-annual variations in the real LOD series, which may *not* be attributed to the deformations of the solid Earth. Peculiar features also exist in the polar motion series. They are observed with space geodetic techniques because the observatories are attached to the solid Earth and therefore describe the rotation of this body (and not of the body formed by the solid Earth, the atmosphere and the oceans). Fortunately, meteorologists and oceanographers are capable of deriving the angular momentum of the atmosphere from their measurements: by comparing the series of AAM (Atmospheric Angular Momentum) emerging from the meteorological global pressure, temperature, and wind fields with the corresponding angular momentum time series of the solid Earth emerging from space geodesy, the "unexplained" features in the space geodetic observation series of Earth rotation are nowadays interpreted by the exchange of angular momentum between solid Earth, atmosphere and oceans – implying that the sum of the angular momenta of the solid Earth and of atmosphere and oceans is nearly constant.

Even after having modelled the Earth as a solid elastic body, partly covered by oceans and surrounded by the atmosphere, it is not yet possible to explain all features of the monitored Earth rotation. Decadal and secular motions in the observed Earth rotation series still await explanation. The explanation of these effects requires even more complex, multi-layer Earth-models, as, e.g., illustrated by Figure 2.55. The development of these complex Earth models is out of the scope of an introductory text. Fortunately, most of their features can already be seen in the simplest generalization, usually referred to as the Poincaré Earth model, consisting of a rigid mantle and a fluid core (see Figure 2.56). It is in particular possible to explain the terms FCN (Free Core Nutation) and NDFW (Nearly-Diurnal Free Wobble). The mathematical deliberations associated with the Poincaré model indicate the degree of complexity associated with the more advanced Earth models. It is expected that such models will be capable of interpreting the as yet unexplained features in the Earth rotation series – provided that Earth rotation is continuously monitored over very long time spans (centuries).

Artificial Earth Satellites. Chapter 3 deals with the orbital motion of artificial Earth satellites. Most illustrations of this chapter stem from program SATORB, which allows it (among other) to generate series of osculating and/or mean elements associated with particular satellite trajectories.

The perturbations of the orbits due to the oblate Earth, more precisely the perturbations due to the term C_{20} of the harmonic expansion of Earth's potential, are discussed first. The pattern of perturbations at first sight seems rather similar to the perturbations due to a third body: No long-period or secular perturbations in the semi-major axis and in the eccentricity, secular perturbations in the right ascension of the ascending node Ω and in the argument ω of perigee. There are, however, remarkable peculiarities of a certain practical relevance. The secular rates of the elements Ω (right ascension of ascending node) and ω (argument of perigee) are functions of the satellite's inclination i w.r.t. the Earth's equatorial plane. The perturbation patterns allow it to establish either sun-synchronous orbital planes or orbits with perigees residing in pre-defined latitudes $\phi \leq \pm i$.

The orbital characteristics are established by simulation techniques (using program SATORB), then explained with first-order general perturbation methods (based on simplified perturbative forces). Higher-order perturbations due to the C_{20}-term and the influence of the higher-order terms of the Earth's potential (which are about three orders of magnitude smaller than C_{20}) are studied subsequently. The attenuating influence of the Earth's oblateness term C_{20} on the perturbations due to the higher-order terms C_{ik} is discussed as well.

If a satellite's revolution period is commensurable with the sidereal revolution period of the Earth, some of the higher-order terms of Earth's potential may produce resonant perturbations, the amplitudes of which may become orders of magnitude larger than ordinary higher-order perturbations. Resonant perturbations are typically of very long periods (years to decades), and the amplitudes may dominate even those caused by the oblateness. Two types of resonances are discussed in more detail, the (1:1)-resonance of geostationary satellites and the (2:1)-resonance of GPS-satellites. In both cases the practical implications are considerable. In the case of GPS-satellites the problem is introduced by a heuristic study, due to my colleague Dr. Urs Hugentobler, which allows it to understand the key aspects of the problem without mathematical developments.

The rest of the chapter is devoted to the discussion of non-gravitational forces, in particular of drag and of solar radiation pressure. As usual in our treatment, the perturbation characteristics are first illustrated by computer simulations, then understood by first-order perturbation methods. Atmospheric drag causes a secular reduction of the semi-major axis (leading eventually to the decay of the satellite orbit) and a secular decrease of the eccentricity (rendering the decaying orbit more and more circular). Solar radiation pressure is (almost) a conservative force (the aspect is addressed explicitly), which (almost) excludes secular perturbations in the semi-major axis. Strong and long-period perturbations occur in the eccentricity, where the period is defined by the periodically changing position of the Sun w.r.t. the satellite's orbital plane.

The essential forces (and the corresponding perturbations) acting on (suffered by) high- and low-orbiting satellites are reviewed at the end of the chapter.

Evolution of the Planetary System. The application part concludes with Chapter 4 pretentiously entitled *Evolution of the Planetary System*. Three major issues are considered: (a) the orbital development of the outer system from Jupiter to Pluto over a time period of two million years (the past million years and the next million years – what makes sure that the illustrations in this chapter will not be outdated in the near future, (b) the orbital development of the complete system (with the exception of the "dwarfs" Mercury and Pluto), where only the development of the inner system from Venus to Mars is considered in detail, and (c) the orbital development of minor planets (mainly of those in the classical asteroid belt between Mars and Jupiter).

The illustrations have three sources, namely (a) computer simulations with program PLASYS, allowing it to numerically integrate any selection of planets of our planetary system with the inclusion of one body of negligible mass (e.g., a minor planet or a comet) with a user-defined set of initial orbital elements (definition in Chapter I- 2), (b) orbital elements obtained through the MPC (Minor Planet Center) in Cambridge, Mass., and (c) spectral analyses of the series of orbital elements (and functions thereof) performed by our program FOURIER.

By far the greatest part of the (mechanical) energy and the angular momentum of our planetary system is contained in the outer system. Jupiter and Saturn are the most massive planets in this subsystem. Computer simulations over relatively short time-spans (of 2000 years) and over the full span of two million years clearly show that even when including the entire outer system the development of the orbital elements of the two giant planets is dominated by the exchange of energy and angular momentum between them. The simulations and the associated spectra reveal much more information.

Venus and Earth are the two dominating masses of the inner system. They exchange energy and angular momentum (documented by the coupling between certain orbital elements) very much like Jupiter and Saturn in the outer system. They are strongly perturbed by the planets of the outer system (by Jupiter in particular). An analysis of the long-term development of the Earth's orbital elements (over half a million years) shows virtually "no long-period structure" for the semi-major axis, whereas the eccentricity varies between $e \approx 0$ and $e \approx 0.5$ (exactly like the orbital eccentricity of Venus).

Such variations might have an impact on the Earth's climate (annual variation of the "solar constant", potential asymmetry between summer- and winter-half-year). The eccentricity is, by the way, approaching a minimum around the year 35'000 A.D., which does not "promise" too much climate-relevant "action" in the near future – at least not from the astronomical point of view. The idea that the Earth's dramatic climatic changes in the past (ice-ages and warm periods) might at least in part be explained by the Earth's orbital motion is due to Milankovitch. Whether or not this correlation is significant cannot be firmly decided (at least not in this book). The long-term changes of the orbital characteristics (of the eccentricity, but also of the inclination of the Earth's orbital plane w.r.t. the so-called invariable plane) are, however, real, noteworthy and of respectable sizes.

Osculating orbital elements of more than 100'000 minor planets are available through the MPC. This data set is inspected to gain some insight into the motion of these celestial objects at present. The classical belt of minor planets is located between Mars and Jupiter. Many objects belonging to the so-called

Edgeworth-Kuiper belt are already known, today. Nevertheless, the emphasis in Chapter 4 is put on the classical belt of asteroids and on the explanation of (some aspects of) its structure. The histogram 4.43 of semi-major axes (or of the associated revolution periods) indicates that the Kirkwood gaps must (somehow) be explained by the commensurabilities of the revolution periods of the minor planets and of Jupiter. After the discussion of the observational basis, the analysis of the orbital motion of minor planets is performed in two steps:

- In a first step the development of the orbital elements of a "normal" planet is studied. This study includes the interpretation of the (amazingly clean) spectra of the minor planet's mean orbital elements. These results lead to the definition of the (well known) so-called proper elements. It is argued that today the definition of these proper elements should in principle be based on numerical analyses, rather than on approximate analytical theories as, e.g., developed by Brouwer and Clemence [27]. A few numerical experiments indicate, however, that the results from the two approaches agree quite well.

- Minor planets in resonant motion are studied thereafter. The Hilda group ((3:2)-resonance) and the (3:1)-resonance are considered in particular. The Ljapunov characteristic exponent is defined as an excellent tool to identify chaotic motion. A very simple and practical method for its establishment (based on the solution of one variational equation associated with the minor planet's orbit) is provided in program PLASYS. The tools of numerical integration of the minor planet's orbit together with one or more variational equations associated with it, allow it to study and to illustrate the development of the orbital elements of minor planets in resonance zones. It is fascinating to see that the revolutionary numerical experiments performed by Jack Wisdom, in the 1980s, using the most advanced computer hardware available at that time, nowadays may be performed with standard PC (Personal Computer) equipment.

1.3 Part III: Program System

The program system, all the procedures, and all the data files necessary to install and to use it on PC-platforms or workstations equipped with a WINDOWS operating system are contained on the CDs accompanying both volumes of this work. The system consists of eight programs, which will be briefly characterized below. Detailed program and output descriptions are available in Part III, consisting of Chapters 5 to 11.

The program system is operated with the help of a menu-system. Figure 1.1 shows a typical panel – actually the panel after having activated the program system *Celestial Mechanics* and then the program PLASYS. The top line of each panel contains the buttons with the program names and the help-key offer real-time information when composing a problem.

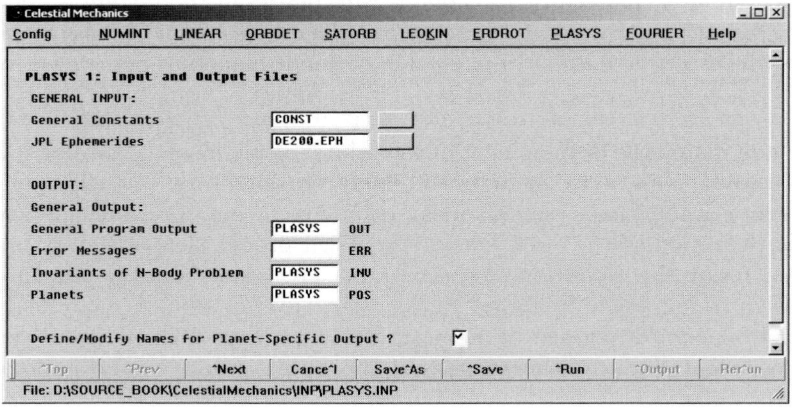

Fig. 1.1. Primary menu for program system *Celestial Mechanics, PLASYS*

The names of input- and output-files may be defined or altered in these panels and input options may be set or changed. By selecting ≪ Next Panel ≫ (bottom line), the next option/input panel of the same program are activated. If all options and file definitions are meeting the user's requirements, the program is activated by selecting ≪ Save and Run ≫ . For CPU (Central Processing Unit) intensive programs, the program informs the user about the remaining estimated CPU-requirements (in %).

The most recent general program output (containing statistical information concerning the corresponding program run and other characteristics) may be inspected by pressing the button ≪ Last Output ≫ . With the exception of LEOKIN all programs allow it to visualize some of the more specific output files using a specially developed graphical tool compatible with the menu-system. The output files may of course also be plotted by the program user with any graphical tool he is acquainted with. All the figures of this book illustrating computer output were, e.g., produced with the so-called "gnu"-graphics package. The gnu-version used here is also contained on the CD. The programs included in the package "Celestial Mechanics" are (in the sequence of the top line of panel 1.1):

1. **NUMINT** is used in the first place to demonstrate or test the mutual benefits and/or deficiencies of different methods for numerical integration. Only two kinds of problems may be addressed, however: either the motion of a minor planet in the gravitational field of Sun and Jupiter

(where the orbits of the latter two bodies are assumed to be circular) or the motion of a satellite in the field of an oblate Earth (only the terms C_{00} and C_{20} of the Earth's potential are assumed to be different from zero).

The mass of Jupiter or the term C_{20} may be set to zero (in the respective program options), in which case a pure two-body problem is solved.

When the orbit of a "minor planet" is integrated, this actually corresponds to a particular solution of the problème restreint. In this program mode it is also possible to generate the well known surfaces of zero velocity (Hill surfaces), as they are shown in Chapter I-4.

2. **LINEAR** is a test program to demonstrate the power of collocation methods to solve linear initial- or boundary-value problems. The program user may select only a limited number of problems. He may test the impact of defining the collocation epochs in three different ways (equidistant, in the roots of the Legendre and the Chebyshev polynomials, respectively).

3. **SATORB** may either be used as a tool to generate satellite ephemerides (in which case the program user has to specify the initial osculating elements), or as an orbit determination tool using *either* astrometric positions of satellites or space debris as observations *or* positions (and possibly position differences) as pseudo-observations. In the latter case SATORB is an ideal instrument to determine a purely dynamical or a reduced-dynamics orbit of a LEO. It may also be used to analyze the GPS and GLONASS ephemerides routinely produced by the IGS (International GPS Service).

The orbit model can be defined by the user, who may, e.g.,

- select the degree and the order for the development of the Earth's gravity potential,

- decide whether or not to include relativistic corrections,

- decide whether or not to include the direct gravitational perturbations due to the Moon and the Sun,

- define the models for drag and radiation pressure, and

- decide whether or not to include the perturbations due to the solid Earth and ocean tides.

Unnecessary to point out that this program was extensively used to illustrate Chapter 3.

When using the program for orbit determination the parameter space (naturally) contains the initial osculating elements, a user-defined selection of dynamical parameters, and possibly so-called pseudo-stochastic pulses (see Chapter I-8).

Programs ORBDET and SATORB were used to illustrate the algorithms presented in Chapter I- 8.

4. **LEOKIN** may be used to generate a file with positions and position differences of a LEO equipped with a spaceborne GPS-receiver. This output file is subsequently used by program SATORB for LEO orbit determination. Apart from the observations in the standard RINEX (Receiver Independent Exchange Format), the program needs to know the orbit and clock information stemming from the IGS.

5. **ORBDET** allows it to determine the (first) orbits of minor planets, comets, artificial Earth satellites, and space debris from a series of astrometric positions. No initial knowledge of the orbit is required, but at least two observations must lie rather close together in time (time interval between the two observations should be significantly shorter than the revolution period of the object considered).

The most important perturbations (planetary perturbations in the case of minor planets and comets, gravitational perturbations due to Moon, Sun, and oblateness of the Earth (term C_{20}) in the case of satellite motion) are included in the final step of the orbit determination. ORBDET is the only interactive program of the entire package.

The program writes the final estimate of the initial orbital element into a file, which may in turn be used subsequently to define the approximate initial orbit, when the same observations are used for orbit determination in program SATORB.

6. **ERDROT** offers four principal options:

- It may be used to study Earth rotation, assuming that the geocentric orbits of Moon and Sun are known. Optionally, the torques exerted by Moon and Sun may be set to zero.

- It may be used to study the rotation of the Moon, assuming that the geocentric orbits of Moon and Sun are known. Optionally, the torques exerted by Earth and Sun may be set to zero.

- The N-body problem Sun, Earth, Moon, plus a selectable list of (other) planets may be studied and solved.

- The program may be used to study the correlation between the angular momenta of the solid Earth (as produced by the IGS or its institutions) and the atmospheric angular momenta as distributed by the IERS (International Earth Rotation and Reference Systems Service).

This program is extensively used in Chapter 2.

7. **PLASYS** numerically integrates (a subset of) our planetary system starting either from initial state vectors taken over from the JPL (Jet Propulsion Laboratory) DE200 (Development Ephemeris 200), or using the approximation found in [72]. A minor planet with user-defined initial

osculating elements may be included in the integration, as well. In this case it is also possible to integrate up to six variational equations simultaneously with the primary equations pertaining to the minor planet. Program PLASYS is extensively used in Chapter 4.

8. **FOURIER** is used to spectrally analyze data provided in tabular form in an input file. The program is named in honour of Jean Baptiste Joseph Fourier (1768–1830), the pioneer of harmonic analysis. In our treatment Fourier analysis is considered as a mathematical tool, which should be generally known. Should this assumption not be (entirely) true, the readers are invited to read the theory provided in Chapter 11, where Fourier analysis is developed starting from the method of least squares. As a matter of fact it is possible to analyze a data set using

- *either* the *least squares* technique – in which case the spacing between subsequent data points may be arbitrary,

- *or* the *classical Fourier analysis*, which is orders of magnitude more efficient than least squares (but requires equal spacing between observations), and where *all* data points are used,

- *or* FFT (Fast Fourier transformation), which is in turn orders of magnitude more efficient than the classical Fourier technique, but where usually the number of data points should be a power of 2 (otherwise a loss of data may occur).

In the FFT-mode the program user is invited to define the decomposition level (maximum power of 2 for the decomposition), which affects the efficiency, but minimizes (controls) loss of data. The general program output contains the information concerning the data loss.

The program may very well be used to demonstrate the efficiency ratio of the three techniques, which should produce identical results. FOURIER is a pure service program.

The computer programs of Part III are used throughout the two volumes of our work. It is considered a minimum set ("starter's kit") of programs that should be available to students entering into the field of Astrodynamics, in particular into one of the applications treated in Part II of this work. Some of the programs, NUMINT, LINEAR, and PLASYS are also excellent tools to study the methods of numerical integration.

2. The Rotation of Earth and Moon

2.1 Basic Facts and Observational Data

2.1.1 Characteristics of the Earth-Moon System

The facts briefly reviewed in this introductory section are taken from the *Explanatory Supplement to the Astronomical Almanac* [107], from the book *Global Earth Physics: A Handbook of Physical Constants* [2], from the *IERS Conventions* [71], and from the textbook by W. Torge [125]. The IERS conventions were used (wherever possible) for the numerical values in Table 2.1. The reference is specified in the last column. If no reference is specified, the numerical value was calculated using other values of the table.

The equations of motion for the generalized three-body-problem Earth-Moon-Sun were derived in section I- 3.3. In order to produce particular solutions of the equations of motion (I- 3.118) describing the geocentric motion of Sun and Moon and of the equations (I- 3.124) together with the kinematic equations (I- 3.68) describing the rotational motion of the Earth and the Moon, we need to know the gravity constant G, the masses of Sun, Earth, Moon, the principal moments of inertia of Earth and Moon, the initial conditions (geocentric position- and velocity-vectors of Sun and Moon at an initial time T_0), the angles describing the initial orientation of Earth and Moon in the inertial space, and the initial conditions for the angular velocity vectors of Earth and Moon.

Table 2.1 recapitulates the relevant facts. It tells that the Earth may be viewed in good approximation as an oblate ellipsoid of rotation with the principal moments of inertia A_δ, B_δ, and C_δ, where $A_\delta \approx B_\delta$ and $C_\delta > A_\delta$. The Moon is oblate, as well, but its oblateness is less pronounced. Also, the two moments of inertia $A_\mathbb{C}$ and $B_\mathbb{C}$ differ significantly and invalidate the assumption of rotational symmetry. In section 2.2 we will see that this difference is the essential condition for the orbital and rotational periods of the Moon to be identical.

The orbital characteristics of the Moon are closely related to the precession and nutation of the Earth's rotation axis. This is why the essential characteristics of the lunar orbital motion will be discussed in the dedicated section 2.2.1.

Table 2.1. Facts associated with the three-body problem Earth-Moon-Sun

Quantity	Value	Unit	Ref.
Mass of the Sun $m_\odot$	$1.9891 \cdot 10^{30}$	kg	[107]
Radius of the Sun	$6.96 \cdot 10^{8}$	m	[107]
Gravity constant G	$6.67259 \cdot 10^{-11}$	$m^3 kg^{-1} s^{-2}$	[107]
Speed of light c	299792458	$m\ s^{-1}$	[107]
$m_\odot : (M + m_{\mathbb{C}})$	328900.55	–	[107]
Mass of the Earth M	$5.9737 \cdot 10^{24}$	kg	–
Geocentric gravity constant GM	$398.6004418 \cdot 10^{12}$	$m^3 s^{-2}$	[71]
Mean distance Earth-Sun $a_\odot$	14959787066	km	[107]
Flattening factor $1/f$	298.25642	–	[71]
Equatorial radius $a_\oplus$	6378136.6	m	[71]
Polar radius $b_\oplus = a_\oplus(1-f)$	6356751.9	m	–
Rate of rotation $\omega_\oplus$	$7.292115 \cdot 10^{-5}$	rad s^{-1}	[71]
Length of sidereal day	0.9972697	days	–
	23.934472	hours	–
Dynamical flattening $(C_\oplus - A_\oplus)/A_\oplus$	1/305.45	–	[125]
Dyn. form factor $J_2 = (C_\oplus - A_\oplus)/a_\oplus^2 M$	$1082.6359 \cdot 10^{-6}$	–	[125]
Maximum moment of inertia $C_\oplus$	$0.3307007 \cdot Ma_\oplus^2$	$m^2 kg$	[2]
(Axis $\approx$ rotation axis)			
Moment of inertia $B_\oplus$	$0.3296181 \cdot Ma_\oplus^2$	$m^2 kg$	[2]
(Geogr. long. of axis at $\lambda = 14.9285°$)			
Minimum moment of inertia $A_\oplus$	$0.3296108 \cdot Ma_\oplus^2$	$m^2 kg$	[2]
$(A_\oplus + B_\oplus)/2$	$0.3296144 \cdot Ma_\oplus^2$	$m^2 kg$	–
Obliquity ε of ecliptic J2000.0	23.43928108	°	[107]
Prec. period around pole of ecliptic	≈ 26500	years	–
Love number h_2	≈ 0.6	–	[125]
Love number l_2	≈ 0.08	–	[125]
Love number k_2	≈ 0.3	–	[125]
$M : m_{\mathbb{C}}$	81.300588	–	[107]
Mass of the Moon $m_{\mathbb{C}}$	$7.3483 \cdot 10^{22}$	kg	[107]
Radius of the Moon $R_{\mathbb{C}}$	1738	km	[107]
Semi-major axis of lunar orbit $a_{\mathbb{C}}$	≈ 384400	km	[107]
Secular increase	≈ 3.8	m/century	[2]
Mean eccentricity of lunar orbit	0.05490	–	[107]
Mean inclination of lunar orbit w.r.t. ecliptic	5.145396	°	[107]
Rev.period of lunar node in ecliptic	18.61	years	[107]
Mean sidereal revolution period	27.32166	days	[107]
Mean sidereal rotation period	27.32166	days	[107]
Inclination $\varepsilon_{\mathbb{C}}$ of rotation axis w.r.t. ecliptic normal	1.542417	°	[107]
Inclination of axis w.r.t. lunar orbital plane	6.683	°	[107]
$\beta = (C_{\mathbb{C}} - A_{\mathbb{C}})/B_{\mathbb{C}}$	$6.313 \cdot 10^{-4}$	–	[107]
$\gamma = (B_{\mathbb{C}} - A_{\mathbb{C}})/C_{\mathbb{C}}$	$2.278 \cdot 10^{-4}$	–	[107]
Maximum moment of inertia $C_{\mathbb{C}}$	$0.39350 \cdot m_{\mathbb{C}} R_{\mathbb{C}}^2$	$m^2 kg$	[2]
(Axis $\approx$ rotation axis)			
Moment of inertia $B_{\mathbb{C}}$	$0.39334 \cdot m_{\mathbb{C}} R_{\mathbb{C}}^2$	$m^2 kg$	[2]
Minimum moment of inertia $A_{\mathbb{C}}$	$0.39325 \cdot m_{\mathbb{C}} R_{\mathbb{C}}^2$	$m^2 kg$	[2]
(Axis on lunar equator facing Earth)			

In section I-3.3 it was pointed out that in principle the differential equations for the orbital and rotational motion are coupled and should be solved together. The coupling is weak, on the other hand, and excellent approximations for Earth and Moon rotation may be also be obtained by assuming the orbits of Sun and Moon to be known and by assuming the Moon as a point mass when studying Earth rotation (and the Earth as a point mass when studying lunar rotation). The orbital motion may in these cases be taken from independent sources, e.g., from the JPL (Jet Propulsion Laboratory) DE200 (Development Ephemerides 200) [111] or even from [72], if a crude approximation is believed to be sufficient.

Subsequently we will often use the terms *Earth rotation parameters*, *length of day*, etc. It seems therefore appropriate to define these terms more precisely here: We will use the term ERP (Earth Rotation Parameters) for the parameters describing

- the position of the Earth's figure axis (or Tissérand axes for a non-rigid Earth, named after Félix Tissérand (1845–1896)) in inertial space,

- the position of the Earth rotation axis $\boldsymbol{\omega}_\delta$ w.r.t. the Earth's figure axis,

- the angular velocity $\omega_\delta(t) \stackrel{\text{def}}{=} |\boldsymbol{\omega}_\delta|$ of the Earth (as a function of time), and

- and the angle $\Omega_\delta(t)$ (e.g., Greenwich sidereal time) describing the position of a conventional meridian in inertial space.

Different parameter sets of ERP may be defined and are in use. Usually precession and nutation, expressed by the angles Ψ_δ and ε_δ, are associated with the orientation of the Earth's figure axis in inertial space. One might also refer to these angles as EOP (Earth Orientation Parameter). In this sense the EOP might be considered as a subset of the ERP.

The ERP in the more restricted sense (and sometimes the term ERP *is* used in this sense) describe the Earth's rotation axis w.r.t. the Earth's figure axis and the angular velocity $\omega_\delta(t)$ of Earth rotation. Usually the PM (Polar Motion) parameters x and y (to be defined more precisely later) and the LOD (Length of Day) are used as parameters to describe the angular velocity vector of the Earth. The length of day is defined as

$$\text{LOD}(t) \stackrel{\text{def}}{=} \frac{2\pi}{\omega_\delta(t)} \ . \tag{2.1}$$

Often, one also uses the term ΔLOD (Excess LOD), which is defined as

$$\Delta\text{LOD} \stackrel{\text{def}}{=} \text{LOD} - 86400\,[\text{s}] \ . \tag{2.2}$$

The sidereal time angle $\Theta_\delta(t)$ simply is the integral of the angular velocity $\omega_\delta(t)$ over time. It emerges naturally as one of the solution components of the Euler or Liouville-Euler equations of Earth rotation (named after Joseph Liouville (1809–1882) and Leonhard(1707–1783)). It is "only" used to orient the Earth (around the figure axis) in inertial space.

Extensive use will be made of numerical solutions of the equations of motion for Earth rotation, lunar rotation, and for the general development of the general three-body problem Earth-Moon-Sun. The computer program ERDROT is described in detail in Chapter 9 of Part III. It allows to study the solution of the equations of motion for

- the rotation of the Earth assuming that the geocentric orbital motions of Sun and Moon are known,
- the rotation of the Moon assuming that the selenocentric orbital motions of the Earth and the Sun are known, and
- the general three-body problem Earth-Moon-Sun as defined in section I- 3.3.

Several solution methods and approximations are implemented in program ERDROT. In all cases it is possible to take the planetary perturbations for the orbital motions of Earth (or Sun) and Moon into account. Several approximations are implemented with the goal to speed up the time of integration in order to allow a numerical solution over several millennia.

2.1.2 Observational Basis

Three observation techniques are relevant today for the determination of the ERP, namely

- *VLBI (Very Long Baseline Interferometry)*,
- *SLR (Satellite Laser Ranging) and LLR (Lunar Laser ranging)*, and
- the U.S. *GPS*.

The three techniques are referred to as *space geodetic techniques*. They all contribute to the determination of the terrestrial reference frame and to the monitoring of Earth rotation.

The contributions by the GPS were introduced in Chapter I- 8, in particular the contributions to polar wobble (Figures I- 8.19 and I- 8.20) and to the determination of the LOD (Figure I- 8.21). SLR was introduced in Chapter I- 8, as well, and needs no further comments.

LLR as a technique is based on the same physical principles as SLR. Thanks to the observed object, the Moon, other objectives can be addressed:

- Of the above mentioned three space geodetic techniques, only LLR contributes to the monitoring of lunar rotation.

- Thanks to the very small area to mass-ratio of $A/m \approx 1.3 \cdot 10^{-10}$ (see Table 3.4 in Chapter I- 3.4), LLR provides an excellent test-field for the theories of general relativity.

- LLR observations are also useful for the determination of UT1-UTC, i.e., the difference between universal time defined by Earth rotation (UT1) and atomic clocks (UTC).

- Last, but not least, the secular increase of the Moon's semi-major axis (see Table 2.1) can be directly measured with LLR.

Laser ranging to the Moon is possible only because Laser reflectors were deployed by the U.S. Apollo missions 11, 14, 15, and the Russian Lunokhod 2 mission. SLR and LLR are coordinated today by the *ILRS*, the International Laser Ranging Service (see http://ilrs.gsfc.nasa.gov).

This leaves us with VLBI as the third essential space geodetic technique contributing to the monitoring of Earth rotation. VLBI is indeed fundamental for this purpose. As no optical observations to artificial satellites are made for geodynamics purposes today, only VLBI measurements allow it to determine all transformation parameters between the terrestrial and the celestial reference frames: VLBI observations allow it to determine the two polar wobble components x and y, UT1-UTC, and precession in longitude and in obliquity. From this point of view VLBI is even *the* central technique for monitoring Earth rotation, because VLBI is the technique defining the official celestial reference frame, the so-called ICRF (International Celestial Reference Frame), which is the realization of the ICRS.

VLBI is a special application of radio astronomy. A global network of radio telescopes simultaneously observes Quasars, i.e., *quasi stellar radio sources*. Quasars are radio galaxies, which, due to their distance, do not show any measurable proper motion. The first Quasars were discovered in the 1960s. The absence of proper motions and their small angular diameters make Quasars ideal objects for the definition and realization of the inertial celestial reference system (see section I- 3.1).

Only the development of radio interferometric methods allowed to exploit the full potential of the Quasars for the definition of the celestial reference frame and the monitoring of Earth rotation. The measurement principle of VLBI is illustrated by Figure 2.1, where the two radio telescopes T_1 and T_2 simultaneously observe one and the same Quasar. The random signals emitted by the Quasar are recorded at T_1 and T_2 as a function of highly accurate clocks (hydrogen masers are used today). By correlating the recorded signals at T_1 and T_2 in time t, one eventually obtains the distance difference d of the two stations T_1 and T_2 as seen from the Quasar. Figure 2.2 shows the VLBI telescope of the Wettzell Observatory in the Bavarian Forest, Germany.

A typical VLBI observation session of one Quasar lasts for a few minutes. The radio signals simultaneously observed by the two telescopes are recorded

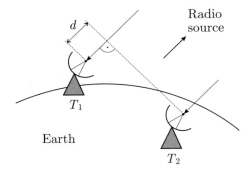

Fig. 2.1. Principle of VLBI, where $d = c \, \Delta t$ is the observed quantity

Fig. 2.2. The Wettzell VLBI-telescope

with broad bandwidth on magnetic tapes together with timing information of very high accuracy – every VLBI telescope is equipped with a hydrogen Maser. The two magnetic tapes are then *correlated* in time, and eventually the distance difference d between the two telescopes, projected onto the direction telescope $\rightarrow$ Quasar, is obtained with a precision of a few millimeters.

VLBI is a microwave observation technique (like GPS), which is why the measured quantity d is also influenced by tropospheric refraction. Due to the long distances between VLBI telescopes it often occurs that observations are made at rather large zenith distances $z > 80°$, which makes tropospheric refraction the ultimate accuracy limiting effect of the VLBI observation technique. As in the GPS observation technique, tropospheric refraction has to be modelled in the analysis by introducing troposphere parameters or by treating tropospheric refraction as a stochastic process. But even with this problem area, all components of Earth rotation may be determined with VLBI on the order of 0.1 mas (milliarcseconds) or better.

The correlation of VLBI signals is very time consuming and, together with data transmission, the real bottleneck of this space geodetic technique. Correlating one day of VLBI observations of a network of about five stations still required in essence one day of CPU of specialized computers in 2000. VLBI observations are today coordinated by the International VLBI Service for Geodesy and Astrometry (IVS). For more information we refer to the internet site http://ivscc.gsfc.nasa.gov.

Figure 2.3 shows the major corrections w.r.t. the IAU nutation model of 1980 (in milliarcseconds) revealed by the VLBI observation technique. The Figures are based on the analysis performed by Herring [55] and others.

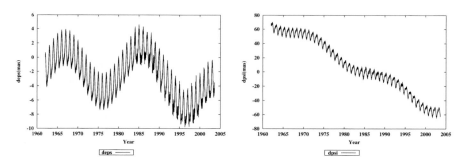

Fig. 2.3. Corrections $\Delta\varepsilon$ and $\Delta\Psi$ in nutation between 1962 and 2003 in obliquity (left) and ecliptical longitude detected by VLBI

2.2 The Rotation of a Rigid Earth and a Rigid Moon

The equations of motion for the three-body problem Earth-Moon-Sun were derived in section I- 3.3. The geocentric orbital motion of Moon and Sun are described by eqns. (I- 3.118) and the rotational motions of Earth and Moon by eqns. (I- 3.124) and (I- 3.68). For very accurate investigations we need to take the perturbations by the planets into account for modeling the geocentric orbital motions of Sun and Moon. The torques on Earth and Moon exerted by the planets are very small (due to the geo- and selenocentric distances to the planets and due to the relatively small masses of these bodies) and are neglected here.

The rotation of the Earth and the Moon is primarily governed by the geocentric orbital motion of the Moon, which is why its essential orbital characteristics are addressed in the next paragraph 2.2.1 before discussing the essential properties of the rotation of the rigid bodies Earth and Moon in sections 2.2.2 and 2.2.3, respectively.

Program ERDROT, as described in Chapter 9 of Part III is used to illustrate the rotational and orbital motions of the Earth, Moon, Sun, and planets.

2.2.1 The Orbit of the Moon

Figures 2.4, 2.5, and 2.6 give an impression of the size and the structure of the perturbing accelerations in radial R, along-track S, and in the out-of-plane direction W. The perturbations are dominated by the gravitational attraction exerted by the Sun.

Figure 2.4 shows the perturbing acceleration in the Moon's $\mathcal{R}$-system (see Table I- 4.3) from node to node approximately in the month of August 1999, Figure 2.5 contains the same information approximately for May 1999. The argument of latitude u was used as independent argument in these Figures. The W-component is about $1-2$ orders of magnitude smaller than the R- and the S-components, which have amplitudes of a few 10^{-5} m/s^2 . These figures have to be compared to the absolute value of the two-body acceleration, which is of the order of $\frac{G(M+m_{\mathbb{C}})}{a_{\mathbb{C}}^2} \approx 2.7 \cdot 10^{-3}$ m/s^2 acting between Moon and Earth. The perturbations are periodic with a period of one draconitic month for the W-component, of half a draconitic month for the R- and the S-components (the draconitic month will be defined below).

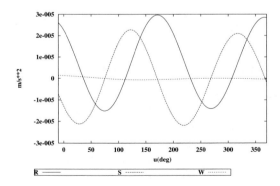

Fig. 2.4. Perturbing accelerations R (radial), S (along-track), and W (out-of-plane) acting on the Moon in August 1999

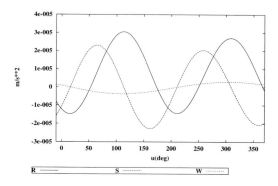

Fig. 2.5. Perturbing accelerations R, S, and W acting on the Moon in May 1999

The W-component always points to the ecliptic and vanishes whenever the Sun lies in the orbital plane of the Moon – which occurs twice per year. The time interval between two subsequent passes of the Sun through the Moon's nodal line is called an *eclipse year*. The W-component is thus negative for $0° < u < 180°$ and positive for $180° < u < 360°$. This pattern also explains the differences of the amplitudes of the W-component in Figures 2.4 and 2.5. The readers of this book (at least the European ones) may remember that a total solar eclipse could be observed in Europe on August 11, 1999. The Sun swept through the Moon's nodal line at this point in time, implying that the W-component of the perturbative acceleration must have been (almost) zero. This is confirmed by Figure 2.6 showing the annual variation of the W-component over the entire year 1999.

The pattern of perturbations has the same structure year after year: the periodic monthly (or semimonthly) perturbations are modulated by the period of a so-called eclipse year. Due to the difference between the tropical and the eclipse year, the minima and maxima occur at different points in time within each year.

We have to make the distinction between different lunar revolution periods. Table 2.2 lists five different types of months: The length of the synodic month differs considerably from that of the other four types. This big difference has nothing to do with perturbations; it is rather a consequence of the fact that the Moon has to sweep an angle of about

$$360° + 360° \, \frac{U_{\mathrm{sid}}}{365.25} \approx 386.9°$$

(where U_{sid} is the sidereal month) in the inertial system in order to complete one revolution w.r.t. the Sun. The synodic month governs by definition the phases of the Moon. Sidereal, anomalistic, and draconitic months would be identical, if the perturbations of the Sun on the Moon were "turned off". The

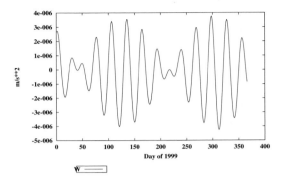

Fig. 2.6. Perturbing acceleration W in 1999

Table 2.2. Lengths of months

Month	Type	Length [days]
Synodic	New Moon to New Moon	29.530589
Tropical	Equinox to Equinox	27.321582
Sidereal	Fixed Star to Fixed Star	27.321661
Anomalistic	Perigee to Perigee	27.554551
Draconitic	Node to Node	27.212220

sidereal and the tropical month only differ, because the true equinox of date (due to precession) moves w.r.t. the inertial space.

Despite the fact that the perturbative accelerations are relatively simple quasi-periodic functions with only two basic periods (the draconitic month and the eclipse year), the resulting perturbations in the Moon's orbital elements and, what is relevant in practice, in its ecliptical longitude are relatively complicated. Thanks to the Moon's proximity, the orbit of the Moon could (and can) be observed rather precisely. This is true for optical observations (where fractions of arcseconds may be achieved) and for distance measurements using the technique of LLR (where the accuracy is of the order of a few centimeters). It is therefore not amazing that the Moon's orbital motion posed a challenge for generations of astronomers and mathematicians.

The list of famous contributions started with the *Principia* [83], [84] of Sir Isaac Newton (1643–1727). Euler, Alexis-Claude Clairaut (1713–1765), and Jean Le Rond d'Alembert (1717–1783), considerably advanced the lunar theory and the methods of analytical mechanics in the second half of the eighteenth century. Lunar ephemerides produced by Johann Tobias Mayer (1723–1762), which were based on Euler's first lunar theory, were even recognized as useful for navigation on sea (method of lunar distances) by the English parliament: In 1762 Mayer and Euler were granted with rewards (3000 English pounds to Mayer and 300 to Euler – showing the mutual weights assigned at that time to practical and theoretical work – in context with the so-called *longitude act* of 1714). Not too much has changed in the estimation of theoretical vs. practical contributions since these days! Many more contributions were made in the eighteenth and nineteenth century by the specialists in the field. Hill's lunar theory was published in 1878. It was innovative in the sense that George William Hill (1838–1914) based his series developments on a rotating "variational" orbit instead of an ellipse fixed in space. Hill's theory was brought to perfection by Ernest William Brown (1866–1938), whose formulae were used extensively for the production of ephemerides.

With the deployment of Laser reflectors on the surface of the Moon and with the availability of LLR-derived distances between Earth and Moon, the analytical theories had to be abandoned for the production of lunar ephemerides and the description of lunar rotation. The lunar ephemerides and the libration data are produced today with numerical integration. The procedure is described, e.g., in [107] or in [111].

Let us now briefly address the perturbations in the Moon's orbital elements. Figure 2.7 shows that the Moon's semi-major axis $a_{\mathbb{C}}$ undergoes periodic variations of about 3500 km with a basic period of half a month over the considered time period of two years. Monthly and semiannual signals modulate this periodic signal. A spectral analysis shows that the principal period actually is half a synodic month.

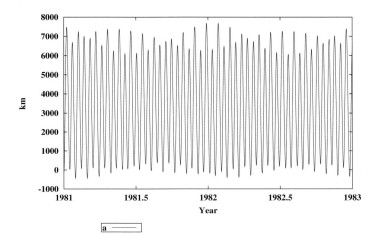

Fig. 2.7. Semi-major axis $a_{\mathbb{C}} - 380'000$ km of lunar orbit 1981-1983

According to Figure 2.8, documenting a time interval of two years, the eccentricity of the lunar orbit varies roughly within the limits $0.025 \leq e \leq 0.075$. The principal period is of the order of one month (according to a spectral analysis of 31.5 days), which is modulated by a semiannual period (more precisely 206 days).

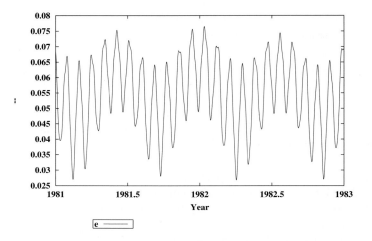

Fig. 2.8. Eccentricity of the lunar orbit 1981-1983

Figures 2.9, 2.10 and 2.11 characterize the motion of the Moon's orbital plane w.r.t. the inertial frame $J2000.0$. As expected from the characteristics of perturbation component W (negative for $0° < u < 180°$ and positive for $180° < u < 360°$) and from the structure of the perturbation equations (I-6.88) for the orbital elements i and Ω, periodic variations in the inclination i and a regression of the node Ω with a period of about 18.6 years are observed. The pole of the Moon's orbital plane therefore precesses approximately on a cone around the pole of the ecliptic. This motion is illustrated by Figure 2.11 showing the projection of the Moon's orbital pole on the ecliptic ($J2000.0$) – scaled by a factor of $180°/\pi$ in order to show directly the angular distance between the poles of the ecliptic and the Moon's orbital plane.

Figure 2.12 shows that the regression of the node causes a very strong periodic variation of the inclination angle $\tilde{i}$ of the Moon's orbital plane w.r.t. the equatorial plane.

This inclination angle $\tilde{i}$ varies within the limits $\varepsilon - i \leq \tilde{i} \leq \varepsilon + i$, i.e., approximately between $23.5° \pm 5.1°$. This variation is responsible for the principal nutation term of 18.6 years. The torque exerted by the Moon is much larger for time periods where $\tilde{i} \approx 28°$ than for time periods where $\tilde{i} \approx 18°$.

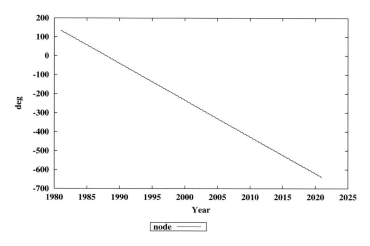

Fig. 2.9. Ecliptical longitude of ascending node of the lunar orbit 1981-2020 in the system $J2000.0$

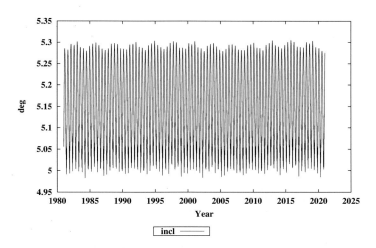

Fig. 2.10. Inclination of the lunar orbit 1981-2020 w.r.t. ecliptic in the system $J2000.0$

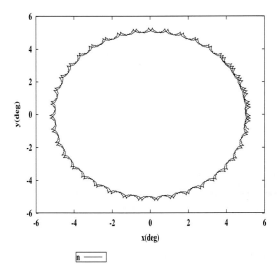

Fig. 2.11. Projection of the normal vector (in degrees) of the lunar orbit 1981-2020 onto the ecliptic J2000.0

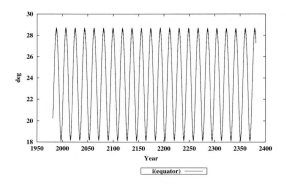

Fig. 2.12. Inclination of the lunar orbit 1981-2380 w.r.t. true equator of date

According to Figure 2.13 the Moon's perigee moves prograde both, w.r.t. the node and w.r.t. the vernal equinox (of a fixed epoch). The period of rotation is of the order of 6.25 years for the motion w.r.t. the node, of 9.0 years for the motion w.r.t. the equinox. The correct explanation of the motion of the Moon's perigee by analytical theories proved to be a challenge for analytical theories.

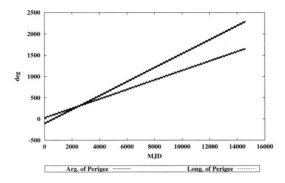

Fig. 2.13. Lunar argument of perigee and longitude of perigee in the system $J2000.0$

All the variations in the semi-major axis imply changes in the mean motion, thus also in the Moon's ecliptical longitude. Figure 2.14 shows the development of the Moon's ecliptical longitude (actually diminished by a mean "mean anomaly" of the Moon), i.e., the quantity $\tilde{l} \overset{\text{def}}{=} \Omega + \omega + v - n_0 (t - t_0)$.

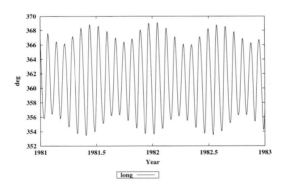

Fig. 2.14. Ecliptical longitude of the Moon (minus mean anomaly) 1981-1982

Figure 2.15 shows the amplitude spectrum (only the ranges between $0 - 400$ days (top) and between $0 - 40$ days (bottom) are reproduced). As described in Chapter 11 of Part III the amplitudes, like the ones in Figure 2.15, of the spectral lines are in general underestimated based on an amplitude spectrum of the type shown in Figure 2.15. The contributing elements of the spectrum to such a line have to be added in a mathematically correct way – as described in Chapter 11 of Part III. The result is contained in Table 2.3.

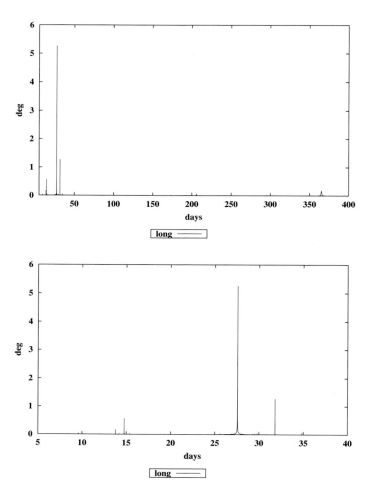

Fig. 2.15. Amplitude spectrum of the ecliptical longitude of the Moon (de-trended) using a 400-years time span

Figure 2.15 and Table 2.3 show that the spectrum is dominated by a few prominent lines, which were discovered empirically a long time ago:

1. **Equation of the Center**. A term with the period of 27.6 days (anomalistic revolution period). It is the dominating term with an amplitude of about 6.3°. It is caused by the Moon's eccentricity.

2. The **Evection** is the second largest term in Figure 2.15. The period is about 31.8 days, the amplitude has a value of about 1°. The effect was already known to Hipparchus (180–125 B.C.). The term is caused by

Table 2.3. Main terms of the amplitude spectrum of the lunar ecliptical longitude*

Name	Period (theor.)	Period (spectrum) [days]	Amp [°]
Equation of Center	$2\pi/n_{\mathbb{C}}$	27.5581	6.2815
Evection	$2\pi\Big/\Big(2\,(n_\omega - n_\odot) + n_{\mathbb{C}}\Big)$	31.8230	1.2759
Variation	$2\pi\Big/\Big(2\,(n_{\mathbb{C}} - n_\odot)\Big)$	14.7670	0.6638
Annual Equation	$2\pi/(2\,n_\odot)$	365.2500	0.1909
	$4\pi\Big/\Big(2\,(n_\omega - n_\odot) + n_{\mathbb{C}}\Big)$	13.7830	0.2150

*($n_{\mathbb{C}}$ and $n_\odot$ are the mean motions of the Moon and the Sun, n_a is the mean motion of the Moon's perigee)

periodic perturbations of the eccentricity e and the argument of perigee ω. The term was at times interpreted as a disturbance of the equation of the center.

3. **Variation.** It is the third largest term in Figure 2.15. It has the period of half a synodic month and an amplitude of about 0.66°. The variation was discovered by Tycho Brahe (1546–1601).

4. **Annual Equation.** With a period of one year and an amplitude of about 0.2°, this term was discovered by Johannes Kepler (1571–1630). It is barely visible in Figure 2.15 (top).

Figure 2.15 shows additional terms in the spectrum of the mean longitude. One largest has a period of half of the period of the equation of the center and might be considered as a modulation of this contribution. The omitted terms are not relevant for the subsequent discussion of the Earth's and the Moon's rotation. The principal perturbation characteristics of the lunar orbit are very more extensively discussed by Danby [31] or by Moulton [77].

Let us conclude this section with a few remarks concerning solar and lunar eclipses. Eclipses only take place if Sun, Moon, and Earth lie approximately on a straight line. Therefore, the Moon has to lie close to its nodal line for an eclipse to take place. Consequently, the length of the draconitic month (see Table 2.2) is the first of essential periods to predict eclipses. Lunar eclipses occur during full Moon, solar eclipses during new Moon, which is why the synodic month is the second important period for eclipse predictions. As the Moon has to be close to the node during eclipses, and as Sun, Moon, and Earth have to be collinear, the Sun must be close to the node of the Moon as well. This makes the *eclipse year*, the time interval between subsequent passes of the Sun through the ascending node of the Moon (with a length of about 346.7 days) the third essential period when predicting eclipses. The following equations hold approximately:

$$19 \text{ eclipse years} \approx 223 \text{ synodic months}$$
$$\approx 242 \text{ draconitic months}$$
$$\approx 18.03 \text{ years} .$$

This fundamental time interval of 19 eclipse years is called a *Saros period*. As its name indicates the Saros period was known already in antique times. Similar eclipses occur after one Saros period.

2.2.2 Rotation of the Rigid Earth

Free Motion. Figure 2.16 shows the projection of the Earth's angular velocity vector onto the plane orthogonal to the axis of maximum moment of inertia (figure axis) in the Earth-fixed system (approximately the equatorial plane) for 1981. The Figure is scaled by $206284.8\ \omega_{\delta}^{-1}$ in order to show the distance between the figure and rotation axis in arcseconds. The following initial conditions were assumed:

$$206264.8\ \frac{\omega_{\delta \mathcal{F}_1}}{\omega_{\delta}} = 0.2''$$
$$206264.8\ \frac{\omega_{\delta \mathcal{F}_2}}{\omega_{\delta}} = 0.0'' . \tag{2.3}$$

What we see in Figure 2.16 is in essence *polar motion* or *polar wobble*. One should, however, keep in mind the conventions of PM: The classical PM components x and y are defined as the rotation angles about the second axis of the Earth-fixed system (x component) and the first axis of the same system (y component) in order to achieve the transformation from the Earth-fixed system to the system of the instantaneous rotation axis. The following equations hold:

$$x = +206264.8\ \frac{\omega_{\delta \mathcal{F}_1}}{\omega_{\delta}}$$
$$y = -206264.8\ \frac{\omega_{\delta \mathcal{F}_2}}{\omega_{\delta}} . \tag{2.4}$$

Subsequently, we will always show the angles $(x, -y)$ which refer to a right-hand system, but, in order to keep the descriptions short, speak of polar wobble (or PM) despite this inconsistency.

Figure 2.16 was generated with program ERDROT (see Chapter 9 of Part III) assuming a rigid, rotationally symmetric rigid Earth and setting the external torques to zero. One easily sees that the rotation axis is moving on a circle around the figure axis. This behavior will be explained rather easily below.

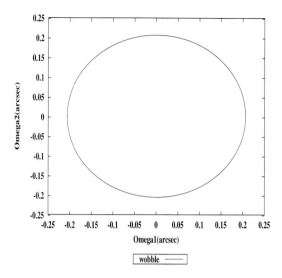

Fig. 2.16. Polar wobble 1981 without torques

Before doing that, let us consider the analogue case for the orbital motion. Equations (I-3.118) for the centers of mass of Earth and Moon relate the accelerations of these centers (left-hand side of the equations) to the sum of the forces acting on all the particles of the rigid body (right-hand sides of the same equations). If the system consists only of one body of finite extension and mass (let us say, a very lonely Earth), the right-hand side becomes zero and the motion is rectilinear (with constant velocity) – exactly as in the case of an isolated point mass. This statement is also true if the original, non-approximated equations of motion (I-3.85) are considered.

The rotation of an isolated body of finite extensions (or the case, where the sum of torques is always zero) may be studied as well. In section I-3.3 we saw that the angular momentum of a finite body is conserved, if the right-hand sides of eqns. (I-3.88) are zero. In the case of the Earth we may therefore write

$$\boldsymbol{h}_{\delta} = \boldsymbol{h}_{\delta_0} \ . \tag{2.5}$$

The inertia tensor w.r.t. the inertial system in eqn. (I-3.73) may be expressed by the diagonal inertia tensor w.r.t. the Earth-fixed system and the transformation matrix $\mathbf{T}_{\delta}$ between the inertial and the Earth-fixed system (see eqns. (I-3.77) and (I-3.56)):

$$\boldsymbol{h}_{\delta_\mathcal{I}} = \mathbf{I}_{\delta_\mathcal{I}} \, \boldsymbol{\omega}_{\delta_\mathcal{I}} = \mathbf{T}_{\delta} \, \mathbf{I}_{\delta_\mathcal{F}} \, \mathbf{T}_{\delta}^T \, \boldsymbol{\omega}_{\delta_\mathcal{I}} = \mathbf{T}_{\delta} \, \{\mathbf{I}_{\delta_\mathcal{F}} \, \boldsymbol{\omega}_{\delta_\mathcal{F}}\} \ . \tag{2.6}$$

The expression in the brackets may be written as a linear combination of three vectors (all of which referring to the Earth-fixed system):

$$\mathbf{I}_{\oplus F}\,\boldsymbol{\omega}_{\oplus F} = \left[A_\oplus\,\mathbf{E} + (B_\oplus - A_\oplus)\begin{pmatrix}0&0&0\\0&1&0\\0&0&0\end{pmatrix} + (C_\oplus - A_\oplus)\begin{pmatrix}0&0&0\\0&0&0\\0&0&1\end{pmatrix}\right]\boldsymbol{\omega}_{\oplus F} \tag{2.7}$$

$$= A_\oplus\,\boldsymbol{\omega}_{\oplus F} + (B_\oplus - A_\oplus)\,\omega_{\oplus F_2}\,\boldsymbol{e}_{2_{\oplus F}} + (C_\oplus - A_\oplus)\,\omega_{\oplus F_3}\,\boldsymbol{e}_{3_{\oplus F}}\ .$$

The column matrices $\boldsymbol{e}_{2_{\oplus F}}$ and $\boldsymbol{e}_{3_{\oplus F}}$ contain the components of the second and third unit vectors of the Earth-fixed coordinate systems as elements. $\omega_{\oplus F_2}$ and $\omega_{\oplus F_3}$ are the second and third components of the angular velocity expressed in the Earth-fixed system. Introducing the previous expression into eqn. (2.6) allows the establishment of the equation

$$\boldsymbol{h}_{\oplus I} = \mathbf{T}_\oplus\left[A_\oplus\,\boldsymbol{\omega}_{\oplus F} + (B_\oplus - A_\oplus)\,\omega_{\oplus F_2}\,\boldsymbol{e}_{2_{\oplus F}} + (C_\oplus - A_\oplus)\,\omega_{\oplus F_3}\,\boldsymbol{e}_{3_{\oplus F}}\right]\ , \tag{2.8}$$

where $\boldsymbol{e}_{2_{\oplus F}}$ and $\boldsymbol{e}_{3_{\oplus F}}$ are the unit vectors lying in the second and third axis of the Earth-fixed system.

Equation (2.8) says that the angular momentum vector of the Earth may be expressed as a linear combination of the angular velocity vector and of two of the unit vectors defining the Earth-fixed system. An inspection of the coefficients shows that for our application the first vector on the right-hand side is dominant ($C_\oplus - A_\oplus \ll A_\oplus$ and $B_\oplus - A_\oplus \ll A_\oplus$). Apart from that the relation is not too informative – after all, in the Euclidean space $\mathbb{E}^3$ it is always possible to express a vector as a linear combination of three other vectors as long as the latter are not linearly dependent. Equation (2.8) is much more interesting if rotational symmetry is assumed, i.e., if $A_\oplus = B_\oplus$. Table 2.1 tells that this is the case in good approximation for the Earth. The above relation then reads as

$$\boldsymbol{h}_{\oplus I} = \mathbf{T}_\oplus\left[A_\oplus\,\boldsymbol{\omega}_{\oplus F} + (C_\oplus - A_\oplus)\,\omega_{\oplus F_3}\,\boldsymbol{e}_{3_{\oplus F}}\right]$$

$$= \mathbf{T}_\oplus\left[C_\oplus\,\boldsymbol{\omega}_{\oplus F} + (C_\oplus - A_\oplus)\,\omega_{\oplus F_3}\,\boldsymbol{e}_{3_{\oplus F}} - (C_\oplus - A_\oplus)\,\boldsymbol{\omega}_{\oplus F}\right]$$

$$= C_\oplus\,\omega_\oplus\,\mathbf{T}_\oplus\left\{\boldsymbol{e}_{\omega_{\oplus F}} + \frac{C_\oplus - A_\oplus}{C_\oplus}\left[\frac{\omega_{\oplus F_3}}{\omega_\oplus}\,\boldsymbol{e}_{3_{\oplus F}} - \boldsymbol{e}_{\omega_{\oplus F}}\right]\right\} \tag{2.9}$$

$$\approx C_\oplus\,\omega_\oplus\,\mathbf{T}_\oplus\left\{\boldsymbol{e}_{\omega_{\oplus F}} + \frac{C_\oplus - A_\oplus}{C_\oplus}\left[\boldsymbol{e}_{3_{\oplus F}} - \boldsymbol{e}_{\omega_{\oplus F}}\right]\right\}\ ,$$

where $\boldsymbol{e}_{\omega_\oplus} \overset{\text{def}}{=} \dfrac{\boldsymbol{\omega}_\oplus}{\omega_\oplus}$.

Equation (2.9) tells that the angular velocity vector, the angular momentum vector, and the figure axis of the Earth (unit vector $\boldsymbol{e}_{3_{\oplus F}}$ of the Earth-fixed system) always lie in one and the same plane. The last line holds, if the angle between the figure axis $\boldsymbol{e}_{3_{\oplus F}}$ and the angular velocity vector is small – which is true in the case of the Earth. Figure 2.17 illustrates the above vector equation. It also shows that the angular momentum and angular velocity vectors are much closer together (by a factor of $(C_{\delta} - A_{\delta})/C_{\delta} \approx \frac{1}{305}$, see Table2.1) than the figure and rotation axes. The latter angle corresponds to the amplitude of PM and is given by the initial state of the system (see developments below).

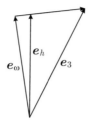

Fig. 2.17. Angular velocity vector e_ω, angular momentum vector e_h, and figure axis e_3 of the rotationally symmetric Earth

From eqn. (2.9) we conclude that the rotation pole moves on a circle around the pole of figure (assuming a rotationally symmetric Earth) in the Earth-fixed system. The precise characteristics of this motion are easily established by setting the right-hand sides of eqns. (I-3.124) to zero (case without external torques):

$$
\begin{pmatrix} \dot{\omega}_{\delta 1} \\ \dot{\omega}_{\delta 2} \\ \dot{\omega}_{\delta 3} \end{pmatrix} + \begin{pmatrix} + \gamma_\delta\, \omega_{\delta 2}\, \omega_{\delta 3} \\ - \gamma_\delta\, \omega_{\delta 3}\, \omega_{\delta 1} \\ 0 \end{pmatrix} = 0
$$

$$
\begin{pmatrix} \dot{\omega}_{\mathbb{C} 1} \\ \dot{\omega}_{\mathbb{C} 2} \\ \dot{\omega}_{\mathbb{C} 3} \end{pmatrix} + \begin{pmatrix} \gamma_{\mathbb{C} 1}\, \omega_{\mathbb{C} 2}\, \omega_{\mathbb{C} 3} \\ \gamma_{\mathbb{C} 2}\, \omega_{\mathbb{C} 3}\, \omega_{\mathbb{C} 1} \\ \gamma_{\mathbb{C} 3}\, \omega_{\mathbb{C} 1}\, \omega_{\mathbb{C} 2} \end{pmatrix} = 0 \ .
$$

(2.10)

Observe that the above equations refer to the Earth fixed system. The index $\mathcal{F}$ was left out in order to reduce the formalism.

In the case of the Earth we assumed rotational symmetry $A_\delta = B_\delta$, which allowed it to eliminate the three parameters γ_{δ_i}, $i = 1, 2, 3$, in favour of only one, namely the dynamical flattening parameter

$$
\gamma_\delta \overset{\text{def}}{=} \frac{C_\delta - A_\delta}{A_\delta} \ .
$$

(2.11)

The third of eqns. (2.10) simply states that the third component of the angular velocity in the body-fixed system is constant:

$$\omega_{\oplus 3} = \omega_{\oplus 0} \ . \tag{2.12}$$

$\omega_{\oplus 0}$ is the constant angular velocity of Earth rotation. It is interesting to note that this quantity is conserved even in the presence of torques, due to the fact that $\gamma_{\oplus 3} = 0$ on the right-hand side of the equations (I- 3.124). In other words: a rigid Earth does not show LOD changes.

Eqns. (2.12) and (2.9) even allow it to conclude that the rotation axis and the figure axis rotate on (straight) cones around the axis of angular momentum.

Introducing the result (2.12) into the equations (2.10) leads to a system of two *linear* differential equations for the first two components of the Earth's angular velocity vector in the Earth-fixed system

$$\begin{aligned}
\dot{\omega}_{\oplus 1} + \gamma_{\oplus}\, \omega_{\oplus 0}\, \omega_{\oplus 2} &= 0 \\
\dot{\omega}_{\oplus 2} - \gamma_{\oplus}\, \omega_{\oplus 0}\, \omega_{\oplus 1} &= 0 \ ,
\end{aligned} \tag{2.13}$$

which are solved by functions of the type

$$\begin{aligned}
\omega_{\oplus 1}(t) &= \rho_{\oplus} \cos(\tilde{\omega}_{\oplus} t + \alpha_{\oplus 0}) \\
\omega_{\oplus 2}(t) &= \rho_{\oplus} \sin(\tilde{\omega}_{\oplus} t + \alpha_{\oplus 0}) \ .
\end{aligned} \tag{2.14}$$

The rotation axis thus precesses prograde with constant angular velocity on a cone around the figure axis of the Earth. The angular velocity of this motion is uniquely a function of the angular velocity of Earth rotation and of the flattening parameter $\gamma_{\oplus}$:

$$\tilde{\omega}_{\oplus 0} = \gamma_{\oplus}\, \omega_{\oplus 0} \ . \tag{2.15}$$

The amplitude $\rho_{\oplus}$ and the phase angle $\alpha_{\oplus 0}$ are defined by the initial state of the system.

The above result is remarkable: in the absence of torques the endpoint of the angular velocity vector moves in the Earth-fixed system with constant angular velocity $\tilde{\omega}_{\oplus 0} = \gamma_{\oplus}\, \omega_{\oplus 0}$ on a circle with radius $\rho_{\oplus}$ in the positive sense of rotation around the figure axis. The period of this rotation is given by

$$P_e = \frac{2\,\pi}{\gamma_{\oplus}\, \omega_{\oplus 0}} \approx 303 \text{ days} \ , \tag{2.16}$$

which is called the *Euler period* of PM. The values in Table 2.1 were used to compute the numerical value for the Euler period of PM.

This important result was first stated by Euler in his famous book [36]. The belief in the Euler period of PM played an essential role in end of the 19th century when PM was confirmed to be real in long time series of latitude observations (see, e.g., [126]). Equations (2.14) explain PM in the absence of torques as shown in Figure 2.16.

How do figure and rotation axes of Earth move in the inertial space in the absence of torques? Figure 2.18 gives the answer for a short time interval of roughly one month: the three Euler angles only show periodic variation with periods of approximately one day. The amplitudes are small: about 0.5″ for the Euler angles Ψ_δ and Θ_δ (Figure 2.18, top, bottom), about 0.2″ for the obliquity ε_δ. Let us now establish the analytical solution corresponding to Figure 2.18.

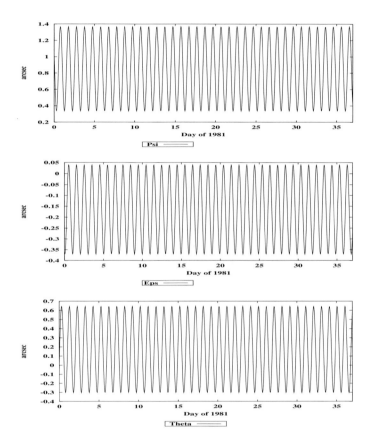

Fig. 2.18. Euler angles Ψ_δ, ε_δ and Θ_δ (minus mean motion) at beginning of 1981 without torques

Having seen that the figure, angular momentum, and rotation axes lie (in this order) in one plane (see Figure 2.17 and eqn. (2.9)) and that the former two axes move on cones around the angular momentum axis (because according to eqn. (2.12) the third component of the Earth's angular velocity vector in the Earth-fixed system is constant), it should be relatively easy to establish the three Euler angles as a function of time. Equation (I- 3.68) provides the relationship between the Earth-fixed components of the angular velocity vector and the time derivatives of the Euler angles. Introducing eqns. (2.12) and (2.14) into Euler's kinematic equations (I- 3.68) we obtain the differential equations for the Euler angles:

$$
\begin{aligned}
\dot{\Psi}_{\text{☾}} &= -\rho_{\text{☾}} \sin(\Theta_{\text{☾}} + \tilde{\omega}_{\text{☾}_0} t + \alpha_{\text{☾}_0})\, \csc \varepsilon_{\text{☾}} \\
\dot{\varepsilon}_{\text{☾}} &= -\rho_{\text{☾}} \cos(\Theta_{\text{☾}} + \tilde{\omega}_{\text{☾}_0} t + \alpha_{\text{☾}_0}) \\
\dot{\Theta}_{\text{☾}} &= +\rho_{\text{☾}} \sin(\Theta_{\text{☾}} + \tilde{\omega}_{\text{☾}_0} t + \alpha_{\text{☾}_0})\, \cot \varepsilon_{\text{☾}} + \omega_{\text{☾}_0}
\end{aligned}
\tag{2.17}
$$

or

$$
\begin{aligned}
\sin \varepsilon_{\text{☾}}\, \dot{\Psi}_{\text{☾}} &= -\rho_{\text{☾}} \sin(\Theta_{\text{☾}} + \tilde{\omega}_{\text{☾}_0} t + \alpha_{\text{☾}_0}) \\
\dot{\varepsilon}_{\text{☾}} &= -\rho_{\text{☾}} \cos(\Theta_{\text{☾}} + \tilde{\omega}_{\text{☾}_0} t + \alpha_{\text{☾}_0}) \\
\sin \varepsilon_{\text{☾}}\, \dot{\Theta}_{\text{☾}} &= +\rho_{\text{☾}} \sin(\Theta_{\text{☾}} + \tilde{\omega}_{\text{☾}_0} t + \alpha_{\text{☾}_0})\, \cos \varepsilon_{\text{☾}} + \omega_{\text{☾}_0} \sin \varepsilon_{\text{☾}} \; .
\end{aligned}
\tag{2.18}
$$

Observe that $\tilde{\omega}_{\text{☾}_0}$ is defined by eqn. (2.15).

These equations may be solved analytically up to (and including) terms of first order in small quantities. Let us introduce for this purpose the two auxiliary angles $\varepsilon_{\text{☾}_h}$ and $\Theta_{\text{☾}_h}$ as the polar coordinates of the vector of angular momentum $\boldsymbol{h}_{\text{☾}}$ in the inertial system:

$$
\boldsymbol{h}_{\text{☾}} = h \begin{pmatrix} -\sin \varepsilon_{\text{☾}_h} \sin \Theta_{\text{☾}_h} \\ +\sin \varepsilon_{\text{☾}_h} \cos \Theta_{\text{☾}_h} \\ \cos \varepsilon_{\text{☾}_h} \end{pmatrix} .
\tag{2.19}
$$

From the above developments one may conclude that $\Delta\varepsilon_{\text{☾}} \overset{\text{def}}{=} \varepsilon_{\text{☾}} - \varepsilon_{\text{☾}_h}$ is a small quantity. Also, in view of the fact that the angular velocity vector moves around the figure axis of the Earth on a small circle of radius $< \rho_{\text{☾}}$, the derivatives $\dot{\Psi}_{\text{☾}}$ and $\dot{\varepsilon}_{\text{☾}}$ must be small quantities as well. This allows us to approximate the non-linear differential equations (2.18) as

$$
\begin{aligned}
\sin \varepsilon_{\text{☾}h}\, \dot{\Psi}_{\text{☾}} &= -\rho_{\text{☾}} \sin(\Theta_{\text{☾}} + \tilde{\omega}_{\text{☾}_0} t + \alpha_{\text{☾}_0}) \\
\dot{\varepsilon}_{\text{☾}} &= -\rho_{\text{☾}} \cos(\Theta_{\text{☾}} + \tilde{\omega}_{\text{☾}_0} t + \alpha_{\text{☾}_0}) \\
\sin \varepsilon_{\text{☾}h}\, \dot{\Theta}_{\text{☾}} &= +\rho_{\text{☾}} \sin(\Theta_{\text{☾}} + \tilde{\omega}_{\text{☾}_0} t + \alpha_{\text{☾}_0})\, \cos \varepsilon_{\text{☾}h} + \omega_{\text{☾}_0} \sin \varepsilon_{\text{☾}h} \; .
\end{aligned}
\tag{2.20}
$$

In this approximation the third equation is separated from the first two. Its zero-order solution in terms of $\rho_{\text{☾}}$ simply is

$$
\Theta_{\text{☾}}(t) \approx \Theta_{\text{☾}_0} + \omega_{\text{☾}_0} t \; .
\tag{2.21}
$$

Using this approximation on the right-hand sides of eqns. (2.20) leads to the following equations and their first-order solution in $\rho_{\text{☾}}$

$$\dot{\Psi}_\leftmoon = - \rho_\leftmoon \sin \left((1 + \gamma_\leftmoon) \omega_{\leftmoon 0} t + \alpha_{\leftmoon 0} + \Theta_{\leftmoon 0} \right) \csc \varepsilon_{\leftmoon h}$$

$$\dot{\varepsilon}_\leftmoon = - \rho_\leftmoon \cos \left((1 + \gamma_\leftmoon) \omega_{\leftmoon 0} t + \alpha_{\leftmoon 0} + \Theta_{\leftmoon 0} \right)$$

$$\dot{\Theta}_\leftmoon = + \rho_\leftmoon \sin \left((1 + \gamma_\leftmoon) \omega_{\leftmoon 0} t + \alpha_{\leftmoon 0} + \Theta_{\leftmoon 0} \right) \cot \varepsilon_{\leftmoon h} + \omega_{\leftmoon 0}$$

$$\Psi_\leftmoon(t) = \Psi_{\leftmoon 0} + \tilde{\rho}_\leftmoon \left\{ \cos \left((1 + \gamma_\leftmoon) \omega_{\leftmoon 0} t + \alpha_{\leftmoon 0} + \Theta_{\leftmoon 0} \right) - \cos \alpha_{\leftmoon 0} \right\} \csc \varepsilon_{\leftmoon h}$$
$$\tag{2.22}$$

$$\varepsilon_\leftmoon(t) = \varepsilon_{\leftmoon 0} - \tilde{\rho}_\leftmoon \left\{ \sin \left((1 + \gamma_\leftmoon) \omega_{\leftmoon 0} t + \alpha_{\leftmoon 0} + \Theta_{\leftmoon 0} \right) - \sin \alpha_{\leftmoon 0} \right\}$$

$$\Theta_\leftmoon(t) = \Theta_{\leftmoon 0} + \omega_{\leftmoon 0} t - \tilde{\rho}_\leftmoon \left\{ \cos \left((1 + \gamma_\leftmoon) \omega_{\leftmoon 0} t + \alpha_{\leftmoon 0} + \Theta_{\leftmoon 0} \right) - \cos \alpha_{\leftmoon 0} \right\} \cot \varepsilon_{\leftmoon h} \, ,$$

where

$$\tilde{\rho}_\leftmoon = \frac{\rho_\leftmoon}{(1 + \gamma_\leftmoon) \omega_{\leftmoon 0}} . \tag{2.23}$$

The quantity $\frac{\rho_\leftmoon}{\omega_{\leftmoon 0}}$ has a simple interpretation: it is the amplitude of PM (angle between rotation axis and figure axis), expressed in radians. Multiplying it by the factor $\frac{180}{\pi} 3600 \approx 206264.8$ gives the amplitude of PM in arcseconds. In the case of the Earth, the amplitude $\tilde{\rho}_\leftmoon$ is only slightly, by a factor of $\frac{1}{1 + \gamma_\leftmoon} \approx 1 - \gamma_\leftmoon$, smaller than the radius $\rho_\leftmoon$ of PM. The results confirm those of the simulation in Figure 2.18.

This result also confirms the expectations raised by Figure 2.17: the amplitude of the variations of the figure axis in inertial space is the angle between the figure axis and the axis of (conserved) angular momentum. The interpretation of the results for $\Psi_\leftmoon(t)$ and $\varepsilon_\leftmoon(t)$ is thus easy: in inertial space the figure axis rotates retrograde (in a distance of $\tilde{\rho}_\leftmoon$) around the axis of angular momentum. The above equation for $\Theta_\leftmoon(t)$ tells that $\Theta_\leftmoon(t)$ contains a periodic component with a period slightly shorter than the sidereal day (remember that $\frac{2\pi}{\omega_{\leftmoon 0}}$ is one sidereal day and that $\gamma_\leftmoon \approx \frac{1}{300}$). The equation for $\Theta_\leftmoon(t)$ also tells, that its angular velocity is not constant during the day but contains small periodic variations (with an amplitude of $\tilde{\rho}_\leftmoon$). The period

$$P_{\mathrm{nd}} = \frac{2\,\pi}{(1 + \gamma_\leftmoon)\,\omega_{\leftmoon 0}} \approx 23^{\mathrm{h}}55^{\mathrm{m}}13^{\mathrm{s}} \tag{2.24}$$

is also called the *nearly-diurnal* or *quasi-diurnal* period of the figure axis in inertial space (e.g., [66]). The period is close to, but not identical with the sidereal day.

With the above results it is also possible to describe the rotation axis of the Earth in inertial space. Figure 2.17 indicates that the amplitude of the motion of the figure axis around the axis of angular momentum will be $\rho_{\leftmoon 0} \gamma_\leftmoon$, thus about a factor of 300 smaller than the amplitude of PM, and that there are phase difference of 180° between the rotation and figure axes in their motion around the axis of angular momentum.

Separation of Rotational and Orbital Motion. If the terms proportional to the principal moments of inertia are neglected in eqns. (I- 3.118), the equations of motion for the centers of mass of Moon and Sun may be solved separately from the equations (I- 3.124) describing the rotation of Earth and Moon. Under this assumption, the equations for the Moon and the Earth in eqns. (I- 3.124) (or the equations (I- 3.125) referring to the inertial system) together with the corresponding kinematic equations may be solved separately, as well. It is thus possible to study the rotational motion of Earth and Moon "independently". In summary this means, that instead of a coupled system of 24 first-order differential equations (or 12 equations of the second order) we only have to deal with one coupled system of second order with $2 \cdot 3$ equations for the geocentric motion of the centers of mass of Moon and Sun and with two first-order systems with 6 equations (or with two second-order systems with 3 equations). This separation of equations considerably simplifies the solution of the original coupled system.

Separation of the Kinematic and Dynamic Equations. Euler's equations for the components of the angular velocity vector of either Earth or Moon and the corresponding kinematic equations together form a coupled, non-linear system of first-order differential equations. The right-hand sides of eqns. (I- 3.124) or (I- 3.125) are small quantities of first order, because they are proportional to the differences between the principal moments of inertia. Note, that this statement is only correct for celestial bodies not departing to much from spherical symmetry.

Under such circumstances, a system of differential equations may be solved using the methods of *perturbation theory* as treated in Chapter I- 6. A good approximation for the true solution therefore results, if on the right-hand sides of the differential equations approximations for the dependent arguments of the differential equations are used instead of the true values as they would emerge from a correct solution.

Let us assume for the moment, that some approximations Ψ_{δ_a}, ε_{δ_a}, and Θ_{δ_a} for Ψ_δ, ε_δ, and Θ_δ are available. In this case eqns. (I- 3.124) or (I- 3.125) may be solved independently from the corresponding kinematic equations, and only after having solved eqns. (I- 3.124) or (I- 3.125) one may wish to solve the corresponding kinematic equations. Moreover, eqns. (I- 3.124) describing the rotation of Earth and Moon are broken up into one trivial equation for the angle $\omega_{...3}$ and one linear, inhomogeneous first order differential equation system of dimension 2 in the first two components of the angular velocity components in the Earth-fixed system. Such problems may be dealt with by standard mathematical methods.

Forced Motion: Simulations. Before making the attempt to derive analytically the main features of the forced motion, we use program ERDROT to illustrate PM and the motion of the Earth's figure axis in inertial space.

Figure 2.19 shows PM for 1981 under the influence of the torques exerted by the Moon and the Sun. Superimposed to the free motion (with an amplitude of 0.2″), which was also present in Figure 2.16, we observe a small circular motion with a slowly varying amplitude of the Earth's rotation pole. The period is approximately one day. The quasi-daily motion is called *Oppolzer motion* in honour of the Austrian Astronomer Freiherr Ritter von Oppolzer (1841–1886).

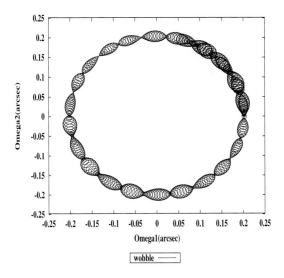

Fig. 2.19. Polar wobble 1981 with torques

Figure 2.20 shows the amplitude spectrum of the two-dimensional Oppolzer motion generated with program FOURIER (see Chapter 11 of Part III). A data-span of forty years with a data sampling of one hour is underlying the spectrum. As opposed to the prograde free motion of the pole, the Oppolzer motion is retrograde (clockwise rotation). The actual motion is relatively complex. Three terms, two of them very close to the solar day, one close to 1.075 days, dominate the spectrum. The amplitudes are of the orde of a few milliarcarcsecond. Figures 2.21 and 2.22 show the development of the Euler angles Ψ_{δ} (approximately precession plus nutation in longitude) and ε_{δ} (approximately nutation in obliquity) in January 1981. The nearly-diurnal terms, as they were already observed in Figure 2.18, are present, as well. Due to the external torques one can see signals of longer periods and – presumably – larger amplitudes.

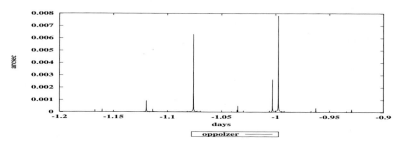

Fig. 2.20. Spectrum of the Oppolzer motion

When studying Earth rotation over years rather than days, the nearly-diurnal terms are usually removed from the time series. In program ERDROT the formulas (2.22) are used for that purpose. The following transformation leads to the Euler angles $\tilde{\Psi}_\delta$, $\tilde{\varepsilon}_\delta$ and $\tilde{\Theta}_\delta$, which are free from quasi-diurnal terms:

$$
\begin{aligned}
\tilde{\Psi}_\delta - \Psi_\delta &= -\frac{1}{\sin\varepsilon_\delta\,(1+\gamma_\delta)\,\omega_{\delta_0}}\left(\omega_{\delta\mathcal{F}_1}\cos\Theta_\delta - \omega_{\delta\mathcal{F}_2}\sin\Theta_\delta\right) \\
\tilde{\varepsilon}_\delta - \varepsilon_\delta &= +\frac{1}{(1+\gamma_\delta)\,\omega_{\delta_0}}\left(\omega_{\delta\mathcal{F}_1}\sin\Theta_\delta + \omega_{\delta\mathcal{F}_2}\cos\Theta_\delta\right) \qquad (2.25) \\
\tilde{\Theta}_\delta - \Theta_\delta &= +\frac{\cos\varepsilon_\delta}{\sin\varepsilon_\delta\,(1+\gamma_\delta)\,\omega_{\delta_0}}\left(\omega_{\delta\mathcal{F}_1}\cos\Theta_\delta - \omega_{\delta\mathcal{F}_1}\sin\Theta_\delta\right) \; .
\end{aligned}
$$

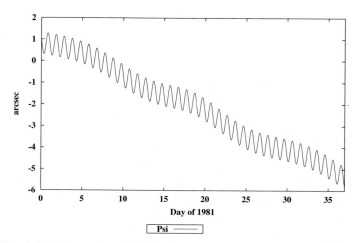

Fig. 2.21. Precession and nutation in longitude at beginning of 1981

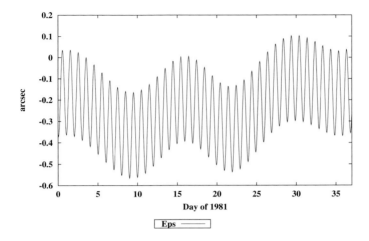

Fig. 2.22. Nutation in obliquity at beginning of 1981

Figure 2.23 shows the PM of a rigid Earth over a time interval of forty years. The rotation pole is contained within a ring centered at the figure axis of the Earth. The thickness of the ring is given by the maximum amplitude of the Oppolzer motion. The Oppolzer terms are not visible, because data were only stored every five days – a clear undersampling, if the nearly-diurnal terms would have been of primary interest.

Figures 2.24, 2.26 and 2.27 show the development of the angles $\tilde{\Psi}_{\delta}$, $\tilde{\varepsilon}_{\delta}$, and $\tilde{\Theta}_{\delta}$, which are related to the corresponding Euler angles by eqn. (2.25), over a time interval of forty years. Figure 2.25 is obtained from Figure 2.24 by removing a linear trend from the former figure. This process shows in essence, how precession (the linear constituent of the Euler angle Ψ_{δ} in Figure 2.26) is separated in astronomy from periodic variations, which are interpreted as nutation terms.

The prominent feature in Figure 2.24 is the linear drift of about $\dot{\Psi}_{\delta} \approx 50.5''$. This is the so-called constant of luni-solar precession, corresponding to a regression of the line of equinoxes on the ecliptic. The effect was presumably discovered by Hipparchus. When subtracting this drift we obtain in essence the nutation in (ecliptical) longitude in Figure 2.25. The main term has a period of about 18.6 years and has an amplitude of about $\Delta\varepsilon(18.6\,\mathrm{y}) \approx 17.3''$.

The nutation in obliquity in Figure 2.26 is dominated by the 18.6 year period, as well. The amplitude is about $\Delta\Psi(18.6\,\mathrm{y}) \approx 9.2''$. The 18.6 year terms in Figures 2.25 and 2.26 are caused by the regression of the node of the lunar orbit which in turn implies the periodic variation of the inclination of the lunar orbit w.r.t. the equator between the limits $\tilde{i} = 23.5° \pm 5.14°$ (see Figure 2.12).

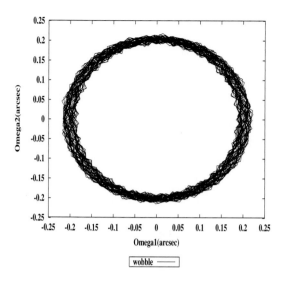

Fig. 2.23. Polar wobble 1981-2020 with torques

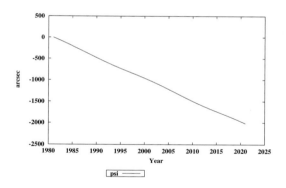

Fig. 2.24. Precession and nutation in longitude 1981-2020 (without daily terms)

Figure 2.27 shows the development of the third Euler angle Θ_{δ} over forty years – after having removed the linear drift $\omega_{\delta_0}(t - t_0)$. The resemblance between Figures 2.27 and 2.25 is striking: When reducing Figure 2.25 by the factor $\cos \varepsilon_{\delta_0}$ one obtains a Figure which is identical with Figure 2.27.

Apart from the 18.6 year period one may clearly observe terms with a semi-annual period in Figures 2.25, 2.26, and 2.27. The resolution is not good enough to detect other periods in these Figures.

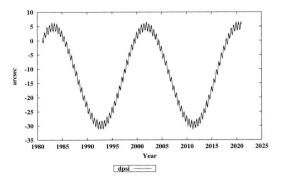

Fig. 2.25. Nutation in longitude 1981-2020 (without daily terms)

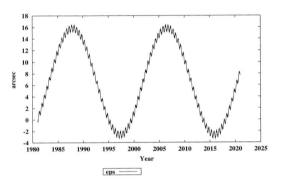

Fig. 2.26. Nutation in obliquity 1981-2020 (without daily terms)

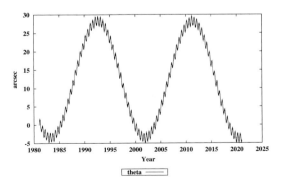

Fig. 2.27. Θ_{δ} 1981-2020 (without daily terms)

A spectral analysis gives more insight. In order to get a fair estimate of the amplitudes of longer periods (in particular the 18.6 year period) we use a time series of 1000 years with a spacing of two days between observation epochs to produce the spectra. Table 2.4 contains the more important terms (up to terms with amplitudes $> 0.01''$) as obtained from the spectral analysis and compares them to the terms officially used in astronomy, the IAU (1980) model of nutation (see Explanatory Supplement [107]). Unnecessary to say that there are many more terms in the spectra. It should be pointed out that periods are found empirically in the spectra (this implies that the numerical value for the longer periods are not so well established in the spectra) whereas the periods follow from the periods of the five so-called fundamental arguments of nutation theory.

Table 2.4. Precession and nutation in longitude and obliquity

Spectral Analysis				Official Values		
Ψ_{δ}		ε_{δ}		Ψ_{δ}		ε_{δ}
P [d]	Amp [$''$]	P [d]	Amp [$''$]	P [d]	Amp [$''$]	Amp [$''$]
∞	50.5731				50.41	
6798.8024	17.3819	6795.6045	9.2673	6798.4	17.1996	9.2025
3397.9567	0.2097	3401.0307	0.0923	3399.2	0.2062	0.0089
365.3278	0.1099	364.9640	0.0089	365.3	0.2426	0.0054
182.6294	1.2722	182.6296	0.5509	182.6	1.3187	0.5736
177.8908	0.0104	177.8869	0.0058	177.8	0.0129	0.0070
121.7441	0.0487	121.7426	0.0211	121.7	0.0517	0.0224
31.8120	0.0149			31.8	0.0158	0.0001
27.5544	0.0677			27.6	0.0712	0.0007
27.0923	0.0114			27.1	0.0123	0.0053
13.6607	0.2044	13.6607	0.0886	13.7	0.2274	0.0977
9.1329	0.0262	9.1329	0.0113	9.1	0.0301	0.0129

It is amazing that the values obtained by our simple simulation are so close to the official values – which were all obtained from analytical series developments. The differences are (must be) mainly due to the differences between the rigid body and elastic models for Earth rotation, an issue to be addressed in section 2.3.

Forced Motion: Analytical Developments. We already presented the analytical solution of the homogeneous equations in the absence of torques. The solution is the circular motion represented by eqns. (2.14). For the subsequent developments it is preferable to write these equations in the form

$$\omega_{\check{o}_1}(t) = \omega_{\check{o}10} \cos\tilde{\omega}_{\check{o}}t - \omega_{\check{o}20} \sin\tilde{\omega}_{\check{o}}t$$
$$\omega_{\check{o}_2}(t) = \omega_{\check{o}10} \sin\tilde{\omega}_{\check{o}}t + \omega_{\check{o}20} \cos\tilde{\omega}_{\check{o}}t \ ,$$

$$(2.26)$$

where

$$\omega_{\check{o}10} \overset{\text{def}}{=} \omega_{\check{o}_1}(0)$$
$$\omega_{\check{o}20} \overset{\text{def}}{=} \omega_{\check{o}_2}(0)$$

$$(2.27)$$

are the initial values of the solution vector at $t = 0$.

It is a standard technique in mathematics to derive the solution of an inhomogeneous system of type (I- 3.124) by starting from the homogeneous solution using the method of *variation of constants*:

$$\omega_{\check{o}_1}(t) = \omega_{\check{o}10}(t) \cos\tilde{\omega}_{\check{o}}t - \omega_{\check{o}20}(t) \sin\tilde{\omega}_{\check{o}}t$$
$$\omega_{\check{o}_2}(t) = \omega_{\check{o}10}(t) \sin\tilde{\omega}_{\check{o}}t + \omega_{\check{o}20}(t) \cos\tilde{\omega}_{\check{o}}t \ .$$

$$(2.28)$$

Introducing the above equations into the inhomogeneous equations (I- 3.124), which may be written in abbreviated form as

$$\begin{pmatrix} \dot{\omega}_{\check{o}_1} \\ \dot{\omega}_{\check{o}_2} \\ \dot{\omega}_{\check{o}_3} \end{pmatrix} + \begin{pmatrix} +\tilde{\omega}_{\check{o}}\,\omega_{\check{o}_2} \\ -\tilde{\omega}_{\check{o}}\,\omega_{\check{o}_1} \\ 0 \end{pmatrix} = \boldsymbol{\ell}_{\mathbb{C}\check{o}} + \boldsymbol{\ell}_{\odot\check{o}} \overset{\text{def}}{=} \boldsymbol{\ell}_{\check{o}} \ ,$$

$$(2.29)$$

where the components of the torques $\boldsymbol{\ell}_{\mathbb{C}\check{o}}$ and $\boldsymbol{\ell}_{\odot\check{o}}$ have been already specified on the right-hand sides of equations (I- 3.124), leads to the following condition equations:

$$\dot{\omega}_{\check{o}10} = + \cos\tilde{\omega}_{\check{o}}t \ \ell_{\check{o}_1} + \sin\tilde{\omega}_{\check{o}}t \ \ell_{\check{o}_2}$$
$$\dot{\omega}_{\check{o}20} = - \sin\tilde{\omega}_{\check{o}}t \ \ell_{\check{o}_1} + \cos\tilde{\omega}_{\check{o}}t \ \ell_{\check{o}_2} \ .$$

$$(2.30)$$

The above equations may be solved easily by (numerical or analytical) quadrature:

$$\omega_{\check{o}10}(t) = \omega_{\check{o}10}(0) + \int_0^t \cos\tilde{\omega}_{\check{o}}t' \ \ell_{\check{o}_1}(t') \ dt' + \int_0^t \sin\tilde{\omega}_{\check{o}}t' \ \ell_{\check{o}_2}(t') \ dt'$$

$$(2.31)$$

$$\omega_{\check{o}20}(t) = \omega_{\check{o}20}(0) - \int_0^t \sin\tilde{\omega}_{\check{o}}t' \ \ell_{\check{o}_1}(t') \ dt' + \int_0^t \cos\tilde{\omega}_{\check{o}}t' \ \ell_{\check{o}_2}(t') \ dt' \ .$$

What remains to be done before approximately solving the above integrals analytically is the calculation of the components $\ell_{\check{o}_i}(t')$, $i = 1, 2$, in the Earth-fixed system. Let us use the following approximations:

- The geocentric orbits of Sun and Moon are assumed to be circles,
- The orbital planes of Sun and Moon are assumed to be the ecliptic,
- The Euler angle Ψ_δ is approximated by $\Psi_\delta \approx 0$, ε_δ is approximated by $\varepsilon_\delta \stackrel{\text{def}}{=} \varepsilon_0 \approx 23.5°$.

With the above simplifying assumptions we will not be able to "see" the main terms of nutation. We should, however, be able to explain the quasi-daily Oppolzer motion (see Figure 2.19) and to calculate the precession constant as observed in Figure 2.24. Obviously it is possible to compute separately the torques for Sun and Moon. For the geocentric coordinates of the Moon we obtain in the Earth-fixed system:

$$
\begin{aligned}
\boldsymbol{r}_{\mathbb{C}} &= a_{\mathbb{C}}\, \mathbf{R}_3(\Theta)\, \mathbf{R}_1(-\varepsilon_0) \begin{pmatrix} \cos u_{\mathbb{C}} \\ \sin u_{\mathbb{C}} \\ 0 \end{pmatrix} \\
&= a_{\mathbb{C}} \begin{pmatrix} +\cos\Theta_\delta \cos u_{\mathbb{C}} + \sin\Theta_\delta \sin u_{\mathbb{C}} \cos\varepsilon_0 \\ -\sin\Theta_\delta \cos u_{\mathbb{C}} + \cos\Theta_\delta \sin u_{\mathbb{C}} \cos\varepsilon_0 \\ \sin\varepsilon_0 \sin u_{\mathbb{C}} \end{pmatrix},
\end{aligned}
\tag{2.32}
$$

where $u_{\mathbb{C}}$ is the (mean and true) ecliptic longitude of the Moon (under the simplified assumptions stated above) and $a_{\mathbb{C}}$ is the radius of the Moon's circular orbit. Designating the Sun's geocentric ecliptical longitude with $u_{\odot}$ and its semi-major axis as $a_{\odot}$ the components of the solar radius vector may be written in the Earth-fixed system as:

$$
\boldsymbol{r}_{\odot} = a_{\odot} \begin{pmatrix} +\cos\Theta_\delta \cos u_{\odot} + \sin\Theta_\delta \sin u_{\odot} \cos\varepsilon_0 \\ -\sin\Theta_\delta \cos u_{\odot} + \cos\Theta_\delta \sin u_{\odot} \cos\varepsilon_0 \\ \sin\varepsilon_0 \sin u_{\odot} \end{pmatrix}.
\tag{2.33}
$$

Using the representation on the right-hand side of eqns. (I- 3.124) the components of the torques in the Earth-fixed system may be written as:

$$
\begin{aligned}
\boldsymbol{l}_{\mathbb{C}\delta} &= -\frac{3\,G\,m_{\mathbb{C}}\,\gamma_\delta \sin\varepsilon_0}{2\,a_{\mathbb{C}}^3} \begin{pmatrix} \sin\Theta_\delta \sin 2u_{\mathbb{C}} - \cos\Theta_\delta(1-\cos 2u_{\mathbb{C}})\cos\varepsilon_0 \\ \cos\Theta_\delta \sin 2u_{\mathbb{C}} + \sin\Theta_\delta(1-\cos 2u_{\mathbb{C}})\cos\varepsilon_0 \\ 0 \end{pmatrix} \\
&\approx -\frac{3\,G\,m_{\mathbb{C}}\,\gamma_\delta \sin\varepsilon_0}{2\,a_{\mathbb{C}}^3} \begin{pmatrix} -\cos\varepsilon_0 \cos\Theta_\delta + \cos(\Theta_\delta - 2u_{\mathbb{C}}) \\ +\cos\varepsilon_0 \sin\Theta_\delta - \sin(\Theta_\delta - 2u_{\mathbb{C}}) \\ 0 \end{pmatrix}.
\end{aligned}
\tag{2.34}
$$

The approximation is relatively crude, as the term $2\sin^2\varepsilon_0 \approx 0.08$ was neglected. In the same approximation, the corresponding result for the torque produced by the Sun reads as

$$
\boldsymbol{l}_{\odot\delta} = -\frac{3\,G\,m_{\odot}\,\gamma_\delta \sin\varepsilon_0}{2\,a_{\odot}^3} \begin{pmatrix} -\cos\varepsilon_0 \cos\Theta_\delta + \cos(\Theta_\delta - 2u_{\odot}) \\ +\cos\varepsilon_0 \sin\Theta_\delta - \sin(\Theta_\delta - 2u_{\odot}) \\ 0 \end{pmatrix}.
\tag{2.35}
$$

If we introduce the latter two results into the equations (2.31) we obtain the final result

$$\omega_{\tilde{o}10}(t) = \omega_{\tilde{o}10}(0) + \int_0^t \cos\tilde{\omega}_{\tilde{o}}t' \, \ell_{\tilde{o}_1}(t') \, dt' + \int_0^t \sin\tilde{\omega}_{\tilde{o}}t' \, \ell_{\tilde{o}_2}(t') \, dt'$$

$$= \omega_{\tilde{o}10}(0) + \frac{3Gm_{\mathbb{C}}\gamma_{\tilde{o}}\sin\varepsilon_0}{2a_{\mathbb{C}}^3\omega^*}\left[-\cos\varepsilon_0\sin\Theta_{\tilde{o}0} + \frac{\omega^*}{\omega^* - 2n_{\mathbb{C}}}\sin(\Theta_{\tilde{o}0} - 2u_{\mathbb{C}0})\right]$$

$$+ \frac{3Gm_{\odot}\gamma_{\tilde{o}}\sin\varepsilon_0}{2a_{\odot}^3\omega^*}\left[-\cos\varepsilon_0\sin\Theta_{\tilde{o}0} + \frac{\omega^*}{\omega^* - 2n_{\odot}}\sin(\Theta_{\tilde{o}0} - 2u_{\odot0})\right]$$

$$- \frac{3Gm_{\mathbb{C}}\gamma_{\tilde{o}}\sin\varepsilon_0}{2a_{\mathbb{C}}^3\omega^*}\left[-\cos\varepsilon_0\sin(\tilde{\omega}_{\tilde{o}}t + \Theta_{\tilde{o}}) + \frac{\omega^*}{\omega^* - 2n_{\mathbb{C}}}\sin(\tilde{\omega}_{\tilde{o}}t + \Theta_{\tilde{o}} - 2u_{\mathbb{C}})\right]$$

$$- \frac{3Gm_{\odot}\gamma_{\tilde{o}}\sin\varepsilon_0}{2a_{\odot}^3\omega^*}\left[-\cos\varepsilon_0\sin(\tilde{\omega}_{\tilde{o}}t + \Theta_{\tilde{o}}) + \frac{\omega^*}{\omega^* - 2n_{\odot}}\sin(\tilde{\omega}_{\tilde{o}}t + \Theta_{\tilde{o}} - 2u_{\odot})\right]$$

$$(2.36)$$

$$\omega_{\tilde{o}20}(t) = \omega_{\tilde{o}20}(0) - \int_0^t \sin\tilde{\omega}_{\tilde{o}}t' \, \ell_{\tilde{o}_1}(t') \, dt' + \int_0^t \cos\tilde{\omega}_{\tilde{o}}t' \, \ell_{\tilde{o}_2}(t') \, dt'$$

$$= \omega_{\tilde{o}20}(0) + \frac{3Gm_{\mathbb{C}}\gamma_{\tilde{o}}\sin\varepsilon_0}{2a_{\mathbb{C}}^3\omega^*}\left[-\cos\varepsilon_0\cos\Theta_{\tilde{o}0} + \frac{\omega^*}{\omega^* - 2n_{\mathbb{C}}}\cos(\Theta_{\tilde{o}0} - 2u_{\mathbb{C}0})\right]$$

$$+ \frac{3Gm_{\odot}\gamma_{\tilde{o}}\sin\varepsilon_0}{2a_{\odot}^3\omega^*}\left[-\cos\varepsilon_0\cos\Theta_{\tilde{o}0} + \frac{\omega^*}{\omega^* - 2n_{\odot}}\cos(\Theta_{\tilde{o}0} - 2u_{\odot0})\right]$$

$$- \frac{3Gm_{\mathbb{C}}\gamma_{\tilde{o}}\sin\varepsilon_0}{2a_{\mathbb{C}}^3\omega^*}\left[-\cos\varepsilon_0\cos(\tilde{\omega}_{\tilde{o}}t + \Theta_{\tilde{o}}) + \frac{\omega^*}{\omega^* - 2n_{\mathbb{C}}}\cos(\tilde{\omega}_{\tilde{o}}t + \Theta_{\tilde{o}} - 2u_{\mathbb{C}})\right]$$

$$- \frac{3Gm_{\odot}\gamma_{\tilde{o}}\sin\varepsilon_0}{2a_{\odot}^3\omega^*}\left[-\cos\varepsilon_0\cos(\tilde{\omega}_{\tilde{o}}t + \Theta_{\tilde{o}}) + \frac{\omega^*}{\omega^* - 2n_{\odot}}\cos(\tilde{\omega}_{\tilde{o}}t + \Theta_{\tilde{o}} - 2u_{\odot})\right],$$

where $\omega^* \stackrel{\text{def}}{=} (1 + \gamma_{\tilde{o}})\,\omega_{\tilde{o}0}$, $u_0 = u(0)$, $u_{\odot0} = u_{\odot}(0)$ and $\Theta_{\tilde{o}0} = \Theta_{\tilde{o}}(0)$.

The final solution of the inhomogeneous equations are obtained by substituting the above equations into eqn. (2.28):

$$\omega_{\tilde{o}1}(t) = \rho_{\tilde{o}}\cos(\gamma_{\tilde{o}}\,\tilde{\omega}_0 t - \alpha_{\tilde{o}0})$$

$$+ \frac{3\,G\,m_{\mathbb{C}}\,\gamma_{\tilde{o}}\,\sin\varepsilon_0}{2\,a_{\mathbb{C}}^3\,\omega^*}\left[\cos\varepsilon_0\sin\Theta_{\tilde{o}} - \frac{\omega^*}{\omega^* - 2\,n_{\mathbb{C}}}\sin(\Theta_{\tilde{o}} - 2\,u_{\mathbb{C}})\right]$$

$$+ \frac{3\,G\,m_{\odot}\,\gamma_{\tilde{o}}\,\sin\varepsilon_0}{2\,a_{\odot}^3\,\omega^*}\left[\cos\varepsilon_0\sin\Theta_{\tilde{o}} - \frac{\omega^*}{\omega^* - 2\,n_{\odot}}\sin(\Theta_{\tilde{o}} - 2\,u_{\odot})\right]$$

$$\omega_{\delta_2}(t) = \rho_\delta \sin(\gamma_\delta \, \omega_0 t - \alpha_{\delta_0})$$
$$+ \frac{3\,G\,m_{\mathbb{C}}\,\gamma_\delta\,\sin\varepsilon_0}{2\,a_{\mathbb{C}}^3\,\omega^*}\left[\cos\varepsilon_0\cos\Theta_\delta - \frac{\omega^*}{\omega^* - 2\,n_{\mathbb{C}}}\cos(\Theta_\delta - 2\,u_{\mathbb{C}})\right]$$
$$+ \frac{3\,G\,m_{\odot}\,\gamma_\delta\,\sin\varepsilon_0}{2\,a_{\odot}^3\,\omega^*}\left[\cos\varepsilon_0\cos\Theta_\delta - \frac{1}{\omega^* - 2\,n_{\odot}}\cos(\Theta_\delta - 2\,u_{\odot})\right].$$
$$(2.37)$$

The above equations show that under the influence of the torques exerted by Moon and Sun the PM consists of the prograde motion with the Euler period of about 303 days which is superposed by three retrograde nearly-daily terms. An inspection of the brackets [...] shows that the amplitude of the lunar Oppolzer contributions vary between (practically) zero and twice the amplitude given by the pre-factor of the brackets within half a synodical month and that the amplitude of the solar Oppolzer contribution varies with an semiannual period between (practically) zero and twice the amplitude given by the pre-factor of the brackets related to the solar term.

The prominent terms of the Oppolzer motion are ω_0 (first terms of solar and lunar contributions), $\omega_0 - 2n_{\mathbb{C}}$ (second lunar term) and $\omega_0 - 2n_{\odot}$ (second solar term). Table 2.5 contains the theoretical values and the values obtained in the spectral analysis. In view of the above approximations the agreement is more than satisfactory. The precession constant is obtained by introducing

Table 2.5. Periods and amplitudes of Oppolzer motion

Spectral Analysis			Theoretical	
P [d]	Amp [$''$]	Ang. vel.	P [d]	Amp [$''$]
-0.9973	0.0087	ω_δ	-0.9973	0.0090
-1.0758	0.0066	$\omega_\delta - 2n_{\mathbb{C}}$	-1.0758	0.0068
-1.0027	0.0029	$\omega_\delta - 2n_{\odot}$	-1.0027	0.0030

the solution (2.37) into Euler's kinematic equation for the Euler angle Ψ_δ (see eqns. (I- 3.68)):

$$\dot{\Psi}_\delta = \left(-\sin\Theta_\delta\,\omega_{\delta_1} - \cos\Theta_\delta\,\omega_{\delta_2}\right)\csc\varepsilon_\delta. \qquad (2.38)$$

Obviously, the contributions $\sim \sin\Theta_\delta$ and $\sim \cos\Theta_\delta$ give rise to a secular term, the lunisolar precession:

$$\dot{\Psi}_\delta = -\frac{3\,G\,\gamma_\delta\,\cos\varepsilon_0}{2\,\omega_\delta\,(1+\gamma_\delta)}\left\{\frac{m_{\mathbb{C}}}{a_{\mathbb{C}}^3} + \frac{m_{\odot}}{a_{\odot}^3}\right\} \approx -50.9''/\text{year}. \qquad (2.39)$$

The value corresponds pretty well with the actual value for the lunisolar precession constant. Mean values were used for $a_{\mathbb{C}}$ and $a_{\odot}$. Unnecessary to say that each planet in the planetary system does correspond to the precession "constant" according to the pattern indicated by formula (2.39). (The semi-major axes would have to be replaced by the mean geocentric distances of the planets).

Observe that all quantities in eqn. (2.39), except for γ_δ, are very accurately known. Equation (2.39) may therefore be viewed as the defining equation for γ_δ.

2.2.3 Rotation of the Moon

History. The basic properties of lunar rotation were detected empirically. Giovanni Domenico Cassini (1625–1712) was the first to state the following laws:

1. The Moon rotates counterclockwise (when seen from the North pole) with constant angular velocity about a fixed axis; the sidereal rotation period is the same as the sidereal revolution period of the Moon in its orbit around the Earth.

2. The inclination of the Moon's orbital plane w.r.t. the ecliptic is constant.

3. The rotation axis of the Moon, the pole of the ecliptic, and the pole of the Moon's orbital plane lie (in this sequence) in one and the same plane.

Cassini, for a long time director of the famous Observatoire de Paris, found these laws towards the end of the 17th century by analyzing long series of observations.

Free Motion. When analyzing the free rotation of the Moon one may no longer assume rotational symmetry of the body (see lunar principal moments of inertia, Table 2.1). Figure 2.28 illustrates the effect of lunar PM over the time interval of 140 years (the figure was generated with program ERDROT using the option of lunar rotation by setting the masses of both, Earth and Sun, to zero). Initially, the rotation pole of the Moon was assumed to be at $\omega_{\mathbb{C}_1} = 0.2''$, $\omega_{\mathbb{C}_2} = 0.0''$ relative to the Moon's axis of maximum moment of inertia. The resulting motion is very similar (at first sight) with the corresponding result of the Earth's rotation (compare Figure 2.16). A closer inspection shows, however, that the Moon's rotation pole is not moving on a circle around the Moon's axis of maximum moment of inertia, but rather on an ellipse with an axis ratio of $4 : 5$, the smaller axis pointing to the positive $\omega_{\mathbb{C}_1}$ direction. Obviously the Euler period of the lunar PM is slightly longer than 140 years.

The free motion of the rotation axis of the Moon may be easily explained analytically: Setting in eqns. (I- 3.124) all torques to zero we may write for the third component of the lunar angular velocity vector:

$$\dot{\omega}_{\mathbb{C}_3} + \gamma_{\mathbb{C}_3}\, \omega_{\mathbb{C}_1}\, \omega_{\mathbb{C}_2} \approx \dot{\omega}_{\mathbb{C}_3} = 0 \;, \tag{2.40}$$

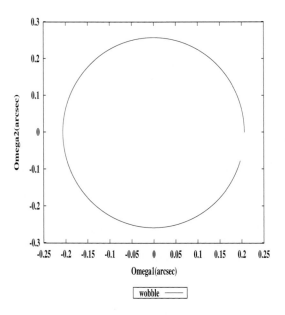

Fig. 2.28. Free motion of the Moon's rotation pole over 140 years (assuming an initial angle of 0.2″ between rotation pole and axis of maximum moment of inertia)

because the two components $\omega_{\mathbb{C}_1}$ and $\omega_{\mathbb{C}_2}$ of PM of the Moon are small quantities. In view of the fact that $\gamma_{\mathbb{C}_3}$ is a small quantity as well (see Table 2.1), the approximation neglects "only" a small quantity of order 3. The above equation is solved by

$$\omega_{\mathbb{C}_3} = \omega_{\mathbb{C}_0} = \text{const.} . \tag{2.41}$$

In the absence of torques the rotation of the Moon therefore would be as regular as that of the Earth. From eqns. (I- 3.124) we obtain the following linear differential equation system for the Moon's PM:

$$\begin{aligned}
\dot{\omega}_{\mathbb{C}_1} + \gamma_{\mathbb{C}_1} \omega_{\mathbb{C}_0} \omega_{\mathbb{C}_2} &= 0 \\
\dot{\omega}_{\mathbb{C}_2} + \gamma_{\mathbb{C}_2} \omega_{\mathbb{C}_0} \omega_{\mathbb{C}_1} &= 0 .
\end{aligned} \tag{2.42}$$

One easily verifies that the above system is solved by

$$\begin{aligned}
\omega_{\mathbb{C}_1}(t) &= \rho_{\mathbb{C}_1} \cos(\tilde{\omega}_{\mathbb{C}} t + \alpha_{\mathbb{C}_0}) \\
\omega_{\mathbb{C}_2}(t) &= \rho_{\mathbb{C}_2} \sin(\tilde{\omega}_{\mathbb{C}} t + \alpha_{\mathbb{C}_0}) ,
\end{aligned} \tag{2.43}$$

where

$$\tilde{\omega}_\mathbb{C} = \sqrt{-\gamma_{\mathbb{C}1}\,\gamma_{\mathbb{C}2}}\ \omega_{\mathbb{C}0}$$

$$\rho_{\mathbb{C}1} = \frac{\rho_\mathbb{C}}{\sqrt{-\gamma_{\mathbb{C}2}}}$$

$$\rho_{\mathbb{C}2} = \frac{\rho_\mathbb{C}}{\sqrt{+\gamma_{\mathbb{C}1}}}\ . \tag{2.44}$$

The actual amplitude $\rho_\mathbb{C}$ and the phase angle $\alpha_{\mathbb{C}0}$ are defined by the initial state of the system.

The above results tell that the rotation pole of the rigid Moon moves with constant angular velocity on an ellipse around the pole of maximum moment of inertia – which corresponds exactly to the result observed in Figure 2.28. The period of the free lunar PM and its constant angular velocity are uniquely defined by the angular velocity of lunar rotation and by the three principal moments of inertia. Using the values in Table 2.1 we obtain the following parameters characterizing the "lunar PM":

$$P_\mathbb{C} = \frac{2\,\pi}{\sqrt{-\gamma_{\mathbb{C}1}\,\gamma_{\mathbb{C}2}}\ \omega_{\mathbb{C}0}} \approx 147 \text{ years}$$

$$\frac{\rho_{\mathbb{C}1}}{\rho_{\mathbb{C}2}} = \sqrt{\frac{+\gamma_{\mathbb{C}1}}{-\gamma_{\mathbb{C}2}}} \approx \frac{4}{5}\ . \tag{2.45}$$

A firm establishment of the Moon's free PM (in particular its period and axes) would tell quite a lot about the satellite's physical properties. Assuming that the amplitude of the free lunar PM is of the order of one arcsecond (or a fraction thereof), the detection of the free motion using observations made from Earth is rather difficult: One arcsecond on the surface of the Moon corresponds to an angle of about $1''\ \frac{R_\mathbb{C}}{a_\mathbb{C}} \approx 0.0045''$ (where $R_\mathbb{C}$ is the radius of the Moon, $a_\mathbb{C}$ its mean distance from Earth) which might be observed from the Earth. It is therefore not amazing that the free motion of the lunar rotation axis could not yet be firmly established – not even using the results of Lunar Laser Ranging (it is extremely difficult to isolate a signal of a small amplitude with a long period of about 147 years with observation series).

How do the Moon's figure and rotation axes move in inertial space? This motion is slightly more complicated than the corresponding motion of the Earth, because rotational symmetry is not an adequate approximation. Euler's kinematic equations (I- 3.68) relating the components of $\boldsymbol{\omega}_\mathbb{C}$ to the Euler angles, when applied to the Moon, provide the differential equations for the Euler angles describing the Moon's figure axis in inertial space:

$$\dot{\Psi}_{\mathbb{C}} = -\{\omega_{\mathbb{C}_1}\sin\Theta_{\mathbb{C}} + \omega_{\mathbb{C}_2}\cos\Theta_{\mathbb{C}}\}\csc\varepsilon_{\mathbb{C}}$$
$$\dot{\varepsilon}_{\mathbb{C}} = -\{\omega_{\mathbb{C}_1}\cos\Theta_{\mathbb{C}} - \omega_{\mathbb{C}_2}\sin\Theta_{\mathbb{C}}\} \tag{2.46}$$
$$\dot{\Theta}_{\mathbb{C}} = +\{\omega_{\mathbb{C}_1}\sin\Theta_{\mathbb{C}} + \omega_{\mathbb{C}_2}\cos\Theta_{\mathbb{C}}\}\cot\varepsilon_{\mathbb{C}} + \omega_{\mathbb{C}_0} .$$

Substituting the results (2.41) and (2.43) into the above equations leads to the following differential equations for the Euler angles related to the Moon:

$$\dot{\Psi}_{\mathbb{C}} = -\rho_{\mathbb{C}_2}\left\{\frac{\rho_{\mathbb{C}_1}}{\rho_{\mathbb{C}_2}}\cos\tilde{\omega}_{\mathbb{C}}t\,\sin\Theta_{\mathbb{C}} + \sin\tilde{\omega}_{\mathbb{C}}t\,\cos\Theta_{\mathbb{C}}\right\}\csc\varepsilon_{\mathbb{C}}$$
$$\dot{\varepsilon}_{\mathbb{C}} = -\rho_{\mathbb{C}_2}\left\{\frac{\rho_{\mathbb{C}_1}}{\rho_{\mathbb{C}_2}}\cos\tilde{\omega}_{\mathbb{C}}t\,\cos\Theta_{\mathbb{C}} - \sin\tilde{\omega}_{\mathbb{C}}t\,\sin\Theta_{\mathbb{C}}\right\} \tag{2.47}$$
$$\dot{\Theta}_{\mathbb{C}} = +\rho_{\mathbb{C}_2}\left\{\frac{\rho_{\mathbb{C}_1}}{\rho_{\mathbb{C}_2}}\cos\tilde{\omega}_{\mathbb{C}}t\,\sin\Theta_{\mathbb{C}} + \sin\tilde{\omega}_{\mathbb{C}}t\,\cos\Theta_{\mathbb{C}}\right\}\cot\varepsilon_{\mathbb{C}} + \omega_{\mathbb{C}_0} .$$

For the sake of simplicity $\alpha_{\mathbb{C}_0}$ was set to zero, which would correspond to a particular initial epoch.

The solution of these equations is slightly complicated by the fact that $\frac{\rho_{\mathbb{C}_1}}{\rho_{\mathbb{C}_2}} \neq 1$. Apart from that, the first order solution in $\rho_{\mathbb{C}_2}$ follows the same pattern as that of eqns. (2.18). The third equation is separated from the first two of the above equations, provided we approximate $\varepsilon_{\mathbb{C}}$ by a suitable mean value on the right-hand side of the third of the above equations. The zero-order solution of this third equation simply is

$$\Theta_{\mathbb{C}}(t) = \omega_{\mathbb{C}_0}t . \tag{2.48}$$

Using this approximation in eqns. (2.47) leads to the first-order approximations $\Psi_{\mathbb{C}}(t)$, $\varepsilon_{\mathbb{C}}(t)$, and $\Theta_{\mathbb{C}}(t)$. As opposed to the corresponding equations for the Earth, we do not only have terms with $(1 + \sqrt{-\gamma_{\mathbb{C}_1}\gamma_{\mathbb{C}_2}})\omega_{\mathbb{C}_0}t$ but also with $(1 - \sqrt{-\gamma_{\mathbb{C}_1}\gamma_{\mathbb{C}_2}})\omega_{\mathbb{C}_0}t$ as time argument in the sin- and cos-functions. The derivation of the exact form of the first-order solution may be left to the interested reader.

Forced Motion: Simulations. Figure 2.29 shows the lunar PM under the inclusion of external torques over the time interval of one year. When comparing this figure to the corresponding Figure 2.19 of Earth rotation one observes significant differences: Whereas one still could see very well the free constituent in Figure 2.19, lunar PM in Figure 2.29 is completely governed by the forced motion. As opposed to effects of the order of fractions of arcseconds in Figure 2.28, we observe effects of the order of about one hundred arcseconds and one gets the impression that there is a secular trend in the lunar PM. The pattern also seems to be rather strange: the periodic part of the signal shows rather abrupt changes.

Figure 2.30 which shows the two components of Figure 2.29 as a function of time explains the strange behavior: whereas the first component of the

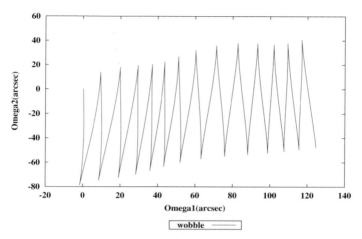

Fig. 2.29. Forced motion of the Moon's rotation pole over one year

Moon's angular velocity vector shows a rather smooth almost secular trend, a strong periodic signal with an amplitude of about $40''$ dominates the second component.

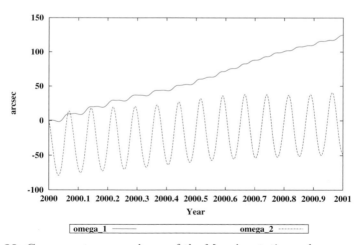

Fig. 2.30. Components $\omega_{\mathbb{C}_1}$ and $\omega_{\mathbb{C}_2}$ of the Moon's rotation pole over one year

Figure 2.31 contains the same information as Figure 2.29, but over the much longer time interval of 140 years. The figure indicates that there is no (obvious) secular drift in the lunar PM. The amplitudes of the variations are about 220″ in the first and about 350″ in the second component.

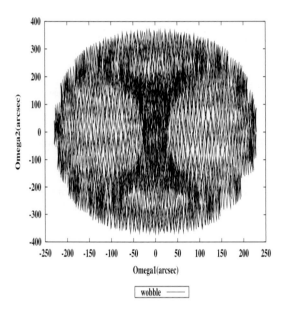

Fig. 2.31. Forced motion of the Moon's rotation pole over 140 years

The example of the rotation of a rotationally symmetric Earth has shown, that the (third component of the) angular velocity is constant. If the rotation axis is close to the figure axis of the body considered, this also implies that the sidereal rotation period of the rigid body is (almost perfectly) constant. Figure 2.32, actually illustrating the variations of the third component of the Moon's angular velocity vector, prove that the rotation period of the Moon shows rather pronounced periodic variations: The rotation period varies by ±0.05%, corresponding to maximum differences about 20 minutes. The period associated with these variations is about 2.8 years.

It is well known that

- due to the (rather small) irregularities of lunar rotation,

- due to the fact that the lunar equator is inclined by $23.5° ± (5.5° + 1.5°)$ w.r.t. Earth's equator, and

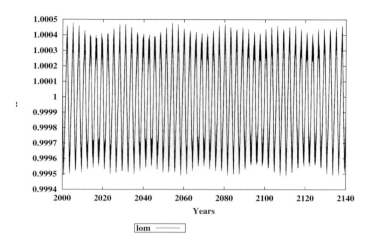

Fig. 2.32. Relative variation of lunar angular velocity

- due to the fact that the Moon revolves with variable velocity around the Earth

we can observe about 58 % of the Moon's surface from the Earth despite the coupling of rotation and revolution periods. Figures 2.33 and 2.34 illustrate the combined effect in a Moon-fixed coordinate system. Figure 2.33 shows

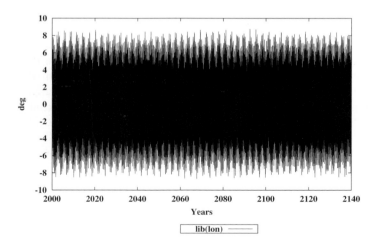

Fig. 2.33. Lunar libration in lunar longitude

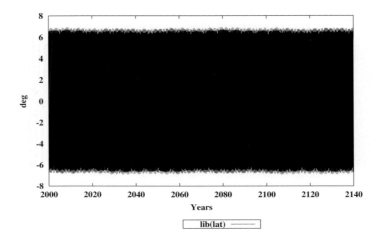

Fig. 2.34. Lunar libration in lunar latitude

the selenocentric longitude difference of the pierce point of the line pointing
from the geo- to the selenocenter w.r.t. the Moon's principal axis of inertia,
Figure 2.34 gives the lunar latitude difference of the same two points. The
amplitude of the variations is in both cases on the average about 6.5°, the
reason is, however, different. The libration in longitude is mainly caused by
the Moon's "irregular" revolution, whereas the libration in latitude is mainly
caused by the inclination of the lunar equator w.r.t. the Earth's equator.

Note that the pierce point used in Figures 2.33 and 2.34 is in principle not
needed to describe either Earth or lunar rotation. The geocentric orbit of the
Moon and the three Euler angles of lunar rotation are needed to calculate the
angles shown in Figure 2.34. It is customary to designate the contribution
caused by the orbital motion of the Moon as *optical libration*, that due to
the rotational motion as *physical libration*. As already mentioned, optical
libration is the larger constituent. *Cum grano salis* one might say that Figures
2.33 and 2.34 illustrate optical libration.

It should be noted that the initial conditions for the three Euler angles $\Psi_{\mathbb{C}_0}$,
$\varepsilon_{\mathbb{C}_0}$, and $\Theta_{\mathbb{C}_0}$ were set only approximately in program ERDROT by assuming
that a "mean Moon" obeys Cassini's laws. The osculating elements of this
mean Moon are identical with the true osculating elements with the exception
of the initial eccentricity which was forced to the value of $e_{\mathbb{C}_0} \stackrel{\text{def}}{=} 0$.

Figure 2.35 shows the development of the lunar nutation angle $\Psi_{\mathbb{C}}(t)$ in (ecliptical) longitude. The drift of $360°/18.6$ years is, as stated by Cassini's laws,

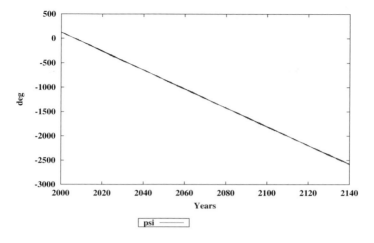

Fig. 2.35. Lunar precession in longitude (angle $\Psi_{\mathbb{C}}$)

identical with the regression period of the lunar node in the ecliptic (see also Figure 2.9).

Figure 2.36 shows the lunar nutation in obliquity. The variations are relatively small (peak to peak about $0.2°$), but they show a relatively complex pattern. A spectral analysis would reveal the relevant periods. This analysis may be left to the reader.

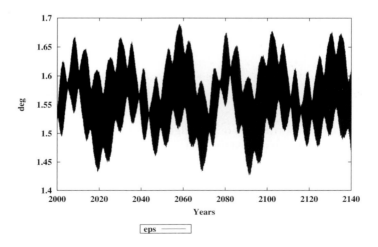

Fig. 2.36. Lunar nutation in obliquity (angle $\varepsilon_{\mathbb{C}}$)

The development of the sum of nutation in longitude and the angle $\Theta_{\mathbb{C}}$ (reduced by the mean angular motion of the orbital motion) illustrates the well known fact that rotation and revolution periods of the Moon agree on the average. The periodic variation with an amplitude of about $0.8°$ and a period of about 2.8 years may be explained by the lunar rotation (see next paragraph).

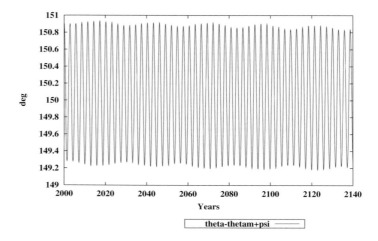

Fig. 2.37. Physical libration in longitude $l_{\mathrm{ph}} \overset{\mathrm{def}}{=} \Psi_{\mathbb{C}} + \Theta_{\mathbb{C}} - n_{\mathbb{C}_0}(t - t_0)$

Forced Motion: Analytical Developments. Considering only the torque exerted by the Earth and neglecting small terms of second and higher order, the equations (I-3.124) for the rotation of the Moon read as

$$
\begin{pmatrix} \dot{\omega}_{\mathbb{C}_1} \\ \dot{\omega}_{\mathbb{C}_2} \\ \dot{\omega}_{\mathbb{C}_3} \end{pmatrix} + \begin{pmatrix} \gamma_{\mathbb{C}_1}\, \omega_{\mathbb{C}_2}\, \omega_{\mathbb{C}_3} \\ \gamma_{\mathbb{C}_2}\, \omega_{\mathbb{C}_3}\, \omega_{\mathbb{C}_1} \\ 0 \end{pmatrix} = \frac{3\,GM}{r_{\mathbb{C}}^5} \begin{pmatrix} \gamma_{\mathbb{C}_1}\, r_{\mathbb{C}\delta_2}\, r_{\mathbb{C}\delta_3} \\ \gamma_{\mathbb{C}_2}\, r_{\mathbb{C}\delta_3}\, r_{\mathbb{C}\delta_1} \\ \gamma_{\mathbb{C}_3}\, r_{\mathbb{C}\delta_1}\, r_{\mathbb{C}\delta_2} \end{pmatrix} , \tag{2.49}
$$

where the Earth's coordinates on the right-hand side of the above equations refer to the selenocentric system of the Moon's principal axes of inertia. Approximating the lunar orbit (the selenocentric Earth orbit) as a circle, the transformation from the inertial to the Moon-fixed coordinate system reads as

$$
\begin{pmatrix} r_{\mathbb{C}\eth_1} \\ r_{\mathbb{C}\eth_2} \\ r_{\mathbb{C}\eth_3} \end{pmatrix} = r_{\mathbb{C}} \, \mathbf{R}_3(\Theta_{\mathbb{C}}) \, \mathbf{R}_1(-\varepsilon_{\mathbb{C}}) \, \mathbf{R}_3(\Psi_{\mathbb{C}} - \Omega_{\eth}) \, \mathbf{R}_1(-i_{\mathbb{C}}) \begin{pmatrix} \cos u_{\eth} \\ \sin u_{\eth} \\ 0 \end{pmatrix} , \quad (2.50)
$$

where $\Omega_{\eth}$ is the ecliptical nodal longitude of the Earth's seleonocentric orbit, $i_{\eth} = i_{\mathbb{C}}$ is its inclination w.r.t. the ecliptic, and $u_{\eth}$ is its argument of latitude. $\Psi_{\mathbb{C}}$, $\varepsilon_{\mathbb{C}}$, and $\Theta_{\mathbb{C}}$ are the three Euler angles of lunar rotation.

Assuming that $\varepsilon_{\mathbb{C}}$, $i_{\mathbb{C}}$ and $\Psi_{\mathbb{C}} + \Theta_{\mathbb{C}} - (u_{\eth} + \Omega_{\eth})$ may be considered as small angles, the above transformation reduces to

$$
\begin{pmatrix} r_{\mathbb{C}\eth_1} \\ r_{\mathbb{C}\eth_2} \\ r_{\mathbb{C}\eth_3} \end{pmatrix} = r_{\mathbb{C}} \begin{pmatrix} 1 \\ u_{\eth} + \Omega_{\eth} - \Theta_{\mathbb{C}} - \Psi_{\mathbb{C}} \\ (\varepsilon_{\mathbb{C}} + i_{\mathbb{C}}) \sin u_{\eth} \end{pmatrix} . \quad (2.51)
$$

Note, that the second and third component of the above vector are small quantities. Up to terms of first order in these small quantities the Earth's torque on the Moon therefore may be written as:

$$
\frac{3\,GM}{r_{\mathbb{C}}^5} \begin{pmatrix} \gamma_{\mathbb{C}_1} r_{\mathbb{C}\eth_2} r_{\mathbb{C}\eth_3} \\ \gamma_{\mathbb{C}_2} r_{\mathbb{C}\eth_3} r_{\mathbb{C}\eth_1} \\ \gamma_{\mathbb{C}_3} r_{\mathbb{C}\eth_1} r_{\mathbb{C}\eth_2} \end{pmatrix} = \frac{3\,GM}{r_{\mathbb{C}}^3} \begin{pmatrix} 0 \\ \gamma_{\mathbb{C}_2} (\varepsilon_{\mathbb{C}} + i_{\mathbb{C}}) \sin u_{\eth} \\ \gamma_{\mathbb{C}_3} (u_{\eth} + \Omega_{\eth} - \Theta_{\mathbb{C}} - \Psi_{\mathbb{C}}) \end{pmatrix} . \quad (2.52)
$$

Substituting eqns. (2.52) into the equations (2.49), we obtain approximate differential equations which may (approximately) be solved analytically. In the approximations considered, the first two of eqns. (2.49) are separated from the third. Considering at present only the third we obtain:

$$
\dot{\omega}_{\mathbb{C}_3} = \frac{3\,\gamma_{\mathbb{C}_3}\,GM}{r_{\mathbb{C}}^3} \{ u_{\eth} + \Omega_{\eth} - (\Theta_{\mathbb{C}} + \Psi_{\mathbb{C}}) \} . \quad (2.53)
$$

Using the approximation $\cos \varepsilon_{\mathbb{C}} \approx 1$, the angle $l_{\eth}(t) \overset{\text{def}}{=} \Omega_{\eth}(t) + u_{\eth}(t)$ is the selenocentric ecliptical longitude of the Earth at time t and the angle $l_{\mathbb{C}_{\mathcal{I}_1}}(t) \overset{\text{def}}{=} \Theta_{\mathbb{C}}(t) + \Psi_{\mathbb{C}}(t)$ is the selenocentric ecliptical longitude of the Moon's first axis of inertia. Both longitudes, $l_{\eth}(t)$ and $l_{\mathbb{C}_{\mathcal{I}_1}}(t)$, may be represented by one and the same linear function and a residual part:

$$
\begin{aligned} l_{\eth}(t) &\overset{\text{def}}{=} n_{\mathbb{C}_0}(t - t_0) + l_{\text{opt}}(t) \\ l_{\mathbb{C}_{\mathcal{I}_1}}(t) &\overset{\text{def}}{=} n_{\mathbb{C}_0}(t - t_0) + l_{\text{ph}}(t) . \end{aligned} \quad (2.54)
$$

$l_{\text{opt}}(t)$ is called the *optical libration* of the Moon, $l_{\text{ph}}(t)$ its *physical libration*. In the approximation considered the third component of the Moon's angular velocity vector is the sum of the angles $\dot{\Psi}_{\mathbb{C}}$ and $\dot{\Theta}_{\mathbb{C}}$:

$$\dot{\omega}_{\mathfrak{C}_3} = \ddot{\Psi}_{\mathfrak{C}} + \ddot{\Theta}_{\mathfrak{C}} = \ddot{l}_{\mathfrak{C}_{I_1}} = \ddot{l}_{\mathrm{ph}} \;. \tag{2.55}$$

Using eqns. (2.54, 2.55), eqn. (2.53) may be written as:

$$\ddot{l}_{\mathrm{ph}} = -\frac{3\,\gamma_{\mathfrak{C}_3}\,GM}{r_{\mathfrak{C}}^3}\,(l_{\mathrm{opt}} + l_{\mathrm{ph}}) \;. \tag{2.56}$$

As the optical libration may be assumed known from the orbital motion, we have obtained a second-order linear differential equation describing the physical libration:

$$\ddot{l}_{\mathrm{ph}} + \frac{3\,\gamma_{\mathfrak{C}_3}\,GM}{r_{\mathfrak{C}}^3}\,l_{\mathrm{ph}} = \frac{3\,GM\,\gamma_{\mathfrak{C}_3}}{r_{\mathfrak{C}}^3}\,l_{\mathrm{opt}} \;. \tag{2.57}$$

The homogeneous part of the equation is that of a pendulum. The basic period is given by:

$$P_{\mathrm{ph}} = \sqrt{\frac{3\,\gamma_{\mathfrak{C}_3}\,GM}{r_{\mathfrak{C}}^3}} = n_{\mathfrak{C}_0}\,\sqrt{3\,\gamma_{\mathfrak{C}_3}} \approx 1045 \text{ days} \;. \tag{2.58}$$

It is exactly this period which is observed in Figure 2.37.

Let us conclude this paragraph by a quick analysis of the first two of the equations (2.49), where we only want to establish amplitude of the short-period variation as they appear in Figures 2.29 and 2.30. For this purpose we may skip the second term on the left-hand side and consider the two separated differential equations:

$$\begin{pmatrix} \dot{\omega}_{\mathfrak{C}_1} \\ \dot{\omega}_{\mathfrak{C}_2} \end{pmatrix} = 3\,\gamma_{\mathfrak{C}_2}\,n_{\mathfrak{C}_0}^2 \begin{pmatrix} 0 \\ (\varepsilon_{\mathfrak{C}} + i_{\mathfrak{C}})\sin u_{\text{\leftmoon}} \end{pmatrix} \;. \tag{2.59}$$

Obviously, these equations are solved by

$$\begin{pmatrix} \omega_{\mathfrak{C}_1}(t) \\ \omega_{\mathfrak{C}_2}(t) \end{pmatrix} = \begin{pmatrix} \omega_{\mathfrak{C}_{10}} \\ -3\,\gamma_{\mathfrak{C}_2}\,n_{\mathfrak{C}_0}\,(\varepsilon_{\mathfrak{C}} + i_{\mathfrak{C}})\cos u_{\text{\leftmoon}} + \mathrm{const} \end{pmatrix} \;, \tag{2.60}$$

implying that the first component of angular momentum is constant, and that the third component is given by a pure sinusoidal term with the period of one draconitic month and an amplitude of

$$a_{\omega_{\mathfrak{C}_2}} = 206264.8 \cdot 3\,\gamma_{\mathfrak{C}_2}\,n_{\mathfrak{C}_0}\,(\varepsilon_{\mathfrak{C}} + i_{\mathfrak{C}}) \approx 45'' \;. \tag{2.61}$$

The above order of magnitude formulae explain the essential features of Figures 2.29 and 2.30.

2.3 Rotation of the Non-Rigid Earth

2.3.1 Proofs for the Non-Rigidity of the Earth

The simulations performed in the previous sections illustrate that many aspects of Earth rotation, in particular the motion of the pole in inertial space (precession and nutation), are explained very well by a rigid body model. This is the main reason why the International Astronomical Union (IAU) switched only in 1980 from a rigid to an elastic Earth model.

There are, on the other hand, quite a few clear proofs that the Earth also has non-rigid properties:

- The Earth's surface is in a good approximation an equipotential surface in its combined gravitational and rotational force field (a detailed discussion will be given below). The Earth's oblateness therefore proves that for (almost) constant forces acting over very long time spans (millennia) the Earth is almost an ideally elastic body.

- The existence of tectonic plates and their relative motion was postulated in the first part of the twentieth century by Alfred Lothar Wegener (1880– 1930). His key article was published in 1915 [129]. As viewed from today Wegener's ideas concerning *continental drift* were mostly speculative. In the age of space geodesy motions of individual observing sites, from which plate motion or refined kinematic models for the Earth's crust may be deduced, are observed "so to speak" in real time (see Figure I- 8.22). Tectonic motions induce very slow changes of the Earth's mass distribution. When considering Earth rotation over time intervals not longer than a few centuries we do not expect observable effects in either PM or LOD variations, which might be attributed to tectonic motions.

- The system Earth consists of the solid Earth, the atmosphere and the oceans. Obviously, the latter two constituents are far from rigid. The question is, whether their contributions to the rotational behavior of the Earth are important enough to be noticed in the observation series. The answer to this question depends on the observational accuracy. Exchange of angular momentum between the solid Earth and the atmosphere were barely noticeable when Earth rotation was monitored with the instruments of optical astrometry (except for the annual term in PM). In the age of VLBI and GPS the correlation between the two types of angular momenta is striking. The correlation is monitored routinely with a daily or even higher time resolution.

- As seen from simulation and predicted from theory a rigid, rotationally symmetric Earth would ask for the Earth's pole to move on the surface of the Earth on a circle around the pole of figure with a period of about 304 days (Euler period). The period is uniquely defined by the length of the sidereal day and the dynamical flattening of the Earth. When spectrally analyzing the actual PM (as, e.g., illustrated by Figures I- 8.19 and I- 8.20) one obtains, however, in the low frequency domain a superposition of an annual signal and one of about 430 days. The latter period is called the *Chandler period* (named after its discoverer Seth Carlo Chandler (1846– 1913)). The annual period is mainly due to the interaction between the atmosphere and the solid Earth. The remarkably big discrepancy between the Chandler and the Euler period is one of the classical (and striking) proofs for the non-rigidity of the Earth. The difference was in part responsible for a long delay in the observational verification of the effect of PM (see Verdun et al. [126]). Figure 2.38 shows the spectrum of the IERS C04-pole series (obtainable through http://hpiers.obspm.fr/), which contains (among other Earth rotation data) the x- and y-coordinates of the pole and the excess LOD values at five day intervals since January 1, 1962. Note that the Chandler wobble is actually a prograde motion – the negative sign arises because the x- and y-coordinates of the pole were analyzed (see eqns. (2.4)).

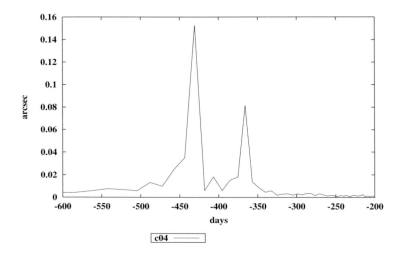

Fig. 2.38. Spectrum of C04 PM (two-dimensional analysis)

The spectrum differs in two aspects from the expected Euler motion of the pole: First of all, we clearly distinguish two lines, one with a period of about 430 days, one with an annual period. As will be shown below, both periods are due to the non-rigidity of the Earth: The annual term is (mainly) due to exchange of angular momentum between the solid Earth and the atmosphere; the Chandlerian motion may be explained as the generalized PM of an elastic body (a detailed analysis will be provided below).

- The length of the sidereal day is constant for a rigid, rotationally symmetric Earth. Figure I-8.21 shows that in reality this is not the case: annual and semiannual terms are clearly visible. Figure 2.39, which is based on the C04-pole series of the IERS, shows that there are also longer-term variations (of the order of decades) in these series. Figure 2.40 shows the

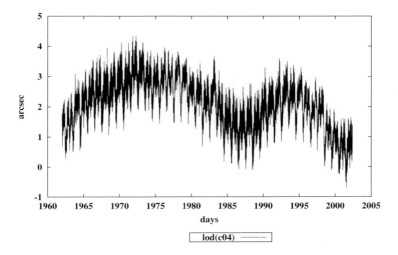

Fig. 2.39. Excess LOD based on C04 series of the IERS

amplitude spectrum (up to 500 days) of the C04 LOD series. As expected from Figures I- 8.21 and 2.39 the annual and semiannual terms are prominent. It is interesting to note that monthly and bi-monthly terms show up as well. We will see below that the latter terms are due to the luni-solar

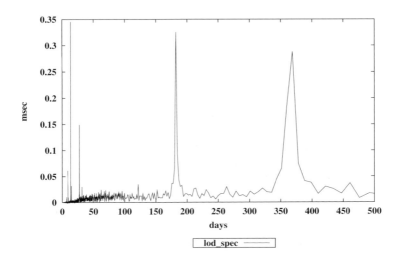

Fig. 2.40. Amplitude spectrum of C04 excess LOD series

tides. As expected, and, as opposed to the spectrum of PM, there is no clear signal near the Chandler period in the spectrum of the excess LOD values.

2.3.2 Hooke's Law and the Earth's Deformations

Let us assume from now on that the Earth is an elastic body, which may be deformed by applied forces. We neglect ocean and atmospheres in this simple approximation, and we do not make the distinction between the different layers of the Earth (crust, mantle, liquid outer and solid inner core). For a perfectly elastic body the deformations are *reversible*, i.e., if the applied forces disappear, the body will assume its original shape – as long as the applied forces are "small".

The relation between the external forces and the resulting deformations is described by *Hooke's law* in an elastic body. The original version of the law is due to Robert Hooke (1635–1703). It simply states that the deformation of an elastic body is proportional to the forces applied. The one-dimensional case is illustrated in Figure 2.41. In order to formulate the one-dimensional

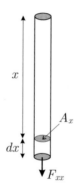

Fig. 2.41. Deformation dx of a cylindrical rod of length x and cross-section A_x due to a force F_{xx} acting along x

version of the law explicitly, we have to introduce the *stress* as the force per cross-section area by

$$\sigma_{xx} \overset{\text{def}}{=} \frac{F_{xx}}{A_x} \tag{2.62}$$

and the resulting deformation per length unit, called *strain*, by

$$\varepsilon_{xx} \overset{\text{def}}{=} \frac{dx}{x} . \tag{2.63}$$

Using the notations (2.62) and (2.63) Hooke's law simply may be written as follows:

$$\varepsilon_{xx} \propto \sigma_{xx} \tag{2.64}$$

or

$$\frac{dx}{x} \propto \frac{F_{xx}}{A_x} \ .$$

Note that the strain is a dimensionless quantity, whereas the stress has the unit of N/m^2 (or Pascal), exactly like pressure.

It is a relatively straightforward task to generalize the one-dimensional version (2.64) of Hooke's law to three dimensions (as it will be needed below). In the three-dimensional case we have to study the deformation of a rectangular prism of side lengths x, y, and z (the corresponding areas are $A_x = yz$, $A_y = xz$, $A_z = xy$) due to the respective applied *normal stress* components $\sigma_{xx} \overset{\text{def}}{=} \frac{F_{xx}}{A_x}$, $\sigma_{yy} \overset{\text{def}}{=} \frac{F_{yy}}{A_y}$, and $\sigma_{zz} \overset{\text{def}}{=} \frac{F_{zz}}{A_z}$. At first sight one would expect that the law (2.64) could simply be transcribed to the components y and z by replacing the symbol x by y and z respectively, for the deformations in the directions y and z. The situation is, however, slightly complicated by the fact that each *longitudinal strain* ε_{xx} also induces *transversal strain* ε_{yy} and ε_{zz} in the directions y and z. Longitudinal and transversal strains are related by

$$\varepsilon_{yy} \propto -\nu \varepsilon_{xx} \qquad \text{and} \qquad \varepsilon_{zz} \propto -\nu \varepsilon_{xx} \ , \tag{2.65}$$

where the dimensionless parameter ν, called *Poisson's ratio* (in honour of Siméon-Denis Poisson (1781–1840)), depends on the material properties of the body. One easily verifies that $\nu = 0.5$ for an incompressible body. The value $\nu = 0$ is assumed when no lateral contractions occur. Obviously, ν must thus lie between the two values for a real body. We are now in a position to write down Hooke's law in three dimensions:

$$\begin{aligned}
\varepsilon_{xx} &\propto \sigma_{xx} - \nu \sigma_{yy} - \nu \sigma_{zz} \\
\varepsilon_{yy} &\propto \sigma_{yy} - \nu \sigma_{zz} - \nu \sigma_{xx} \\
\varepsilon_{zz} &\propto \sigma_{zz} - \nu \sigma_{xx} - \nu \sigma_{yy} \ .
\end{aligned} \tag{2.66}$$

These equations complete our excursion into the theory of elasticity. Let us mention in conclusion that σ_{xx}, σ_{yy}, and σ_{zz} are the diagonal elements of the *stress matrix*, and that ε_{xx}, ε_{yy}, and ε_{zz} are the diagonal elements of the resulting *strain matrix*. We will not need the off-diagonal elements of the two matrices in our context. For a more complete treatment of the theory of elasticity, as used in geodynamics, we refer to the elementary treatment [69] or to any other standard textbook on geophysics.

In the simplest case an elastic body may thus be characterized by one proportionality coefficient in Hooke's law (2.66) *and* by a value for the Poisson ratio ν.

Subsequently, we will apply Hooke's law to the Earth as a whole. Such a treatment must be a gross simplification. The Earth should be treated more correctly as a composite body, consisting (at least) of different layers (as, e.g., specified in Figure 2.55), where each layer is governed by a specific set of elastic parameters. Such treatments must be left to special treatments in geophysics. For our purpose it is sufficient to assume that the elastic Earth is governed by eqns. (2.66); we will even assume that no lateral contraction occur, i.e., that

$$\nu \stackrel{\text{def}}{=} 0 \ . \tag{2.67}$$

This assumption certainly is not *very* realistic. It has, however, the advantage of significantly reducing the formalism – and it is justified by the fact that the errors due to this assumption are of second or higher order in small quantities.

The "external" forces which have to be considered in the case of an elastic Earth model are:

- Centrifugal force due to the rotation of the Earth around its figure axis.

- Differential centrifugal force due to the fact that the Earth's figure and rotation axes do *not* coincide. Because the angle between the two axes is the angle of PM, one usually refers to this effect as *polar tides*.

- Tidal deformations due to Sun and Moon. (The tidal torques due to the planets exist as well, but they are very small and usually ignored.)

Permanent Deformation. The combined gravitational and centrifugal potential $W(\boldsymbol{r})$ of the Earth, called *gravity potential* in geodesy, may be written approximately as (see (I- 3.157))

$$
\begin{aligned}
W(r,\phi) &= V(r,\phi) + \tfrac{1}{2}\,\omega_\delta^2\,r^2\cos^2\beta \\
&= \frac{GM}{r}\left\{1 + \frac{C_\delta - A_\delta}{M\,r^2}\left(\frac{3}{2}\cos^2\beta - 1\right)\right\} + \frac{1}{2}\,\omega_\delta^2\,r^2\cos^2\beta \ .
\end{aligned} \tag{2.68}
$$

Rotational symmetry, i.e., $A_\delta = B_\delta$ and $I_{\delta_{12}} = 0$ was assumed. Moreover, it was assumed that the angular velocity vector coincides with the Earth's figure axis. We thus assume that the Earth rotates about the figure axis.

A particular equipotential surface of the potential (2.68) which is tangential to the Earth at the equator is defined by

$$W(r,\phi) \stackrel{\text{def}}{=} W(a_\delta,0°) \ , \tag{2.69}$$

where a_δ is the equatorial radius of the Earth and ϕ is the geocentric latitude. The surface defined by the above equation is approximately that of a flattened, rotationally symmetric ellipsoid with its semi-minor axis coinciding with the rotation axis (the proof may be left to the reader). The polar radius b_δ (semi-minor axis of the ellipsoid) is obtained by the condition equation

$$W(b_\delta, 90°) = W(a_\delta, 0°) \ . \tag{2.70}$$

This equation is approximately solved by

$$\frac{a_\delta - b_\delta}{a_\delta} = \frac{3}{2}\frac{C_\delta - A_\delta}{M\, a_\delta^2} + \frac{1}{2}\frac{\omega_\delta^2\, a_\delta^3}{GM} \approx \frac{1}{297.6} \ , \tag{2.71}$$

which gives the following value for the difference between the equatorial and polar radius of the Earth:

$$a_\delta - b_\delta \approx 21.431 \text{ km} \ . \tag{2.72}$$

These values agree to within about 0.2% with the actual values for the geometric flattening and the polar radius of the Earth, respectively. The agreement proves that the Earth's surface is in an excellent approximation an equipotential surface of the Earth's gravity field. This agreement indicates in turn that the Earth reacts nearly like an ideally elastic body to constant forces acting over millions of years.

The second term on the right-hand side of formula (2.71) may be interpreted in terms of the fraction ξ of the centrifugal force and of the gravitational force at the equator

$$\xi \stackrel{\text{def}}{=} \frac{\omega_\delta^2\, a_\delta}{\frac{GM}{a_\delta^2}} = \frac{\omega_\delta^2\, a_\delta^3}{GM} \approx \frac{1}{288.9} \ . \tag{2.73}$$

Polar Tides. The effect of the so-called *polar tides* is illustrated by Figure 2.42. The polar tides are a consequence of the fact that the Earth does not rotate about its figure axis (unit vector e_{3_δ} in Figure 2.42) but about the actual rotation axis (represented by the vector ω_δ in the same Figure). Due to the differential force caused by this mismatch of axes the Earth's surface tries to make its surface (solid ellipsoid in Figure 2.42) an equipotential surface (dashed line) in the gravity field of an Earth rotating about the axis ω_δ .

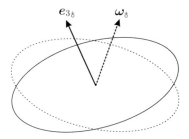

Fig. 2.42. Polar tides: deformation of Earth due to its rotation

Let us now assume that the Earth is an elastic body and explain the resulting deformation due to the polar tides as a global application of Hooke's law. Let us calculate the resulting differential forces, etc. in the Earth-fixed PAI-system. A mass element $dM \stackrel{\text{def}}{=} \rho(\boldsymbol{r}_{\delta\mathcal{F}}) \, dV$ with the geocentric coordinates $\boldsymbol{r}_{\delta\mathcal{F}}$ (in the rigid, rotating coordinate system) rotating with an angular velocity

$$\boldsymbol{\omega}_{\delta\mathcal{F}} \stackrel{\text{def}}{=} \boldsymbol{\omega}_{\delta\mathcal{F}_0} + \begin{pmatrix} \delta\omega_{\delta\mathcal{F}_1} \\ \delta\omega_{\delta\mathcal{F}_2} \\ \delta\omega_{\delta\mathcal{F}_3} \end{pmatrix} = \omega_{\delta\mathcal{F}_0} \begin{pmatrix} 0 \\ 0 \\ 1 \end{pmatrix} + \begin{pmatrix} \delta\omega_{\delta\mathcal{F}_1} \\ \delta\omega_{\delta\mathcal{F}_2} \\ \delta\omega_{\delta\mathcal{F}_3} \end{pmatrix} \tag{2.74}$$

experiences the centrifugal force

$$\boldsymbol{f}_{\delta\mathcal{F}} = (\boldsymbol{\omega}_{\delta\mathcal{F}} \times \boldsymbol{r}_{\delta\mathcal{F}}) \times \boldsymbol{\omega}_{\delta\mathcal{F}} \, dM = \omega_{\delta\mathcal{F}}^2 \, \boldsymbol{r}_{\delta\mathcal{F}} - (\boldsymbol{\omega}_{\delta\mathcal{F}} \cdot \boldsymbol{r}_{\delta\mathcal{F}}) \, \boldsymbol{\omega}_{\delta\mathcal{F}} \; . \tag{2.75}$$

The same mass element would suffer from a centrifugal force

$$\boldsymbol{f}_{\delta\mathcal{F}_0} = \omega_{\delta\mathcal{F}_3}^2 (\boldsymbol{e}_{3_{\delta\mathcal{F}}} \times \boldsymbol{r}_{\delta\mathcal{F}}) \times \boldsymbol{e}_{3_{\delta\mathcal{F}}} \, dM = \omega_{\delta\mathcal{F}_0}^2 \left(\boldsymbol{r}_{\delta\mathcal{F}} - (\boldsymbol{e}_{3_{\delta\mathcal{F}}} \cdot \boldsymbol{r}_{\delta\mathcal{F}}) \, \boldsymbol{e}_{3_{\delta\mathcal{F}}} \right) , \tag{2.76}$$

when rotating with angular velocity $\omega_{\delta\mathcal{F}_0}$ around the figure axis $\boldsymbol{e}_{3_{\delta}}$.

Neglecting the terms of the second and higher orders in the small quantities $\delta\omega_{\delta\mathcal{F}_i}$, $i = 1, 2, 3$, the differential centrifugal acceleration is obtained as the difference of the above two forces

$$d\boldsymbol{f}_{\delta\mathcal{F}} \stackrel{\text{def}}{=} \boldsymbol{f}_{\delta\mathcal{F}} - \boldsymbol{f}_{\delta\mathcal{F}_0} = \omega_{\delta\mathcal{F}_3} \begin{pmatrix} 2\,\delta\omega_{\delta\mathcal{F}_3}\, r_{\delta\mathcal{F}_1} - \delta\omega_{\delta\mathcal{F}_1}\, r_{\delta\mathcal{F}_3} \\ 2\,\delta\omega_{\delta\mathcal{F}_3}\, r_{\delta\mathcal{F}_2} - \delta\omega_{\delta\mathcal{F}_2}\, r_{\delta\mathcal{F}_3} \\ -\delta\omega_{\delta\mathcal{F}_2}\, r_{\delta\mathcal{F}_2} - \delta\omega_{\delta\mathcal{F}_1}\, r_{\delta\mathcal{F}_1} \end{pmatrix} dM \; . \tag{2.77}$$

One easily verifies that this differential force may be expressed as the gradient of the following scalar potential:

$$dV = dM\, \omega_{\delta\mathcal{F}_3} \left\{ \delta\omega_{\delta\mathcal{F}_3} \left(r_{\delta\mathcal{F}_1}^2 + r_{\delta\mathcal{F}_2}^2 \right) - r_{\delta\mathcal{F}_3} \left(\delta\omega_{\delta\mathcal{F}_1} r_{\delta\mathcal{F}_1} + \delta\omega_{\delta\mathcal{F}_2} r_{\delta\mathcal{F}_2} \right) \right\} \; . \tag{2.78}$$

Hooke's law states that the deformation suffered by an elastic body as a consequence of an applied force is proportional to the force. Obviously this deformation also must be inversely proportional to the mass. The displacement of the mass element dM considered may therefore be written as

$$\delta\boldsymbol{r}_{\delta\mathcal{F}} \stackrel{\text{def}}{=} \frac{k\,\xi}{\omega_{\delta\mathcal{F}_3}} \begin{pmatrix} 2\,\delta\omega_{\delta\mathcal{F}_3}\, r_{\delta\mathcal{F}_1} - \delta\omega_{\delta\mathcal{F}_1}\, r_{\delta\mathcal{F}_3} \\ 2\,\delta\omega_{\delta\mathcal{F}_3}\, r_{\delta\mathcal{F}_2} - \delta\omega_{\delta\mathcal{F}_2}\, r_{\delta\mathcal{F}_3} \\ -\delta\omega_{\delta\mathcal{F}_1}\, r_{\delta\mathcal{F}_1} - \delta\omega_{\delta\mathcal{F}_2}\, r_{\delta\mathcal{F}_2} \end{pmatrix} , \tag{2.79}$$

where k is a dimensionless constant and ξ is the fraction of the centrifugal and the gravitational forces at the equator (as defined by eqn. (2.73)).

k is (one of) the well-known *Love number(s)* (named in honour of Augustus Edward Hough Love (1863–1940)), providing a measure for the elasticity of a planet. Obviously, $k = 0$ corresponds to the rigid-body model of the Earth. As we will see below, $k \approx 1$ characterizes an ideally elastic Earth – which justifies the use of the dimensionless "constant" ξ in the expression (2.79) for the displacement of the mass element dM.

Let us now explicitly calculate the elements of the inertia tensor (I- 3.75) in a coordinate system with its third axis lying in the figure axis of the unperturbed body. For the minimum moment of inertia we obtain, e.g.:

$$A_{\delta} \stackrel{\text{def}}{=} \int_V \rho(\boldsymbol{r}_{\delta\mathcal{F}}) \left\{ \left(r_{\delta\mathcal{F}_2} + \delta r_{\delta\mathcal{F}_2} \right)^2 + \left(r_{\delta\mathcal{F}_3} + \delta r_{\delta\mathcal{F}_3} \right)^2 \right\} dV , \qquad (2.80)$$

where the integral has to be extended over the unperturbed volume of the Earth. Substituting the deformations (2.79) as expected by Hooke's law, making use of the relations (I- 3.79) and (I- 3.80), and denoting the unperturbed moments of inertia of the Earth rotating about the figure axis by A_0, B_0, and C_0, one easily verifies the following relations:

$$A_{\delta} = A_0 + 2\,\frac{k\,\xi\,\delta\omega_{\delta\mathcal{F}_3}}{\omega_{\delta\mathcal{F}_3}}\,(C_0 + A_0 - B_0)$$

$$B_{\delta} = B_0 + 2\,\frac{k\,\xi\,\delta\omega_{\delta\mathcal{F}_3}}{\omega_{\delta\mathcal{F}_3}}\,(C_0 - A_0 + B_0) \qquad (2.81)$$

$$C_{\delta} = C_0 + 4\,\frac{k\,\xi\,\delta\omega_{\delta\mathcal{F}_3}}{\omega_{\delta\mathcal{F}_3}}\,C_0 .$$

For a rotationally symmetric Earth one obtains in particular

$$A_{\delta} = A_0 + 2\,\frac{k\,\xi\,\delta\omega_{\delta\mathcal{F}_3}}{\omega_{\delta\mathcal{F}_3}}\,C_0$$

$$B_{\delta} = A_0 + 2\,\frac{k\,\xi\,\delta\omega_{\delta\mathcal{F}_3}}{\omega_{\delta\mathcal{F}_3}}\,C_0 \qquad (2.82)$$

$$C_{\delta} = C_0 + 4\,\frac{k\,\xi\,\delta\omega_{\delta\mathcal{F}_3}}{\omega_{\delta\mathcal{F}_3}}\,C_0 .$$

The deformations exerted by the polar tides also induce changes in the elements $I_{\delta\mathcal{F}_{13}}$ and $I_{\delta\mathcal{F}_{23}}$ of the inertia tensor. Their computation follows the same pattern as that of the principal moments of inertia. Using the definition (I- 3.76) for the inertia tensor and the deformations (2.79) we obtain, e.g.:

$$I_{\delta\mathcal{F}_{13}} \stackrel{\text{def}}{=} -\int_V \rho(\boldsymbol{r}_{\delta\mathcal{F}}) \left(r_{\delta\mathcal{F}_1} + \delta r_{\delta\mathcal{F}_1} \right) \left(r_{\delta\mathcal{F}_3} + \delta r_{\delta\mathcal{F}_3} \right) dV = \frac{k\,\xi\,\delta\omega_{\delta\mathcal{F}_1}}{\omega_{\delta\mathcal{F}_3}}\,B_0 .$$

$$(2.83)$$

Along the same lines one obtains:

$$I_{\delta\mathcal{F}23} \stackrel{\text{def}}{=} \frac{k\,\xi\,\delta\omega_{\delta\mathcal{F}_2}}{\omega_{\delta\mathcal{F}_3}}\,A_0\;. \tag{2.84}$$

In conclusion, the inertia tensor, expressed in the PAI-system of the unperturbed body, may be written as

$$\mathbf{I}_{\delta\mathcal{F}} = \mathbf{I}_0 + \delta\mathbf{I}_{\delta\mathcal{F}} = \begin{pmatrix} A_0 & 0 & 0 \\ 0 & B_0 & 0 \\ 0 & 0 & C_0 \end{pmatrix} + \delta\mathbf{I}_{\delta\mathcal{F}}\;, \tag{2.85}$$

where

$$\delta\mathbf{I}_{\delta\mathcal{F}} = \frac{k\,\xi}{\omega_{\delta\mathcal{F}_3}} \begin{pmatrix} 2\,\delta\omega_{\delta\mathcal{F}_3}(C_0 + A_0 - B_0) & 0 & \delta\omega_{\delta\mathcal{F}_1} B_0 \\ 0 & 2\,\delta\omega_{\delta\mathcal{F}_3}(C_0 - A_0 + B_0) & \delta\omega_{\delta\mathcal{F}_2} A_0 \\ \delta\omega_{\delta\mathcal{F}_1} B_0 & \delta\omega_{\delta\mathcal{F}_2} A_0 & 4\,\delta\omega_{\delta\mathcal{F}_3} C_0 \end{pmatrix}\;. \tag{2.86}$$

For a rotationally symmetric Earth one obtains in particular

$$\delta\mathbf{I}_{\delta\mathcal{F}} = \frac{k\,\xi}{\omega_{\delta\mathcal{F}_3}} \begin{pmatrix} 2\,\delta\omega_{\delta\mathcal{F}_3} C_0 & 0 & \delta\omega_{\delta\mathcal{F}_1} A_0 \\ 0 & 2\,\delta\omega_{\delta\mathcal{F}_3} C_0 & \delta\omega_{\delta\mathcal{F}_2} A_0 \\ \delta\omega_{\delta\mathcal{F}_1} A_0 & \delta\omega_{\delta\mathcal{F}_2} A_0 & 4\,\delta\omega_{\delta\mathcal{F}_3} C_0 \end{pmatrix}\;. \tag{2.87}$$

In order to solve the Liouville-Euler equations, we also need the first time derivatives of the inertia tensor. Obviously, one may write (for a rotationally symmetric Earth):

$$\dot{\mathbf{I}}_{\delta\mathcal{F}} = \delta\dot{\mathbf{I}}_{\delta\mathcal{F}} = \frac{k\,\xi}{\omega_{\delta\mathcal{F}_3}} \begin{pmatrix} 2\,\delta\dot\omega_{\delta\mathcal{F}_3} C_0 & 0 & \delta\dot\omega_{\delta\mathcal{F}_1} A_0 \\ 0 & 2\,\delta\dot\omega_{\delta\mathcal{F}_3} C_0 & \delta\dot\omega_{\delta\mathcal{F}_2} A_0 \\ \delta\dot\omega_{\delta\mathcal{F}_1} A_0 & \delta\dot\omega_{\delta\mathcal{F}_2} A_0 & 4\,\delta\dot\omega_{\delta\mathcal{F}_3} C_0 \end{pmatrix}\;. \tag{2.88}$$

It is important to note that both, the elements of matrices $\delta\mathbf{I}_{\delta\mathcal{F}}$ and $\dot{\mathbf{I}}_{\delta\mathcal{F}}$ are small quantities of the first order (because they are proportional to $\delta\omega_{\delta F_i}$ or their time derivatives).

With the expressions for the inertia tensor and its time derivative the Liouville-Euler equations may be set up for an elastic Earth rotating about an axis (slightly) different from the figure axis. We assume that it is possible to realize a Tissérand system using all (space-) geodetic and geophysical informations – not an entirely unproblematic assumption. Under these circumstances the Liouville-Euler equations may be used in the form (I- 3.137). Neglecting all second- and higher-order terms in the small quantities $\delta\dot\omega_{\delta\mathcal{F}_2}$, the individual terms may be approximated by

$$\dot{\mathbf{I}}_{\delta\mathcal{F}}\,\boldsymbol{\omega}_{\delta\mathcal{F}} = k\,\xi \begin{pmatrix} B_0\,\dot\omega_{\delta\mathcal{F}_1} \\ A_0\,\dot\omega_{\delta\mathcal{F}_2} \\ 4\,C_0\,\dot\omega_{\delta\mathcal{F}_3} \end{pmatrix}\;, \tag{2.89}$$

$$\mathbf{I}_{\mathring{\delta}\mathcal{F}}\,\dot{\boldsymbol{\omega}}_{\mathring{\delta}\mathcal{F}} = \begin{pmatrix} A_0\,\dot{\omega}_{\mathring{\delta}\mathcal{F}_1} \\ B_0\,\dot{\omega}_{\mathring{\delta}\mathcal{F}_2} \\ C_0\,\dot{\omega}_{\mathring{\delta}\mathcal{F}_3} \end{pmatrix}, \tag{2.90}$$

and

$$\boldsymbol{\omega}_{\mathring{\delta}\mathcal{F}} \times (\mathbf{I}_{\mathring{\delta}\mathcal{F}}\,\boldsymbol{\omega}_{\mathring{\delta}\mathcal{F}}) = \begin{pmatrix} (C_0 - B_0 - k\,\xi\,A_0)\,\omega_{\mathring{\delta}\mathcal{F}_2}\,\omega_{\mathring{\delta}\mathcal{F}_3} \\ (A_0 - C_0 + k\,\xi\,B_0)\,\omega_{\mathring{\delta}\mathcal{F}_1}\,\omega_{\mathring{\delta}\mathcal{F}_3} \\ 0 \end{pmatrix}. \tag{2.91}$$

By introducing the above relations into the Liouville-Euler equations one obtains (after the division of the first equation by A_0, of the second by B_0, and of the third by C_0) the Liouville-Euler equations for the free rotation of the elastic Earth

$$\left(1 + k\,\xi\,\frac{B_0}{A_0}\right)\dot{\omega}_{\mathring{\delta}\mathcal{F}_1} + (\gamma_{\mathring{\delta}_1} - k\,\xi)\,\omega_{\mathring{\delta}\mathcal{F}_2}\,\omega_{\mathring{\delta}\mathcal{F}_3} = 0$$

$$\left(1 + k\,\xi\,\frac{A_0}{B_0}\right)\dot{\omega}_{\mathring{\delta}\mathcal{F}_2} + (\gamma_{\mathring{\delta}_2} + k\,\xi)\,\omega_{\mathring{\delta}\mathcal{F}_1}\,\omega_{\mathring{\delta}\mathcal{F}_3} = 0 \tag{2.92}$$

$$\dot{\omega}_{\mathring{\delta}\mathcal{F}_3} \qquad\qquad\qquad\qquad = 0\,,$$

where

$$\gamma_{\mathring{\delta}_1} = \frac{C_0 - B_0}{A_0}\,, \qquad \gamma_{\mathring{\delta}_2} = \frac{A_0 - C_0}{B_0}\,, \qquad \gamma_{\mathring{\delta}_3} = \frac{B_0 - A_0}{C_0} \tag{2.93}$$

exactly as in the case of the rigid body.

Assuming rotational symmetry one obtains the somewhat simpler relations

$$(1 + k\,\xi)\,\dot{\omega}_{\mathring{\delta}\mathcal{F}_1} + \gamma_{\mathring{\delta}}\left(1 - \frac{k\,\xi}{\gamma_{\mathring{\delta}}}\right)\omega_{\mathring{\delta}\mathcal{F}_2}\,\omega_{\mathring{\delta}\mathcal{F}_3} = 0$$

$$(1 + k\,\xi)\,\dot{\omega}_{\mathring{\delta}\mathcal{F}_2} - \gamma_{\mathring{\delta}}\left(1 - \frac{k\,\xi}{\gamma_{\mathring{\delta}}}\right)\omega_{\mathring{\delta}\mathcal{F}_1}\,\omega_{\mathring{\delta}\mathcal{F}_3} = 0 \tag{2.94}$$

$$\dot{\omega}_{\mathring{\delta}\mathcal{F}_3} \qquad\qquad\qquad\qquad = 0\,,$$

where

$$\gamma_{\mathring{\delta}} = \frac{C_0 - A_0}{A_0}\,. \tag{2.95}$$

The equations (2.92) and (2.94) are very closely related to the corresponding equations of the rigid body (see eqns. (2.10)). The above equations are, as a matter of fact, identical with the equations of the rigid body for $k = 0$. The mathematical structure of the equations for the rigid and the elastic body is the same: The equation for the third component of the angular velocity is separated from the equations for PM and has the trivial solution $\omega_{\mathring{\delta}\mathcal{F}_3}(t) = \omega_{\mathring{\delta}_0} = $ const. As a consequence the first two components form a coupled system of linear, homogeneous differential equations with constant coefficients which may solved with the same methods as in the case of the rigid body. For a rotationally symmetric Earth, the free rotation of which is defined by eqns. (2.94), the pole moves with constant angular velocity on a circle around the pole of figure. The amplitude is given by the initial conditions, the period by

$$P = \frac{2\,\pi}{\gamma_\delta\,\omega_{\delta 0}}\left(1 - k\,\frac{\xi}{\gamma_\delta}\right)^{-1}, \qquad (2.96)$$

(where $1 + k\,\xi \approx 1$ was used as an approximation for the coefficients of the terms $\omega_{\delta\mathcal{F}_i}$, $i = 1, 2, 3$ in the Liouville-Euler equations).

The term

$$k_s \stackrel{\text{def}}{=} \frac{\gamma_\delta}{\xi} = \frac{GM\,(C_0 - A_0)}{A_0\,\omega_{\delta 0}^2\,a_\delta^3} \approx \frac{3\,G\,(C_0 - A_0)}{\omega_0^2\,R_\delta^5} \approx 0.95, \qquad (2.97)$$

where R_δ is the mean radius of the Earth, often is referred to as *secular Love number*. The designation is somewhat unfortunate. It may be justified by the fact that for an ideally elastic Earth, one would expect $k_s \stackrel{\text{def}}{=} 1$. For a geophysical discussion of the secular Love number k_s we refer to the standard textbook by Munk and MacDonald [78].

Formula (2.96) tells that the period P of the Chandler motion (Euler motion for $k = 0$) heavily depends on the Love number k, which characterizes the ability of the Earth's surface to assume the form of an equipotential surface in the actual gravity potential. For $k = k_s$, the period becomes infinite. This corresponds to the case of an ideally elastic Earth: The surface of the Earth is always an ellipsoid of rotation with its semi-minor axis coinciding with the angular velocity axis. For values $0 < k < k_s$ the Earth's response to forces with periods of about one day is "too slow" to behave like an ideally elastic body. The actual value for k may, e.g., be extracted from a spectral analysis of PM. From the spectrum in Figure 2.38 of the IERS C04 series of PM we find a period of about $P = 435^\text{d}$, corresponding to a Love number k of about $k \approx 0.3$.

The most prominent feature in the PM series in Figure 2.38 could thus be explained as a consequence of the so-called *polar tides*. Figure 2.43 illustrates the free PM over a time interval of 300 days using program ERDROT. Only the free Earth rotation was considered and the Love number was set to the values $k = 0$, $k = 0.3$, and $k = 0.7$. The second value characterizes the real situation.

Lunisolar Tides. Figure 2.44 gives a first insight into the characteristics of the lunisolar tides: The upper part illustrates the gravitational forces acting on a mass element in the inertial system, the lower part the resulting tidal forces, i.e., the difference between the gravitational forces at a particular point on the Earth surface (or in the Earth's interior) and the gravitational force in the geocenter. From Figure 2.44 (bottom) we expect, that the Earth

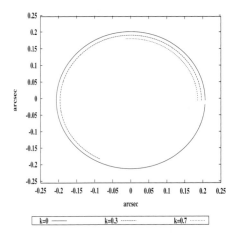

Fig. 2.43. Free PM for Love numbers $k = 0,\ 0.3,\ 0.7$, over a time interval of 300 days

is elongated along the line *geocenter* → *selenocenter*. Somewhat exaggerating one might say that the Earth assumes the shape of an American football pointing to the Moon due to the tidal force exerted by the Moon. As will be seen below, the deformed Earth assumes the shape of an elongated (as opposed to oblate) ellipsoid. Note that the four points in Figure 2.44 were

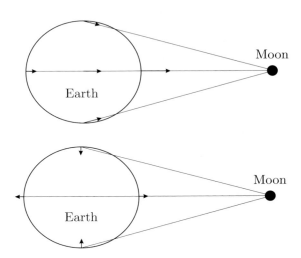

Fig. 2.44. Gravitational forces (top) and tidal forces (bottom) acting on mass unit at four points in the equatorial plane

selected in a special way as the tidal acceleration always points into the vertical direction. At locations between these four special cases horizontal accelerations do occur as well.

Using equations of type (I-3.21) (where geocentric instead of heliocentric vectors have to be used and the constant k^2 has to be replaced by $Gm_{\mathbb{C}}$) the tidal acceleration $\boldsymbol{a}_{\mathrm{tid}}$ due to Sun and Moon acting on a mass element at point P with geocentric position $\boldsymbol{r}$ (on the Earth's surface or in the Earth interior) may be written as

$$\boldsymbol{a}_{\mathrm{tid}} = - G\, m_{\mathbb{C}} \left\{ \frac{\boldsymbol{r} - \boldsymbol{r}_{\mathbb{C}}}{|\,\boldsymbol{r} - \boldsymbol{r}_{\mathbb{C}}\,|^3} + \frac{\boldsymbol{r}_{\mathbb{C}}}{r_{\mathbb{C}}^3} \right\} - G\, m_{\odot} \left\{ \frac{\boldsymbol{r} - \boldsymbol{r}_{\odot}}{|\,\boldsymbol{r} - \boldsymbol{r}_{\odot}\,|^3} + \frac{\boldsymbol{r}_{\odot}}{r_{\odot}^3} \right\} . \quad (2.98)$$

This acceleration may be written as the gradient of a potential function

$$\boldsymbol{a}_{\mathrm{tid}} = \boldsymbol{\nabla} V_{\mathrm{tid}} , \quad (2.99)$$

where the tide-generating potential may be written in the form (compare eqn. (I-3.28))

$$V_{\mathrm{tid}} = \frac{G\, m_{\mathbb{C}}}{|\,\boldsymbol{r} - \boldsymbol{r}_{\mathbb{C}}\,|} - G\, m_{\mathbb{C}} \frac{\boldsymbol{r} \cdot \boldsymbol{r}_{\mathbb{C}}}{r_{\mathbb{C}}^3} + \frac{G\, m_{\odot}}{|\,\boldsymbol{r} - \boldsymbol{r}_{\odot}\,|} - G\, m_{\odot} \frac{\boldsymbol{r} \cdot \boldsymbol{r}_{\odot}}{r_{\odot}^3} . \quad (2.100)$$

The tidal potential (2.100) is the sum of two terms, the tidal potential $V_{tid\mathbb{C}}$ due to the Moon and that due to the Sun $V_{tid\odot}$. Let us consider only the term due to the Moon (the tidal potential due to the Sun may be treated in an analogous way):

$$V_{tid\mathbb{C}} = \frac{G\, m_{\mathbb{C}}}{|\,\boldsymbol{r} - \boldsymbol{r}_{\mathbb{C}}\,|} - G\, m_{\mathbb{C}} \frac{\boldsymbol{r} \cdot \boldsymbol{r}_{\mathbb{C}}}{r_{\mathbb{C}}^3} . \quad (2.101)$$

Note that $|\,\boldsymbol{r}\,| \ll |\,\boldsymbol{r}_{\mathbb{C}}\,|$, because $\boldsymbol{r}$ is the geocentric position vector of a point on the Earth's surface or interior. We may therefore approximate the first term by a series of Legendre polynomials with the cosine of the angle between the vectors $\boldsymbol{r}$ and $\boldsymbol{r}_{\mathbb{C}}$ as argument

$$\cos \alpha \overset{\mathrm{def}}{=} \frac{\boldsymbol{r} \cdot \boldsymbol{r}_{\mathbb{C}}}{r\, r_{\mathbb{C}}} . \quad (2.102)$$

The procedure is analogous to that performed in section (see eqn. (3.102) in section I-3.3.5). The result reads as:

$$\begin{aligned}
V_{tid\mathbb{C}} &= G\, m_{\mathbb{C}} \left\{ \frac{1}{|\,\boldsymbol{r} - \boldsymbol{r}_{\mathbb{C}}\,|} - \frac{\boldsymbol{r} \cdot \boldsymbol{r}_{\mathbb{C}}}{r_{\mathbb{C}}^3} \right\} \\
&= \frac{G\, m_{\mathbb{C}}}{r_{\mathbb{C}}} \left\{ 1 + (1 - 1) \frac{\boldsymbol{r} \cdot \boldsymbol{r}_{\mathbb{C}}}{r_{\mathbb{C}}^2} - \frac{1}{2} \frac{r^2}{r_{\mathbb{C}}^2} + \frac{3}{2} \left(\frac{\boldsymbol{r} \cdot \boldsymbol{r}_{\mathbb{C}}}{r_{\mathbb{C}}^2} \right)^2 \right\} \quad (2.103) \\
&\overset{\mathrm{def}}{=} + \frac{1}{2} \frac{G\, m_{\mathbb{C}}}{r_{\mathbb{C}}^3} r^2 \left\{ 3 \cos^2 \alpha - 1 \right\} ,
\end{aligned}$$

where the term $\dfrac{G\, m_{\mathbb{C}}}{r_{\mathbb{C}}}$ was skipped in the transition from the second to the last line, because it does not depend on the vector $\boldsymbol{r}$ (this does of course mean that mathematical equality is not preserved in this transition).

The deformations caused by the tide-generating potential (2.103) are best studied in a geocentric Cartesian coordinate system with coordinates x, y and z, the third axis of which coincides with the line geocenter $\rightarrow$ selenocenter. From the last line of eqns. (2.103) one immediately obtains:

$$V_{\text{tid}\mathbb{C}} = \frac{1}{2} \frac{G\, m_{\mathbb{C}}}{r_{\mathbb{C}}^3} \left\{ 2\, z^2 - x^2 - y^2 \right\}\ , \tag{2.104}$$

from where the accelerations follow as

$$\boldsymbol{a}_{\text{tid}\mathbb{C}} = \boldsymbol{\nabla} V_{\text{tid}\mathbb{C}} = \frac{G\, m_{\mathbb{C}}}{r_{\mathbb{C}}^3} \begin{pmatrix} -x \\ -y \\ +2\, z \end{pmatrix}\ . \tag{2.105}$$

According to Hooke's law the deformations are proportional to the applied accelerations. Consequently, one may write

$$\begin{pmatrix} \delta x \\ \delta y \\ \delta z \end{pmatrix} = \frac{k\, \xi}{\omega_{\delta 0}^2} \boldsymbol{\nabla} V_{\text{tid}\mathbb{C}} = \frac{k\, \xi}{\omega_{\delta 0}^2} \frac{G\, m_{\mathbb{C}}}{r_{\mathbb{C}}^3} \begin{pmatrix} -x \\ -y \\ +2\, z \end{pmatrix}\ . \tag{2.106}$$

The dimensionless constant k is again the Love number. The factor $\frac{\xi}{\omega_{\delta 0}^2}$ was introduced for easy comparisons of lunisolar tide results with those of the polar tides.

How do the principal moments of inertia change due to the tidal deformation (2.106)? In order to isolate the effect we assume that the unperturbed Earth is of spherical symmetry with three identical principal moments of inertia (see Table 2.1):

$$A_0 = B_0 = C_0 = 0.3296144\, M R_{\delta}^2\ . \tag{2.107}$$

The new principal moments of inertia $A_{\mathbb{C}}$, $B_{\mathbb{C}}$, and $C_{\mathbb{C}}$ due to the lunar tides are calculated as

$$\begin{aligned} A_{\mathbb{C}} &= \int_V \rho(\boldsymbol{r}_{\delta\mathcal{F}}) \left\{ (y_{\delta\mathcal{F}} + \delta y_{\delta\mathcal{F}})^2 + (z_{\delta\mathcal{F}} + \delta z_{\delta\mathcal{F}})^2 \right\} dV \\[2mm] &= A_0 + 2 \int_V \rho(\boldsymbol{r}_{\delta\mathcal{F}}) \left\{ y_{\delta\mathcal{F}}\, \delta y_{\delta\mathcal{F}} + z_{\delta\mathcal{F}}\, \delta z_{\delta\mathcal{F}} \right\} dV \\[2mm] &= A_0 + \frac{k\, \xi}{\omega_{\delta 0}^2} \frac{G\, m_{\mathbb{C}}}{r_{\mathbb{C}}^3} \left\{ -A_0 + 2\, A_0 \right\} \\[2mm] &= \left(1 + \frac{k\, \xi}{\omega_{\delta 0}^2} \frac{G\, m_{\mathbb{C}}}{r_{\mathbb{C}}^3} \right) A_0\ . \end{aligned} \tag{2.108}$$

For symmetry reasons the same result is obtained for $B_{\mathbb{C}}$, whereas the third moment of inertia is smaller than $A_{\mathbb{C}} = B_{\mathbb{C}}$. The result may be summarized as follows:

$$A_{\mathbb{C}} = \left(1 + \frac{k\,\xi}{\omega_{\mathfrak{d}o}^2}\,\frac{G\,m_{\mathbb{C}}}{r_{\mathbb{C}}^3}\right) A_0$$

$$B_{\mathbb{C}} = \left(1 + \frac{k\,\xi}{\omega_{\mathfrak{d}o}^2}\,\frac{G\,m_{\mathbb{C}}}{r_{\mathbb{C}}^3}\right) A_0 \tag{2.109}$$

$$C_{\mathbb{C}} = \left(1 - \frac{2\,k\,\xi}{\omega_{\mathfrak{d}o}^2}\,\frac{G\,m_{\mathbb{C}}}{r_{\mathbb{C}}^3}\right) A_0 \ .$$

In the coordinate system used (third axis pointing into the direction geocenter $\rightarrow$ selenocenter) a body of mass M and principal moments of inertia $A_{\mathbb{C}} = B_{\mathbb{C}}$ and $C_{\mathbb{C}}$ gives rise to a potential of the form (see eqn. (I-3.157))

$$V_{\mathbb{C}}(r,\alpha) = \frac{GM}{r} + \frac{G}{2\,r^3}\,(A_{\mathbb{C}} - C_{\mathbb{C}})\,(3\cos^2\alpha - 1) \overset{\text{def}}{=} \frac{GM}{r} + \Delta V_{\mathbb{C}}(r,\alpha) \ , \tag{2.110}$$

where in view of the expressions (2.109) the term $A_{\mathbb{C}} - C_{\mathbb{C}}$ reads as

$$A_{\mathbb{C}} - C_{\mathbb{C}} = +\frac{3\,k\,\xi}{\omega_{\mathfrak{d}o}^2}\,\frac{G\,m_{\mathbb{C}}}{r_{\mathbb{C}}^3}\,A_0 \ . \tag{2.111}$$

The tide-induced modification $\Delta V_{\mathbb{C}}(r,\alpha)$ of the external potential of the Earth therefore may be written as

$$\Delta V_{\mathbb{C}}(r,\alpha) = \left\{3\,k\,\frac{A_0\,a_{\mathfrak{d}}^3}{M r^5}\right\}\left\{\frac{1}{2}\,\frac{G\,m_{\mathbb{C}}}{r_{\mathbb{C}}^3}\,r^2\,(3\cos^2\alpha - 1)\right\} \ . \tag{2.112}$$

In view of the definition of the tide-generating potential (2.103) and in view of the fact that $A_{\mathfrak{d}} \approx \frac{1}{3}\,M a_{\mathfrak{d}}^2$ we may further simplify the above result:

$$\Delta V_{\mathbb{C}}(r,\alpha) = k\,\left(\frac{a_{\mathfrak{d}}}{r}\right)^5 V_{\text{tid}\mathbb{C}}(r,\alpha) \ , \quad r \geq a_{\mathfrak{d}} \ . \tag{2.113}$$

On the sphere with radius $r \overset{\text{def}}{=} a_{\mathfrak{d}}$ the tide-induced external potential thus reads as

$$\Delta V_{\mathbb{C}}(r,\alpha) = k\,V_{\text{tid}\mathbb{C}}(r,\alpha) \ , \quad r = a_{\mathfrak{d}} \ . \tag{2.114}$$

Equation (2.114) may be used as the defining relation for the Love number k. Under these conditions we have:

$$0 \leq k \leq 1 \ , \tag{2.115}$$

where $k = 0$ holds for a rigid Earth, $k = 1$ for an ideally elastic Earth. According to the IERS conventions [71]

$$k = 0.3 \ , \tag{2.116}$$

which means that the Earth in its current state of development is far from being a rigid, but also far from being an ideally elastic body.

In order to solve the Liouville-Euler equations (I- 3.135) we have to transform the inertia tensor from the particular system used in eqns. (2.109) into the rigid, rotating coordinate system to which the eqns. (I- 3.135) refer to. If we designate with $\lambda_{\mathbb{C}}$ and $\beta_{\mathbb{C}}$ the longitude and latitude of the Moon for the time t considered, the transformation between the rigid rotating system and the system underlying eqns. (2.109) may be written as follows:

$$\begin{pmatrix} x \\ y \\ z \end{pmatrix} = \mathbf{R}_3 \left(\frac{\pi}{2} + \lambda_{\mathbb{C}} \right) \mathbf{R}_1 \left(\frac{\pi}{2} - \beta_{\mathbb{C}} \right) \mathbf{r}_{\delta\mathcal{F}} \overset{\text{def}}{=} \tilde{\mathbf{T}}_{\mathbb{C}}^T \mathbf{r}_{\delta\mathcal{F}} . \tag{2.117}$$

In view of the transformation equations (I- 3.77) and eqns. (2.109) for the inertia tensors the change of the inertia tensor due to lunar tides may be written as:

$$\delta \mathbf{I}_{\text{tid}\mathbb{C}\,\mathcal{F}} = \frac{k\,\xi}{\omega_{\delta 0}^2} \frac{G\,m_{\mathbb{C}}}{r_{\mathbb{C}}^3} \mathbf{A}_0 \, \tilde{\mathbf{T}}_{\mathbb{C}} \begin{pmatrix} 1 & 0 & 0 \\ 0 & 1 & 0 \\ 0 & 0 & -2 \end{pmatrix} \tilde{\mathbf{T}}_{\mathbb{C}}^T . \tag{2.118}$$

As mentioned earlier, the change of the inertia tensor due to the solar tides may be computed in an analogous manner:

$$\delta \mathbf{I}_{\text{tid}\odot\mathcal{F}} = \frac{k\,\xi}{\omega_{\delta 0}^2} \frac{G\,m_{\odot}}{r_{\odot}^3} \mathbf{A}_0 \, \tilde{\mathbf{T}}_{\odot} \begin{pmatrix} 1 & 0 & 0 \\ 0 & 1 & 0 \\ 0 & 0 & -2 \end{pmatrix} \tilde{\mathbf{T}}_{\odot}^T . \tag{2.119}$$

In order to solve the Liouville-Euler equations, the expressions (2.118, 2.119) and their time derivatives have to be taken into account when computing the terms $\mathbf{I}$ and $\dot{\mathbf{I}}$. In the program ERDROT the Liouville-Euler equations are used directly in the form (I- 3.137).

In order to gain additional insight into the structure of the solution of the Liouville-Euler equations for the elastic, rotationally symmetric Earth we also provide an approximate version of these equations. Up to terms of first order in the small quantities $\delta\boldsymbol{\omega}_{\delta\mathcal{F}}$ and $\dot{\boldsymbol{\omega}}_{\delta\mathcal{F}}$ these equations read as (compare eqns. (2.94)):

$$(1+k\xi)\,\dot{\omega}_{\delta\mathcal{F}_1} + \gamma_\delta \left(1 - \frac{k\xi}{\gamma_\delta} \right) \omega_{\delta\mathcal{F}_2} \omega_{\delta\mathcal{F}_3} = -\frac{\delta\dot{I}_{\text{tid}\delta\mathcal{F}_{13}}\,\omega_{\delta 0}}{A_0} + \frac{\delta I_{\text{tid}\delta\mathcal{F}_{23}}\,\omega_{\delta 0}^2}{A_0} + \frac{\ell_{\delta\mathcal{F}_1}}{A_0}$$

$$(1+k\xi)\,\dot{\omega}_{\delta\mathcal{F}_1} - \gamma_\delta \left(1 - \frac{k\xi}{\gamma_\delta} \right) \omega_{\delta\mathcal{F}_2} \omega_{\delta\mathcal{F}_3} = -\frac{\delta\dot{I}_{\text{tid}\delta\mathcal{F}_{23}}\,\omega_{\delta 0}}{A_0} - \frac{\delta I_{\text{tid}\delta\mathcal{F}_{13}}\,\omega_{\delta 0}^2}{A_0} + \frac{\ell_{\delta\mathcal{F}_2}}{A_0}$$

$$\dot{\omega}_{\delta\mathcal{F}_3} = -\frac{\delta\dot{I}_{\text{tid}\delta\mathcal{F}_{33}}\,\omega_{\delta 0}}{C_0} , \tag{2.120}$$

where the lunisolar torques may be dealt with exactly like in the case of the rigid body (see eqns. (I-3.124)) in our approximation. The system of equations (2.120) is an inhomogeneous, linear system of differential equations. The underlying homogeneous part is given by eqns. (2.94) and therefore needs no further discussion.

As opposed to the rigid body the inhomogeneous part does not only consist of the lunisolar torques, but in addition of the terms due to the lunisolar tides. Note in particular, that the third component of the angular velocity is no longer constant, but that the changes of the angular velocity are proportional to the changes of the third component of the inertia tensor in the rigid, rotating coordinate system. The solution of the third of equations (2.120) is

$$\omega_{\delta\mathcal{F}_3} = \omega_{\delta\mathcal{F}_0} - \frac{\delta I_{\text{tid}\delta\mathcal{F}33}}{C_0} \omega_{\delta 0} \overset{\text{def}}{=} \omega_{\delta\mathcal{F}_0} + \delta\omega_{\delta\mathcal{F}_3} . \tag{2.121}$$

This in turn gives rise to LOD changes of

$$\delta P_{\text{tid}} = -\frac{\delta\omega_{\delta\mathcal{F}_3}}{\omega_{\delta\mathcal{F}_0}} P_{\delta 0} = +\frac{\delta I_{\text{tid}\delta\mathcal{F}33}}{C_0} P_{\delta 0} , \tag{2.122}$$

where

$$P_{\delta 0} = \frac{2\pi}{\omega_{\delta 0}} \tag{2.123}$$

is the LOD of a rigid Earth with principal moments of inertia $A_0 = B_0$ and C_0.

With respect to a rigid Earth one expects an increase in the LOD. In the case of the lunar tides this increase is expected to assume the maximum value when the Moon lies in the equatorial plane, the minimum value when the Moon's argument of latitude is $u_{\mathbb{C}} \pm 90°$. In the case of the solar tides the maximum impact on the length of day is expected when the Sun crosses the equatorial plane, i.e., in spring and in fall.

Figure 2.45 shows the LOD variations in milliseconds as expected for an elastic Earth model with characteristics according to Table 2.1 (solid line) and the values corresponding to the C04-values of the IERS. The latter values are uniquely based on measurements as provided by the space geodetic techniques. Obviously, the elastic Earth model explains very well the monthly and bimonthly terms: Frequency and amplitude agree very well. It should be mentioned that only the variation and not the general offset between the two series should be interpreted – the LOD as emerging from our simulation is given by the initial condition as opposed to the C04-values characterizing the true length of day. Figure 2.46 contains the same type of information as that of Figure 2.45, but for a time interval of five years. One clearly sees that the actually measured values show much stronger LOD variations than those expected from a purely elastic model of the Earth. One notices in particular strong annual and semiannual variations. These effects will be explained in the next paragraph.

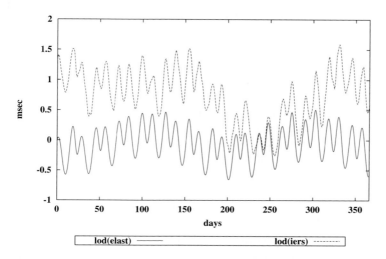

Fig. 2.45. LOD variations in year 2000 (elastic Earth (solid line) and according to C04 (dotted line))

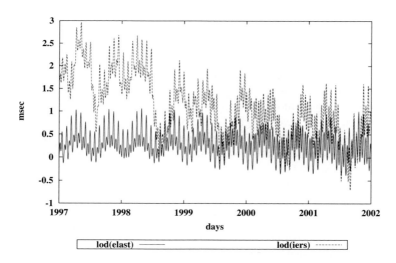

Fig. 2.46. LOD variations 1997-2001 (elastic Earth (solid line) and according to C04 (dotted line))

2.3.3 Atmosphere and Oceans

Introduction. Figures of type 2.38 may have served as early proofs that the Earth's atmosphere, with its pronounced annual variations of temperature, pressure, and wind, plays a key role to explain at least some of the short-period excursions (with periods from days to months) of PM. The relation between atmosphere and Earth rotation was firmly established in the era of space geodesy. VLBI and satellite geodetic results did not only explain the annual and semiannual terms in PM as a consequence of atmospheric effects, they did also reveal the strong correlation between LOD variations as observed, e.g., in Figures I- 8.21 and 2.39 and atmospheric variations. The GARP (Global Atmospheric Research Program) Experiments, which were performed in the 1980s, were instrumental to study this relationship. The fundamental reference for investigations of this kind is due to Barnes [5].

In order to study the impact of the atmosphere on Earth rotation it is of vital importance that meteorological observations with a high temporal and spacial resolution are available on a global basis. Institutions coordinating this work are the *U.S. National Meteorological Center* of NOAA (National Oceanic and Atmospheric Administration), the ECMWF (European Center for Medium-Range Weather Forecasts), the Japan Meteorological Agency, and the U.K. Meteorological Office. Their products relevant for Earth rotation are available through the IERS Special Bureau for Atmospheric Angular Momentum (see [96]).

It is intuitively clear that the Ocean tides influence Earth rotation as well. The tides are also of considerable importance in practice, e.g., for navigation on sea. It is therefore not amazing that a considerable number of so-called *tide gauges* was deployed along the coast lines world-wide. Tide gauges register local changes of sea level height. A significant number of these tide gauges are operational for centuries. When analyzing the data of tide gauges, it becomes clear that the elevation $\zeta(\lambda, \phi, t)$ of the actual sea surface above the equipotential surface at sea level is a very complicated function of the *tide-generating potential* (2.100).

Since the early days of satellite geodesy tide gauge measurements are complemented by altimeter measurements. Altimeter missions allow it to establish the sea surface topography with unprecedented spacial and temporal resolution through radar measurements establishing the radial distance between the satellite and the sea surface. The TOPEX/Poseidon Mission and its follow-up mission *Jason* should be mentioned in this context (see, e.g., [43]).

The establishment of an accurate equipotential surface (few cm accuracy is required), the reference surface for the altimeter measurements, is one of the most demanding problems in geodesy today. This surface is determined by analyzing terrestrial gravity measurements, the orbits of artificial satellites, and by dedicated gravity missions. Modern dedicated gravity missions have two key elements: *On-board GPS receivers* for precise orbit determination (cm-accuracy) and gravity meter and gradiometer measurements to derive in situ the first and second derivative (in space) of the gravitational potential. Gravity meters may, e.g., be realized by "twin satellites" (where the relative motion of two neighbored satellites is the basic observed quantity) or by sets of accelerometers within the space vehicles. The gravity missions CHAMP, GRACE (Gravity Recover and Climate Experiment) and GOCE (Gravity field and steady-state Ocean Circulation Experiment) will provide the observational basis for a new equipotential surface of unprecedented accuracy. Gravity and altimeter missions together allow it to monitor the sea surface and the ocean currents with a very high accuracy. A variable sea surface and ocean currents induce changes of the Earth's tensor of inertia, leading in turn to variations in Earth rotation.

For a rigid Earth with oceans the variations of the inertia tensor are uniquely given by the changes of the sea surface height. For a deformable Earth so-called *ocean loading effects* have to be considered in addition. Ocean loading depresses and tilts the continents due to the ocean tides. Loading effects are small when compared to the direct tidal effects. We will not discuss them in this introductory text. We have to keep in mind, however, that they show up in VLBI and satellite geodetic results.

Liouville-Euler Equations for a Rigid Earth with Atmosphere and Oceans. The rotation of a rigid Earth surrounded by an atmosphere and partly covered by oceans may be described by the Liouville-Euler equations in the form (I-3.133). The rigid rotating coordinate system may be associated with the Earth-fixed PAI-system. In view of the fact that the equations (I-3.133) turned out to be linear in first approximation we may study the effect of the atmosphere and the oceans separately form the external torques (i.e., by ignoring them). Assuming rotational symmetry for the rigid part of the Earth, the Liouville-Euler equations for the Earth consisting of the rigid part, atmosphere, and oceans, may be written in the form

$$\frac{d}{dt}\left\{\mathbf{I}_{\delta_{\mathcal{F}}}\,\boldsymbol{\omega}_{\delta}+\boldsymbol{\kappa}_{ao}\right\}+\boldsymbol{\omega}_{\delta}\times\left\{\mathbf{I}_{\delta_{\mathcal{F}}}\,\boldsymbol{\omega}_{\delta}+\boldsymbol{\kappa}_{ao}\right\}=\mathbf{0}\;,\qquad(2.124)$$

where

$$\mathbf{I}_{\delta_{\mathcal{F}}}=\begin{pmatrix}A_0 & 0 & 0\\ 0 & A_0 & 0\\ 0 & 0 & C_0\end{pmatrix}+\delta\mathbf{I}_{ao}\qquad(2.125)$$

is the sum of the rigid Earth's inertia tensor and the contribution of the atmosphere and the oceans. According the general definition (I-3.74) of the inertia tensor the contribution of the atmosphere and the oceans may be written as

$$\delta\mathbf{I}_{ao}\stackrel{\text{def}}{=}\int\limits_{V}\rho(\boldsymbol{r}_p)\left(r_p^2\,\boldsymbol{E}-\boldsymbol{r}_p\otimes\boldsymbol{r}_p\right)dV\;,\qquad(2.126)$$

where the integration has to be extended over the volume occupied by the atmosphere and the oceans.

When adopting a rigid Earth model, the inner angular momentum $\boldsymbol{\kappa}_\delta$ in eqn. (I-3.130) is uniquely defined by $\boldsymbol{\kappa}_{ao}$ representing the angular momentum vector due to the atmosphere and the oceans relative to the PAI-system:

$$\boldsymbol{\kappa}_{ao} = \int_V \rho(\boldsymbol{r}_p)\, \boldsymbol{r}_p \times \delta\dot{\boldsymbol{r}}\; dV \; . \tag{2.127}$$

The velocities $\delta\dot{\boldsymbol{r}}$ are those of the particles p (of atmosphere and/or oceans) relative to Earth-fixed PAI-system.

The integration has to be extended over the volume occupied by the atmosphere and the oceans (the contributions of the two non-rigid constituents will be studied separately below).

It is important to note that the atmospheric and oceanic contributions (2.126) to the inertia tensor and to the angular momentum (2.127) may be computed from meteorological measurements (atmospheric contribution) and ocean-related measurements (ocean contribution). If known, they may be used as a priori information when solving the Liouville-Euler equations. One might then associate changes in PM and in LOD with these quantities.

When monitoring PM and LOD with space geodetic methods, one usually faces the situation, however, that such information is not available with sufficient precision in due time. Therefore, one has to introduce the current position (and velocity) of PM and LOD as parameters in space geodetic analyses. Usually, space geodetic estimates of these quantities are established with a daily resolution. Long series of LOD and PM are thus available from space geodesy with a time resolution of (at least) one day.

A posteriori, it is therefore possible to compare LOD and PM changes as computed from meteorological and/or oceanic measurements with the corresponding results emerging from space geodesy. Following Barnes [5] such comparisons are not performed directly for PM and LOD, but for the so-called *AMF (angular momentum functions)*. These functions may be defined quite naturally by re-arranging the Liouville-Euler equations (2.124). As usual, the terms of higher than the first order in small quantities are ignored. The result simply reads as (compare with eqns. (2.10)):

$$\dot{\omega}_{\delta\mathcal{F}_1} + \gamma_\delta\, \omega_{\delta 0}\, \omega_{\delta\mathcal{F}_2} = +\frac{1}{A_0}\left(\delta I_{23}\, \omega_{\delta 0}^2 - \delta\dot{I}_{13}\, \omega_{\delta 0} - \dot{\kappa}_1 + \kappa_2\, \omega_{\delta 0}\right)$$

$$\dot{\omega}_{\delta\mathcal{F}_2} - \gamma_\delta\, \omega_{\delta 0}\, \omega_{\delta\mathcal{F}_1} = -\frac{1}{A_0}\left(\delta I_{13}\, \omega_{\delta 0}^2 + \delta\dot{I}_{23}\, \omega_{\delta 0} + \dot{\kappa}_2 + \kappa_1\, \omega_{\delta 0}\right) \tag{2.128}$$

$$\dot{\omega}_{\delta\mathcal{F}_3} = -\frac{1}{C_0}\left(\delta\dot{I}_{33}\, \omega_{\delta 0} + \dot{\kappa}_3\right) \; ,$$

where the index ao was skipped to simplify the notation. Not surprisingly, eqns. (2.128) are a linear system of first-order differential equations. To the approximation needed only the third column of matrix $\delta\mathbf{I}_{ao}$ needs to be known (the other components of the tensor only give rise to second order contributions because they are multiplied with the PM components).

The third of eqns. (2.128) may be separated from the first two. Its solution is readily obtained by

$$\omega_{\delta\mathcal{F}_3} = \omega_{\delta 0} - \frac{1}{C_0}\left(\delta I_{33}\,\omega_{\delta 0} + \kappa_3\right) . \tag{2.129}$$

The relative change of the Earth's angular velocity induced by the atmosphere and the oceans is thus given by

$$\frac{\delta\omega_{\delta\mathcal{F}_3}}{\omega_{\delta 0}} = -\frac{\delta I_{33}\,\omega_{\delta 0} + \kappa_3}{\omega_{\delta 0},C_0} \stackrel{\text{def}}{=} -\chi_3 . \tag{2.130}$$

χ_3 is usually referred to as the *axial angular momentum function*. The term seems to have been coined by Barnes [5]. Obviously, the axial angular momentum function consists of two terms, the first due to the current mass distribution and the second due to the motion of the atmosphere relative to the Earth-fixed system.

The actual LOD (modified by the atmosphere and the oceans) is then given by

$$\frac{2\,\pi}{\omega_{\delta 0} + \delta\omega_{\delta\mathcal{F}_3}} = \frac{2\,\pi}{\omega_{\delta 0}}\left(1 - \frac{\delta\omega_{\delta\mathcal{F}_3}}{\omega_{\delta 0}}\right) = \frac{2\,\pi}{\omega_{\delta 0}}\left(1 + \chi_3\right) , \tag{2.131}$$

implying that the excess day length ΔLOD due to the atmosphere and the oceans may be written as

$$\frac{\Delta\text{LOD}}{\text{LOD}} = \chi_3 , \tag{2.132}$$

because $P_{\text{sid}} \stackrel{\text{def}}{=} \text{LOD} = \frac{2\,\pi}{\omega_{\delta 0}}$ is the nominal length of the sidereal day.

Equation (2.132) tells that the relative LOD change is given by the axial angular momentum function χ_3. Obviously, the above equation may also be used to compute χ_3 associated with LOD changes estimated from space geodetic time series (after having subtracted the tidal contributions). As χ_3 also may be computed from meteorological and/or oceanographic measurements alone using eqns. (2.126) and (2.127), χ_3 is perfectly suited to compare space geodetic axial angular momentum functions with those resulting from meteorology and/or oceanography.

In analogy to the axial angular momentum function two *equatorial angular momentum functions* may be defined. The definition is somewhat arbitrary. According to Barnes [5] the two functions are defined by

$$\chi_1 \overset{\text{def}}{=} \frac{\delta I_{13}\, \omega_{\delta 0} + \kappa_1}{\omega_{\delta 0}\,(C_0 - A_0)}$$

$$\chi_2 \overset{\text{def}}{=} \frac{\delta I_{23}\, \omega_{\delta 0} + \kappa_2}{\omega_{\delta 0}\,(C_0 - A_0)} \ . \tag{2.133}$$

With these definitions the first two of the Liouville-Euler equations may be written as

$$\dot{\omega}_{\delta \mathcal{F}_1} + \gamma_\delta\, \omega_{\delta 0}\, \omega_{\delta \mathcal{F}_2} = +\gamma_\delta\, \omega_{\delta 0}^2 \left(\chi_2 - \frac{\dot{\chi}_1}{\omega_{\delta 0}} \right)$$

$$\dot{\omega}_{\delta \mathcal{F}_2} - \gamma_\delta\, \omega_{\delta 0}\, \omega_{\delta \mathcal{F}_1} = -\gamma_\delta\, \omega_{\delta 0}^2 \left(\chi_1 + \frac{\dot{\chi}_{\delta 2}}{\omega_{\delta 0}} \right) \ . \tag{2.134}$$

Assuming that PM (i.e., the components of the angular velocity vector $\boldsymbol{\omega}_{\delta \mathcal{F}}$ and its first derivative) has been determined as a time series using space geodetic methods, eqns. (2.134) may be interpreted as a system of differential equations for the first two components of the angular momentum function. If the initial values are given for an (arbitrary) initial epoch, the angular momentum components may be reconstructed unambiguously. It is a straightforward procedure to show that in general

$$| \chi_1\, \omega_{\delta 0} | \gg | \dot{\chi}_2 | \quad \text{and} \quad | \chi_2\, \omega_{\delta 0} | \gg | \dot{\chi}_1 | \tag{2.135}$$

which is why a simplified relationship between PM and the equatorial angular momentum functions may be written as:

$$\dot{\omega}_{\delta \mathcal{F}_1} + \gamma_\delta\, \omega_{\delta 0}\, \omega_{\delta \mathcal{F}_2} \approx +\gamma_\delta\, \omega_{\delta 0}^2\, \chi_2$$

$$\dot{\omega}_{\delta \mathcal{F}_2} - \gamma_\delta\, \omega_{\delta 0}\, \omega_{\delta \mathcal{F}_1} \approx -\gamma_\delta\, \omega_{\delta 0}^2\, \chi_1 \ . \tag{2.136}$$

Keeping in mind the relation (2.4) between the PM components and the angular velocity components, one obtains a simplified relationship between the equatorial angular momentum functions and the PM components:

$$\chi_2 \approx -y + \frac{\dot{x}}{\gamma_\delta\, \omega_{\delta 0}}$$

$$\chi_1 \approx +x + \frac{\dot{y}}{\gamma_\delta\, \omega_{\delta 0}} \ , \tag{2.137}$$

where the PM and its first derivatives have to be measured in radian (and radian per time unit). The above equations thus may be used to calculate approximately the equatorial angular momentum function as a function of the PM components.

Earth Rotation and Atmosphere. The atmosphere-induced contributions to the inertia tensor and the inner angular momentum due to the atmosphere have to be calculated now. According to the Liouville-Euler equations (2.128) only three terms are needed, namely:

$$\delta I_{13} = -\int_V \rho(\mathbf{r}_{\delta \mathcal{F}}) \, r_{\delta \mathcal{F}_1} r_{\delta \mathcal{F}_3} \, dV$$

$$\delta I_{23} = -\int_V \rho(\mathbf{r}_{\delta \mathcal{F}}) \, r_{\delta \mathcal{F}_2} r_{\delta \mathcal{F}_3} \, dV \tag{2.138}$$

$$\delta I_{33} = +\int_V \rho(\mathbf{r}_{\delta \mathcal{F}}) \left(r_{\delta \mathcal{F}_1}^2 + r_{\delta \mathcal{F}_2}^3 \right) dV \ ,$$

where the integration has to be extended over the volume V occupied by the atmosphere. $dM = \rho(\mathbf{r}_{\delta \mathcal{F}}) \, dV$ is the atmospheric mass contained in the volume element. This is why the above terms also are referred to as *mass terms* or *pressure terms* of the atmospheric contribution to Earth rotation. Using the well-known relation of hydrostatic equilibrium,

$$\frac{dp}{dr} = -g\rho \ , \tag{2.139}$$

between pressure p and density ρ, the above volume integrals may be transformed into integrals over geographical longitude λ, latitude ϕ, and pressure p. The transformation matters, because global pressure fields are available from meteorology. Using the relation

$$dV = r^2 \cos\phi \, d\lambda \, d\phi \, dr = -\frac{r^2 \cos\phi}{g\rho} \, dp \, d\lambda \, d\phi \tag{2.140}$$

we may transform the above volume integrals as follows:

$$\delta I_{13} = +\int\int\int r_{\delta \mathcal{F}_1} r_{\delta \mathcal{F}_3} \frac{r^2 \cos\phi}{g} \, dp \, d\lambda \, d\phi$$

$$\delta I_{23} = +\int\int\int r_{\delta \mathcal{F}_2} r_{\delta \mathcal{F}_3} \frac{r^2 \cos\phi}{g} \, dp \, d\lambda \, d\phi \tag{2.141}$$

$$\delta I_{33} = -\int\int\int \left(r_{\delta \mathcal{F}_1}^2 + r_{\delta \mathcal{F}_2}^2 \right) \frac{r^2 \cos\phi}{g} \, dp \, d\lambda \, d\phi \ .$$

The integration over the pressure p must be extended from the Earth's surface, where the pressure is $p = p_s$, to the upper atmosphere boundary, where $p = 0$. Adopting a spherical model for the rigid Earth with radius $r = R_\delta$, making use of the fact that the atmosphere is thin (a few km), and neglecting the dependency of the gravity constant g on height, i.e., using $g \stackrel{\text{def}}{=} g_\delta$, where g_δ is the gravitational acceleration at sea level, the above triple integral may be reduced to a surface integral:

$$\delta I_{13} = - \frac{R_\delta^4}{g_\delta} \iint p_s \cos^2 \phi \sin \phi \cos \lambda \, d\lambda \, d\phi$$

$$\delta I_{23} = - \frac{R_\delta^4}{g_\delta} \iint p_s \cos^2 \phi \sin \phi \sin \lambda \, d\lambda \, d\phi \qquad (2.142)$$

$$\delta I_{33} = + \frac{R_\delta^4}{g_\delta} \iint p_s \cos^3 \phi \, d\lambda \, d\phi \, .$$

The above integrals may be solved with the methods of numerical quadrature using the global surface pressure fields $p_s(\lambda, \phi)$ available from meteorology.

The inner angular momentum of the atmosphere and the oceans is defined by eqn. (2.127). Assuming a spherical Earth, using the approximation $g \stackrel{\text{def}}{=} g_\delta$ within the entire atmosphere, replacing the density by the pressure (using relation (2.139)), and neglecting the contribution of the oceans, these integrals may be transformed as follows (due to the dependence of pressure on height, the integration over the pressure is non-trivial in this case):

$$\kappa_{\text{atm}_1} = - \frac{R_\delta^3}{g_\delta} \iiint \left(\cos \phi \sin \lambda \, \dot{r}_{\delta \mathcal{F}_3} - \sin \phi \, \dot{r}_{\delta \mathcal{F}_2} \right) \cos^2 \phi \, dp \, d\lambda \, d\phi$$

$$\kappa_{\text{atm}_2} = - \frac{R_\delta^3}{g_\delta} \iiint \left(\sin \phi \, \dot{r}_{\delta \mathcal{F}_1} - \cos \phi \cos \lambda \, \dot{r}_{\delta \mathcal{F}_3} \right) \cos^2 \phi \, dp \, d\lambda \, d\phi$$

$$\kappa_{\text{atm}_3} = - \frac{R_\delta^3}{g_\delta} \iiint \left(\cos \phi \cos \lambda \, \dot{r}_{\delta \mathcal{F}_2} - \cos \phi \sin \lambda \, \dot{r}_{\delta \mathcal{F}_1} \right) \cos^2 \phi \, dp \, d\lambda \, d\phi \, ,$$

$$(2.143)$$

where the integration over p has to be performed from $p = p_s$ to $p = 0$. Obviously three-dimensional wind profiles are required to evaluate the above integrals. This is why the above terms are also referred to as the *wind terms* of the atmospheric contribution to Earth rotation.

If the integrals (2.142) and (2.143) are available from meteorology, we are now in a position to compare the angular momentum functions as resulting from meteorology with those emerging from space geodetic analyses for a rigid Earth.

It turns out that the required modifications of the above formulae when replacing the rigid by an elastic Earth are minor. First, we have to replace eqns. (2.137) relating the PM components with the two axial angular momentum functions by

$$\chi_2 \approx -y + \frac{\dot{x}}{\gamma_{\delta}\left(1 - \frac{k\,\xi}{\gamma_{\delta}}\right)\omega_{\delta 0}}$$

$$\chi_1 \approx +x + \frac{\dot{y}}{\gamma_{\delta}\left(1 - \frac{k\,\xi}{\gamma_{\delta}}\right)\omega_{\delta 0}} \quad . \tag{2.144}$$

This means in essence that the Euler period was replaced by the Chandler period. Then, the angular momentum functions have to be replaced by the so-called *effective angular momentum functions*. According to Barnes [5] these functions are defined by

$$\chi_{\text{eff}_1} = 1.43 \cdot (0.7 \cdot \chi_{p_1} + \chi_{w_1})$$

$$\chi_{\text{eff}_2} = 1.43 \cdot (0.7 \cdot \chi_{p_2} + \chi_{w_2}) \tag{2.145}$$

$$\chi_{\text{eff}_3} = 0.7 \cdot \chi_{p_3} + \chi_{w_3} \, ,$$

where the χ_{p_i}, $i = 1, 2, 3$, are the angular momentum functions due to the *pressure* or *mass term*, χ_{w_i}, $i = 1, 2, 3$, those due to the *wind term*. The different behavior of the two types of contributions is caused by the loading effects associated with the pressure terms (i.e., due to the fact that an elastic Earth is slightly deformed by the applied atmospheric pressure).

We are now in a position to compare the atmospheric angular momentum functions as computed from meteorological data and the angular momentum functions computed from PM and LOD data series as emerging from space geodesy. We use two different data series from space geodesy and one from the analysis of meteorological data, namely

- the series x, y, and ΔLOD of PM with a daily time resolution as established by the CODE Analysis Center of the IGS between 1993 and early June 2002.

- C04 PM and ΔLOD series with a time resolution of one day between 1963 and early June 2002 computed by the IERS (series available at http://hpiers.obspm.fr).

- Reanalysis time series produced by the IERS Special Bureau for the Atmosphere (SBA), which is a joint effort of Atmospheric and Environmental Research, Inc. (AER) and the NCEP (U.S. National Centers for Environmental Prediction) to provide atmospheric data relevant to the study of the Earth's variable rotation. The spacing between subsequent sets of data is six hours. For the purpose of comparison plain daily mean values were produced. More elaborate schemes may be found in [65].

Let us first focus on comparisons based on the GPS-derived series (which at the time of writing this text have a length of about nine years). Figure 2.47 shows the third component of the axial angular momentum function, once computed with the CODE time series (solid line), once with the meteorological series from NCEP. For specialists in the field we mention that pressure and wind data were used; the so-called inverted barometer option was used to take the pressure contribution into account (consult [5] or [65] for more information). Only a mean value was removed from the two series in Figure

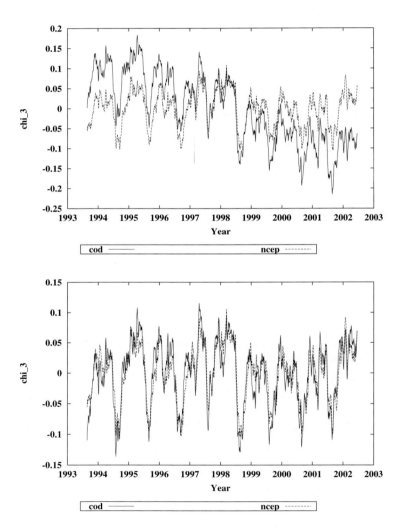

Fig. 2.47. Axial angular momentum function χ_3 from GPS analysis (solid line) and from meteorological data (NCEP reanalysis)

2.47 (top), whereas second-degree polynomials, fitted to the individual series, were subtracted from the original series in Figure 2.47 (bottom).

The two series are highly correlated when considering annual or shorter period variations. The correlation coefficient related to the top figure is $r = 0.60$, it is $r = 0.98$ when analyzing Figure 2.47 (bottom). Remember, that the correlation coefficient r between two time series x_i, y_i, $i = 1, 2, \ldots, n$, is defined by

$$r = \frac{\sum_{i=1}^{n}(x_i - \bar{x})(y_i - \bar{y})}{\sqrt{\sum_{i=1}^{n}(x_i - \bar{x})^2}\sqrt{\sum_{i=1}^{n}(y_i - \bar{y})^2}} \stackrel{\text{def}}{=} \sum_{i=1}^{n}\tilde{x}_i\,\tilde{y}_i \ . \qquad (2.146)$$

A correlation coefficient of $r = 0.98$ implies that most of the short period (annual periods or shorter) are explained by interactions between the solid part of the Earth (to which the space geodetic observatories are attached) and the atmosphere.

Figure 2.48 contains the amplitude spectra of the time series illustrated in Figure 2.47. With the exception of the semiannual term the two spectra match almost perfectly – at least in the range of periods shown. The correlation is even somewhat higher when taking the oceanic angular momentum into account (see [65]).

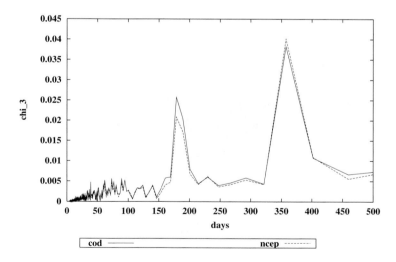

Fig. 2.48. Amplitude spectra of CODE-derived (solid line) and NCEP-derived χ_3-series

Figure 2.49 shows the first (top) and second (bottom) equatorial angular momentum functions χ_1 and χ_2 resulting from the CODE- and the NCEP-time series (after removing the best fitting polynomials of degree 2 from the contributing series). The correlation is not as striking as in the case of the axial angular momentum function χ_3. Nevertheless, the correlation coefficients are $r(\chi_1) = 0.59$ and $r(\chi_2) = 0.74$. The correlation between the

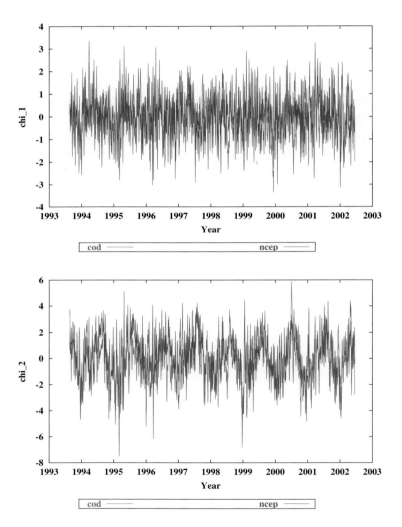

Fig. 2.49. Equatorial angular momentum function χ_1 (top) and χ_2 (bottom) from GPS analysis (solid line) and from meteorological data (NCEP reanalysis)

two series is perhaps better illustrated by Figures 2.50, containing the same type of information as in Figure 2.49 but only for the year 2001.

It is interesting to note that χ_2 shows more structure than χ_1. Interpreting χ_1 and χ_2 as the two components of one vector (what they actually are) one

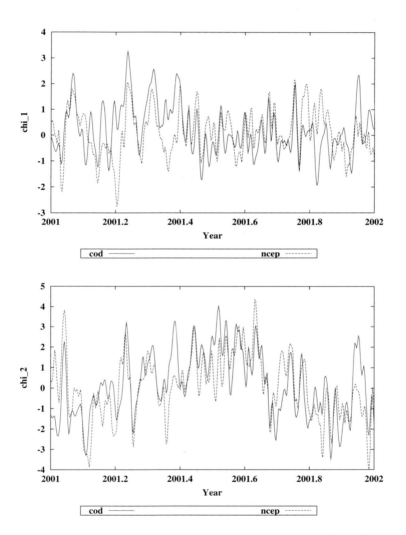

Fig. 2.50. Equatorial angular momentum function χ_1 (top) and χ_2 (bottom) in 2001 from GPS analysis (solid line) and from meteorological data (NCEP reanalysis)

obtains the spectra shown in Figure 2.51. The agreement is not too bad in the case of the terms with an annual period. The shorter periods would need more attention. Exactly as in the case of χ_3 the tide-induced terms should be removed before the comparison.

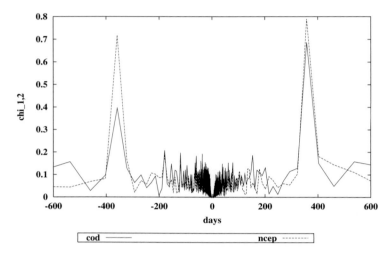

Fig. 2.51. Amplitude spectra of equatorial angular momentum functions χ_1 and χ_2 from GPS analysis (solid line) and from meteorological data (NCEP reanalysis)

Earth Rotation and Oceans. When studying the effects of ocean tides in PM and LOD one may assume the generating frequencies and periods to agree with those of the tide-generating potential. These periods may be classified according to a scheme used by Arthur T. Doodson (1890–1968) (see [32]), who introduced six "independent" variables. Table 2.6 defines the so-called *Doodson variables.* Alternatively, the fundamental arguments of precession, nutation (see [107]) and the Greenwich mean sidereal time, may be used. The fundamental arguments are included in Table 2.6 as well. Both sets of arguments are equivalent. Recently, the fundamental arguments (also called *Delaunay variables*, named after Charles Eugène Delaunay (1816–1872)) were used almost exclusively. This is why this second set will be referred to subsequently.

In the frequency domain the tidal potential may be expressed as a series of discrete tides. Table 2.7 contains the most important constituents of this decomposition of the tide-generating potential. A more extensive list may, e.g., be found in [76].

Table 2.6. Doodson-variables (top) and fundamental nutation arguments (bottom)

Argument	Period	Definition
τ	1.03505 days	Mean Lunar Time
s	27.32158 days	Mean Ecliptical Longitude of Moon
h	1 year	Mean Ecliptical Longitude of Sun
p	8.85 years	Mean Ecliptical Longitude of Lunar Perigee
$N' = -\Omega_{\mathbb{C}}$	18.61 years	Negative Mean Longitude of Lunar Ascending Node
p_s	$20'940$ years	Mean Ecliptical Longitude of Solar Perigee
$F_1 = l$	27.53 days	Mean Anomaly of Moon
$F_2 = l'$	365.26 days	Mean Anomaly of Sun
$F_3 = F$	27.21 days	Moon's Argument of Latitude
$F_4 = D$	29.53 days	Moon's Mean Elongation from the Sun
$F_5 = \Omega$	18.61 years	Ecliptical Longitude of Lunar Ascending Node
$F_6 = \Theta$	$23^{\mathrm{h}}56^{\mathrm{m}}$	Greenwich Mean Sidereal Time

Table 2.7. Leading terms of the tide-generating potential

Tide	Multiple of						Period	V
	l	l'	F	D	Ω	$\Theta + \pi$	[hours]	$[\mathrm{mm}^2/\mathrm{s}^2]$
K_1	0	0	0	0	0	1	23.93	-930
O_1	0	0	2	0	2	1	25.82	$+660$
P_1	0	0	-2	2	-2	1	24.07	$+308$
M_2	0	0	-2	0	-2	2	12.42	$+795$
S_2	0	0	-2	2	-2	2	12.00	$+370$
N_2	-1	0	-2	0	-2	2	12.66	$+152$
K_2	0	0	0	0	0	2	11.97	$+101$

Naturally, one expects to find the periods given in Table 2.7 in the tidally driven variations of the Earth potential. These variations may be expressed as follows (e.g., Gipson [46]):

$$\Delta x(t) = \sum_{j=1}^{n} \left(-p_j^c \cos \phi_j(t) + p_j^s \sin \phi_j(t) \right)$$

$$\Delta y(t) = \sum_{j=1}^{n} \left(+p_j^s \sin \phi_j(t) + p_j^c \cos \phi_j(t) \right) \tag{2.147}$$

$$\Delta \mathrm{UT1}(t) = \sum_{j=1}^{n} \left(+u_j^c \cos \phi_j(t) + u_j^s \sin \phi_j(t) \right) ,$$

where Δx, Δy, and $\Delta \mathrm{UT1}$ are the tidal variations in the x- and y-component of PM and in UT1, respectively. n is the number of tides considered. p_j^c and

p_j^s are the cosine and sine amplitudes of the tidal variations in PM, u_j^c and u_j^s the corresponding amplitudes in UT1. The angle argument $\phi_j(t)$ denotes a linear combination of the six following arguments:

$$\phi_j(t) = \sum_{i=1}^{6} N_{ij} \, F_i(t) \;, \qquad (2.148)$$

where N_{ij} are the integer multipliers of the five fundamental nutation arguments $F_i \, (i = 1, 2, \ldots, 5)$, and of $F_6 = \Theta + \pi$. They uniquely characterize the tide with index j. The circular frequency of tide j is given by $\omega_j \overset{\text{def}}{=} d\phi_j/dt$. The above defined use of the numbers N_{ij} to characterize the tides follows the convention introduced by Woolard [132].

Tidal terms with the multiplier $N_6 \in \{1, 2, -1, -2\}$ are called prograde diurnal (1), prograde semidiurnal (2), retrograde diurnal (−1), and retrograde semi-diurnal tides (−2), respectively. The terms of the generating potential are ordered in this way in Table 2.7.

In order to obtain the tidally-induced LOD variations, one has to take the *time derivative* of the third of eqns. (2.147):

$$\frac{d}{dt} \left(\Delta \text{UT1}\right)(t) = \sum_{j=1}^{n} \omega_j \left(- u_j^c \sin \phi_j(t) + u_j^s \cos \phi_j(t)\right) \;. \qquad (2.149)$$

From the daily drift values one may easily calculate the associated LOD variations associated with the individual tide constituents. This technique was used by Rothacher et al. [93] to estimate the parameters u_j^c, u_j^s in the above equation.

Figures 2.52 and 2.53 show the power spectrum of a Fourier analysis of a CODE PM- and LOD-solution with a two hours time resolution. Only the portions relevant for ocean tides, i.e., the diurnal and semidiurnal terms, are reproduced here. Note, that the power associated with these terms is small (the major contributions are at longer periods). All the periods expected from Table 2.7 are actually found in the figures mentioned, however. The reader is invited to identify the individual terms. As expected from the above short excursion into theory, prograde and retrograde terms associated with the tide constituents of Table 2.7 are found in Figure 2.53, except for the retrograde diurnal terms, where essentially no signal is observed. Retrograde diurnal terms do exist. They are, however, not observable using the technique outlined here if "only" satellite-geodetic observations are available: the retrograde diurnal terms are perfectly correlated with the nodes of the satellite orbits.

The retrograde diurnal band of PM would contain in essence the first derivative of the nutation time series (i.e., of the motion of the Earth pole in space). Euler's kinematic equations (I- 3.67) or (I- 3.68) are but another form of this

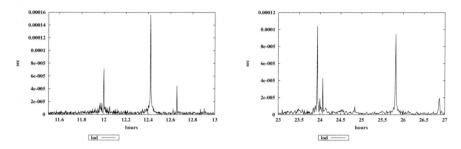

Fig. 2.52. Power spectrum of CODE LOD series with two hours time resolution

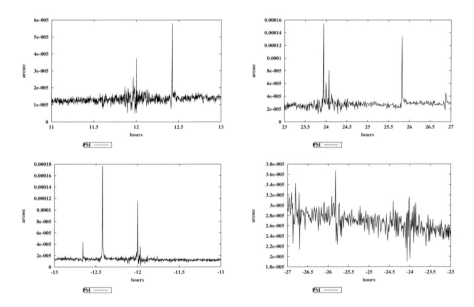

Fig. 2.53. Power spectrum of CODE PM series with two hours time resolution

statement. It is nevertheless possible to access this information (to some extent) using the GPS: It is possible to solve for the first derivatives of nutation in longitude and obliquity when processing the GPS data of a global network of observatories (together with all the other parameters necessary to accomplish this job). Time series of such nutation drift estimates may then be used to solve directly for the nutation parameters. We refer to Rothacher et al. [92] for a comprehensive discussion of this technique. VLBI is of course in a better position in this respect. This is why the nutation parameters are today (almost uniquely) determined with this technique. We refer to Herring [55].

The question naturally arises whether the spectral lines observed are actually due to the ocean tides. This relationship could be firmly established in the series of articles [93], [19], [128]. Figure 2.54 supports this interpretation, as well: If the effects due to official IERS ocean tide model are subtracted from the measured CODE series, there is only very little power left in the relevant bands. The interpretation of these figures is left to the careful readers. A similar result is obtained when comparing the subdaily results for PM. For a complete documentation of the exploitation of the CODE high-frequency PM series, the reader is referred to [93].

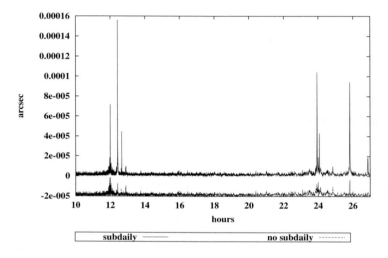

Fig. 2.54. Power spectrum of CODE LOD series with two hours time resolution; solid line: "normal" series, dotted line (shifted by $-0.00002''$) after subtraction of IERS Ocean tide model)

2.3.4 The Poincaré Earth Model

Composite Earth Models. So far, we treated the Earth either as a rigid body, an elastic body, or as a rigid body surrounded by oceans and atmosphere. In the latter case the Earth rotation was studied as that of a rigid (or solid) body, allowing for "small, known" motions of the non-rigid parts w.r.t. the rigid (solid) body expressed by a time-dependent inertia tensor and known "inner" angular momentum of the non-rigid parts. Today's knowledge of the Earth's interior in principle forbids such simplifying treatments. Figure 2.55 illustrates a more realistic model of the Earth (resembling somewhat the layers of an onion). From the center to the surface one has to distinguish

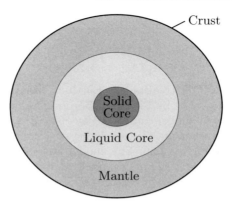

Fig. 2.55. Realistic Earth model

- a *solid inner core*,
- a *liquid outer core*,
- an *elastic mantle*,
- an *elastic crust*, and
- *atmosphere* and *oceans* (not shown in Figure 2.55).

With a thickness of about 30 km the crust is extremely thin. Note that all space geodetic measurements refer to the crust, because all relevant observatories are located at the outer border of this layer. The mantle is assumed to have a thickness of about 2860 km. The boundary between the mantle and the liquid core thus is located at a geocentric distance of about 3485 km. The radius of the solid inner core is estimated to be about 1220 km. Most of the knowledge of the Earth's interior is due to seismic measurements.

More precisely, the boundary between the core and the mantle is assumed to be an ellipsoid with rotational symmetry with a flattening of

$$\varepsilon_{\delta_c} = \frac{a_{\delta_c} - b_{\delta_c}}{a_{\delta_c}} \approx \frac{C_{\delta_c} - A_{\delta_c}}{C_{\delta_c}} \approx \frac{1}{391} \; . \tag{2.150}$$

a_{δ_c} and b_{δ_c} are the semi-major and semi-minor axes of the ellipsoid, C_{δ_c} and A_{δ_c} are the polar (maximum) and equatorial (minimum) principal moments of inertia of the (inner plus outer) core. Note that the estimated oblateness of the core is significantly smaller than the value measured at the Earth's surface.

Subsequently we will need approximate numerical values for the mass and the moments of inertia of the core. In this section the indices "δ_c", "δ_m" and "δ" will refer to the core, to the mantle (plus crust), and to the entire Earth, respectively. The numerical values were taken from [76]:

$$
\begin{aligned}
M_{\delta_c} &\approx 0.37\, M_\delta \\
A_{\delta_c} &\approx 0.11\, A_\delta \approx 9.1 \cdot 10^{36}\ \text{kg m}^2 \\
C_{\delta_c} &\approx 0.11\, C_\delta \\
C_\delta/C_{\delta_m} &\approx A_\delta/A_{\delta_m} \approx 1.13 \ .
\end{aligned}
\tag{2.151}
$$

The first Earth models *not* treating the Earth as *one* solid body but resembling the model represented by Figure 2.55 were developed towards the end of the 19th century. The model considered by Henri Poincaré (1854–1912) in 1910 in the famous treatment [87], consisting of a *rigid mantle* and a *liquid core*, is illustrated by Figure 2.56. This model was subsequently generalized to cope with the more general Earth models as represented by Figure 2.55. The Earth model represented by Figure 2.56, which usually is referred to

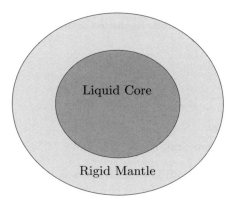

Fig. 2.56. Poincaré Earth model

as *Poincaré model*, was already considered earlier on by Sir William Kelvin (1824–1907), Hough and Sloudskii. The treatment by Poincaré is, however, of exceptional conciseness and elegance, justifying the naming convention.

From now on we will uniquely deal with the Poincaré model for the Earth's interior. The Poincaré model treats the elastic mantle plus the crust of the actual Earth model as a *rigid mantle* and the liquid plus solid core as one *liquid, incompressible core*.

Obviously, the motion of the rigid mantle formally may be described by one angular velocity vector $\boldsymbol{\omega}_\delta$, which may also be considered as the angular velocity vector of the entire Earth. The geocentric velocity of a particular volume element of the mantle thus is uniquely given by its position vector and the mentioned angular velocity vector:

$$
\dot{\boldsymbol{r}}_{\delta m} = \boldsymbol{\omega}_\delta \times \boldsymbol{r}_{\delta m} \ .
\tag{2.152}
$$

In order to describe the motion of a volume element within the liquid, incompressible core we have to recall two equations from hydrodynamics.

Two Results from Hydrodynamics. Following intuition, one may assume that the motion in a volume element within the core does not differ much from the motion it would have, if the entire Earth were rigid. This justifies to calculate the velocity at the location r_{δ_c} within the core as

$$\dot{r}_{\delta_c} = \omega_\delta \times r_{\delta_c} + v_{\delta_c} \ . \tag{2.153}$$

According to this definition v_{δ_c} is the relative velocity at the given location w.r.t. a mantle-fixed coordinate system.

The velocity $\dot{r}_{\delta_c}$ w.r.t. the inertial space must satisfy the basic equations of hydrodynamics. Moreover, the velocity component normal to the boundary between core and mantle of the relative velocity w.r.t. the rigid mantle must be zero at each point of the surface of this boundary (the ellipsoid characterized by eqn. (2.150)).

Having made the simplifying assumption of an incompressible fluid, we know that the density of the fluid is constant within liquid core. This reduces the equation of continuity of hydrodynamics to the requirement that the divergence of the velocity vector is zero within the core:

$$\nabla \cdot \dot{r}_{\delta_c} = 0 \ . \tag{2.154}$$

Equation (2.154) is linear in the components of the velocity.

The equations of motion for a fluid were first formulated by Euler in [35]. In modern formalism (see, e.g., [67]) they read as:

$$\frac{\partial \dot{r}_{\delta_c}}{\partial t} + \left(\dot{r}_{\delta_c} \cdot \nabla \right) \dot{r}_{\delta_c} = - \frac{\nabla p}{\rho} \ , \tag{2.155}$$

where ρ is the density (constant in the case considered), p is the pressure at the location considered.

For our purpose there is a more convenient version of the equations of motion (2.155) containing only the velocity $\dot{r}_{\delta_c}$. If the density is constant (as assumed above) this alternative version of the equations of motion of hydrodynamics simply results by taking the curl of the above equation. Observing that the curl of a gradient is always zero one obtains (after a few transformations) the "new" equations of motion

$$\frac{\partial}{\partial t} \left(\nabla \times \dot{r}_{\delta_c} \right) - \nabla \times \left(\dot{r}_{\delta_c} \times \left(\nabla \times \dot{r}_{\delta_c} \right) \right) = 0 \ . \tag{2.156}$$

This concludes our excursion into the theory of hydrodynamics. Basically, the theory of hydrodynamics asks, that the velocities in the core have to observe the equation of continuity (2.154) and the equations of motion of hydrodynamics (2.156).

Parametrization of the Velocities in the Liquid Core. Poincaré came up with an ingenious, yet simple assumption:

$$\dot{\boldsymbol{r}}_{\delta c} = \boldsymbol{\omega}_{\delta} \times \boldsymbol{r}_{\delta c} + \boldsymbol{v}_{\delta c}$$

$$\stackrel{\text{def}}{=} \begin{pmatrix} r_{\delta c_3}\,\omega_{\delta 2} - r_{\delta c_2}\,\omega_{\delta 3} \\ r_{\delta c_1}\,\omega_{\delta 3} - r_{\delta c_3}\,\omega_{\delta 1} \\ r_{\delta c_2}\,\omega_{\delta 1} - r_{\delta c_1}\,\omega_{\delta 2} \end{pmatrix} + \begin{pmatrix} \dfrac{a_{\delta c}}{c_{\delta c}}\,r_{\delta c_3}\,\chi_{\delta 2} - \dfrac{a_{\delta c}}{b_{\delta c}}\,r_{\delta c_2}\,\chi_{\delta 3} \\[2mm] \dfrac{b_{\delta c}}{a_{\delta c}}\,r_{\delta c_1}\,\chi_{\delta 3} - \dfrac{b_{\delta c}}{c_{\delta c}}\,r_{\delta c_3}\,\chi_{\delta 1} \\[2mm] \dfrac{c_{\delta c}}{b_{\delta c}}\,r_{\delta c_2}\,\chi_{\delta 1} - \dfrac{c_{\delta c}}{a_{\delta c}}\,r_{\delta c_1}\,\chi_{\delta 2} \end{pmatrix}.$$

$$(2.157)$$

$a_{\delta c}$, $b_{\delta c}$, and $c_{\delta c}$ are the three semi-axes of the core ellipsoid. As these parameters are assumed known, the relative velocities $\boldsymbol{v}_{\delta c}$ at different locations within the core (at one and the same time) are characterized by only three scalar quantities χ_{δ_i}, $i = 1, 2, 3$, which will be interpreted subsequently as the components of a vector $\boldsymbol{\chi}_{\delta}$ in the inertial system. Because $a_{\delta c} \approx b_{\delta c} \approx c_{\delta c}$ for the real Earth, $\boldsymbol{\chi}_{\delta}$ may be interpreted *approximately* as the angular velocity vector of the rotation of the core relative to the mantle.

One may easily verify that the velocity (2.157) observes the continuity equation (2.154). One may also verify that the velocity normal to the boundary between core and mantle vanishes at each surface point of the core. This is why we only have to make sure that eqn. (2.157) also observes the equations of motion of hydrodynamics.

The Equations of Motion of Hydrodynamics in the Inertial System. By introducing eqn. (2.157) into the equations of motion (2.156) of hydrodynamics, one eventually obtains the differential equations for vector $\boldsymbol{\chi}_{\delta}$. We first observe that:

$$\boldsymbol{\nabla} \times \dot{\boldsymbol{r}}_{\delta c} = 2\,\boldsymbol{\omega}_{\delta} + \begin{pmatrix} \dfrac{c_{\delta c}^2 + b_{\delta c}^2}{b_{\delta c}\,c_{\delta c}}\,\chi_{\delta 1} \\[2mm] \dfrac{a_{\delta c}^2 + c_{\delta c}^2}{a_{\delta c}\,c_{\delta c}}\,\chi_{\delta 2} \\[2mm] \dfrac{a_{\delta c}^2 + b_{\delta c}^2}{a_{\delta c}\,b_{\delta c}}\,\chi_{\delta 3} \end{pmatrix} \stackrel{\text{def}}{=} 2\,\tilde{\boldsymbol{\omega}}\,. \qquad (2.158)$$

$\tilde{\boldsymbol{\omega}}$ thus in essence represents the angular velocity vector of the liquid core w.r.t. the inertial system. This allows us to calculate the second term of eqns. (2.156). In a first step we obtain:

$$[\dot{\boldsymbol{r}}_{\delta_c} \times (\boldsymbol{\nabla} \times \dot{\boldsymbol{r}}_{\delta_c})]_1 = + \left(r_{\delta c_1}\omega_{\delta 3} - r_{\delta c_3}\omega_{\delta 1} + \frac{b_{\delta c}}{a_{\delta c}}r_{\delta c_1}\chi_{\delta 3} - \frac{b_{\delta c}}{c_{\delta c}}r_{\delta c_3}\chi_{\delta 1} \right) 2\tilde{\omega}_3$$

$$- \left(r_{\delta c_2}\omega_{\delta 1} - r_{\delta c_1}\omega_{\delta 2} + \frac{c_{\delta c}}{b_{\delta c}}r_{\delta c_2}\chi_{\delta 1} - \frac{c_{\delta c}}{a_{\delta c}}r_{\delta c_1}\chi_{\delta 2} \right) 2\tilde{\omega}_2$$

$$[\dot{\boldsymbol{r}}_{\delta_c} \times (\boldsymbol{\nabla} \times \dot{\boldsymbol{r}}_{\delta_c})]_2 = + \left(r_{\delta c_2}\omega_{\delta 1} - r_{\delta c_1}\omega_{\delta 2} + \frac{c_{\delta c}}{b_{\delta c}}r_{\delta c_2}\chi_{\delta 1} - \frac{c_{\delta c}}{a_{\delta c}}r_{\delta c_1}\chi_{\delta 2} \right) 2\tilde{\omega}_1$$

$$- \left(r_{\delta c_3}\omega_{\delta 2} - r_{\delta c_2}\omega_{\delta 3} + \frac{a_{\delta c}}{c_{\delta c}}r_{\delta c_3}\chi_{\delta 2} - \frac{a_{\delta c}}{b_{\delta c}}r_{\delta c_2}\chi_{\delta 3} \right) 2\tilde{\omega}_3$$

$$[\dot{\boldsymbol{r}}_{\delta_c} \times (\boldsymbol{\nabla} \times \dot{\boldsymbol{r}}_{\delta_c})]_3 = + \left(r_{\delta c_3}\omega_{\delta 2} - r_{\delta c_2}\omega_{\delta 3} + \frac{a_{\delta c}}{c_{\delta c}}r_{\delta c_3}\chi_{\delta 2} - \frac{a_{\delta c}}{b_{\delta c}}r_{\delta c_2}\chi_{\delta 3} \right) 2\tilde{\omega}_2$$

$$- \left(r_{\delta c_1}\omega_{\delta 3} - r_{\delta c_3}\omega_{\delta 1} + \frac{b_{\delta c}}{a_{\delta c}}r_{\delta c_1}\chi_{\delta 3} - \frac{b_{\delta c}}{c_{\delta c}}r_{\delta c_3}\chi_{\delta 1} \right) 2\tilde{\omega}_1 \ .$$

$$(2.159)$$

Taking the curl of this vector equation we obtain:

$$\boldsymbol{\nabla} \times (\dot{\boldsymbol{r}}_{\delta_c} \times (\boldsymbol{\nabla} \times \dot{\boldsymbol{r}}_{\delta_c})) = - \begin{pmatrix} \left(\omega_{\delta 3} + \dfrac{a_{\delta c}}{b_{\delta c}}\chi_{\delta 3} \right) 2\tilde{\omega}_2 - \left(\omega_{\delta 2} + \dfrac{a_{\delta c}}{c_{\delta c}}\chi_{\delta 2} \right) 2\tilde{\omega}_3 \\[2mm] \left(\omega_{\delta 1} + \dfrac{b_{\delta c}}{c_{\delta c}}\chi_{\delta 1} \right) 2\tilde{\omega}_3 - \left(\omega_{\delta 3} + \dfrac{b_{\delta c}}{a_{\delta c}}\chi_{\delta 3} \right) 2\tilde{\omega}_1 \\[2mm] \left(\omega_{\delta 2} + \dfrac{c_{\delta c}}{a_{\delta c}}\chi_{\delta 2} \right) 2\tilde{\omega}_1 - \left(\omega_{\delta 1} + \dfrac{c_{\delta c}}{b_{\delta c}}\chi_{\delta 1} \right) 2\tilde{\omega}_2 \end{pmatrix} .$$

$$(2.160)$$

Introducing the expressions (2.158) and (2.160) into the equation of motion (2.156) one obtains after few transformation steps the equations of motion of hydrodynamics referring to the inertial system for the special motion (2.157) within the core:

$$2\dot{\omega}_{\delta 1} + \frac{c_{\delta c}^2 + b_{\delta c}^2}{b_{\delta c}c_{\delta c}}\dot{\chi}_{\delta 1} + \frac{a_{\delta c}^2 - b_{\delta c}^2}{a_{\delta c}b_{\delta c}}\omega_{\delta 2}\chi_{\delta 3} + \frac{c_{\delta c}^2 - a_{\delta c}^2}{a_{\delta c}c_{\delta c}}\omega_{\delta 3}\chi_{\delta 2} + \frac{c_{\delta c}^2 - b_{\delta c}^2}{b_{\delta c}c_{\delta c}}\chi_{\delta 2}\chi_{\delta 3} = 0$$

$$2\dot{\omega}_{\delta 2} + \frac{a_{\delta c}^2 + c_{\delta c}^2}{a_{\delta c}c_{\delta c}}\dot{\chi}_{\delta 2} + \frac{b_{\delta c}^2 - c_{\delta c}^2}{b_{\delta c}c_{\delta c}}\omega_{\delta 3}\chi_{\delta 1} + \frac{a_{\delta c}^2 - b_{\delta c}^2}{a_{\delta c}b_{\delta c}}\omega_{\delta 1}\chi_{\delta 3} + \frac{a_{\delta c}^2 - c_{\delta c}^2}{a_{\delta c}c_{\delta c}}\chi_{\delta 3}\chi_{\delta 1} = 0$$

$$2\dot{\omega}_{\delta 3} + \frac{a_{\delta c}^2 + b_{\delta c}^2}{a_{\delta c}b_{\delta c}}\dot{\chi}_{\delta 3} + \frac{c_{\delta c}^2 - a_{\delta c}^2}{a_{\delta c}c_{\delta c}}\omega_{\delta 1}\chi_{\delta 2} + \frac{b_{\delta c}^2 - c_{\delta c}^2}{b_{\delta c}c_{\delta c}}\omega_{\delta 2}\chi_{\delta 1} + \frac{b_{\delta c}^2 - a_{\delta c}^2}{a_{\delta c}b_{\delta c}}\chi_{\delta 1}\chi_{\delta 2} = 0.$$

$$(2.161)$$

We may assume that these equations refer to an inertial coordinate system coinciding with the PAI-system at the time considered.

Transformation into the Mantle-fixed System. When transforming eqns. (2.161) into the mantle-fixed system we first have to distinguish formally between the coordinates in the inertial and the mantle-fixed system (what we did not do till now in order not to overload the formalism). As a matter of fact, we should have explicitly written $\omega_{\delta_{\mathcal{I}_i}}$ and not only $\omega_{\delta i}$, etc. in the above equations to indicate that they refer to the inertial system. Following the transformation pattern of section I-3.3.4 one easily verifies that the transformation equations for vectors $\dot{\omega}$ and $\dot{\chi}$ between the special inertial system $\mathcal{I}$ introduced above and the mantle-fixed system $\mathcal{F}$ read as

$$\dot{\omega}_{\delta_{\mathcal{I}_1}} = \dot{\omega}_{\delta_{\mathcal{F}_1}}$$
$$\dot{\omega}_{\delta_{\mathcal{I}_2}} = \dot{\omega}_{\delta_{\mathcal{F}_2}} \tag{2.162}$$
$$\dot{\omega}_{\delta_{\mathcal{I}_3}} = \dot{\omega}_{\delta_{\mathcal{F}_3}}$$

and

$$\frac{c_{\delta_c}^2 + b_{\delta_c}^2}{b_{\delta_c} c_{\delta_c}} \dot{\chi}_{\mathcal{I}_1} = \frac{c_{\delta_c}^2 + b_{\delta_c}^2}{b_{\delta_c} c_{\delta_c}} \dot{\chi}_{\delta_{\mathcal{F}_1}} + \frac{a_{\delta_c}^2 + b_{\delta_c}^2}{a_{\delta_c} b_{\delta_c}} \omega_{\delta_{\mathcal{F}_2}} \chi_{\delta_{\mathcal{F}_3}} - \frac{a_{\delta_c}^2 + c_{\delta_c}^2}{a_{\delta_c} c_{\delta_c}} \omega_{\delta_{\mathcal{F}_3}} \chi_{\delta_{\mathcal{F}_2}}$$

$$\frac{a_{\delta_c}^2 + c_{\delta_c}^2}{a_{\delta_c} c_{\delta_c}} \dot{\chi}_{\mathcal{I}_2} = \frac{a_{\delta_c}^2 + c_{\delta_c}^2}{a_{\delta_c} c_{\delta_c}} \dot{\chi}_{\delta_{\mathcal{F}_2}} + \frac{b_{\delta_c}^2 + c_{\delta_c}^2}{b_{\delta_c} c_{\delta_c}} \omega_{\delta_{\mathcal{F}_3}} \chi_{\delta_{\mathcal{F}_1}} - \frac{a_{\delta_c}^2 + b_{\delta_c}^2}{a_{\delta_c} b_{\delta_c}} \omega_{\delta_{\mathcal{F}_1}} \chi_{\delta_{\mathcal{F}_3}}$$

$$\frac{a_{\delta_c}^2 + b_{\delta_c}^2}{a_{\delta_c} b_{\delta_c}} \dot{\chi}_{\mathcal{I}_3} = \frac{a_{\delta_c}^2 + b_{\delta_c}^2}{a_{\delta_c} b_{\delta_c}} \dot{\chi}_{\delta_{\mathcal{F}_3}} + \frac{a_{\delta_c}^2 + c_{\delta_c}^2}{a_{\delta_c} c_{\delta_c}} \omega_{\delta_{\mathcal{F}_1}} \chi_{\delta_{\mathcal{F}_2}} - \frac{b_{\delta_c}^2 + c_{\delta_c}^2}{b_{\delta_c} c_{\delta_c}} \omega_{\delta_{\mathcal{F}_2}} \chi_{\delta_{\mathcal{F}_1}} . \tag{2.163}$$

Using this latter relation in eqns. (2.161) the equations of motion of hydrodynamics referring to the rotating, mantle-fixed system read as follows:

$$2\dot{\omega}_{\delta_{\mathcal{F}_1}} + \frac{c_{\delta_c}^2 + b_{\delta_c}^2}{b_{\delta_c} c_{\delta_c}} \dot{\chi}_{\delta_{\mathcal{F}_1}} + 2\frac{a_{\delta_c}}{b_{\delta_c}} \omega_{\delta_{\mathcal{F}_2}} \chi_{\delta_{\mathcal{F}_3}} - 2\frac{a_{\delta_c}}{c_{\delta_c}} \omega_{\delta_{\mathcal{F}_3}} \chi_{\delta_{\mathcal{F}_2}} + \frac{c_{\delta_c}^2 - b_{\delta_c}^2}{b_{\delta_c} c_{\delta_c}} \chi_{\delta_{\mathcal{F}_2}} \chi_{\delta_{\mathcal{F}_3}} = 0$$

$$2\dot{\omega}_{\delta_{\mathcal{F}_2}} + \frac{a_{\delta_c}^2 + c_{\delta_c}^2}{a_{\delta_c} c_{\delta_c}} \dot{\chi}_{\delta_{\mathcal{F}_2}} + 2\frac{b_{\delta_c}}{c_{\delta_c}} \omega_{\delta_{\mathcal{F}_3}} \chi_{\delta_{\mathcal{F}_1}} - 2\frac{b_{\delta_c}}{a_{\delta_c}} \omega_{\delta_{\mathcal{F}_1}} \chi_{\delta_{\mathcal{F}_3}} + \frac{a_{\delta_c}^2 - c_{\delta_c}^2}{a_{\delta_c} c_{\delta_c}} \chi_{\delta_{\mathcal{F}_3}} \chi_{\delta_{\mathcal{F}_1}} = 0$$

$$2\dot{\omega}_{\delta_{\mathcal{F}_3}} + \frac{a_{\delta_c}^2 + b_{\delta_c}^2}{a_{\delta_c} b_{\delta_c}} \dot{\chi}_{\delta_{\mathcal{F}_3}} + 2\frac{c_{\delta_c}}{a_{\delta_c}} \omega_{\delta_{\mathcal{F}_1}} \chi_{\delta_{\mathcal{F}_2}} - 2\frac{c_{\delta_c}}{b_{\delta_c}} \omega_{\delta_{\mathcal{F}_2}} \chi_{\delta_{\mathcal{F}_1}} + \frac{b_{\delta_c}^2 - a_{\delta_c}^2}{a_{\delta_c} b_{\delta_c}} \chi_{\delta_{\mathcal{F}_1}} \chi_{\delta_{\mathcal{F}_2}} = 0. \tag{2.164}$$

In a final step the above equations of motion are transformed to replace the semi-axes a_{δ_c}, b_{δ_c}, and c_{δ_c} of the core-mantle boundary by the principal moments of inertia of the core. Keeping in mind that the fluid in the core was assumed to be incompressible, implying a constant density of the fluid, we may calculate these moments simply as

$$A_{\delta_c} = \tfrac{1}{5} M_{\delta_c} \left(b_{\delta_c}^2 + c_{\delta_c}^2 \right)$$
$$B_{\delta_c} = \tfrac{1}{5} M_{\delta_c} \left(a_{\delta_c}^2 + c_{\delta_c}^2 \right) \tag{2.165}$$
$$C_{\delta_c} = \tfrac{1}{5} M_{\delta_c} \left(a_{\delta_c}^2 + b_{\delta_c}^2 \right) ,$$

from where one also obtains

$$\frac{1}{5} M_{\delta_c}\left(c_{\delta_c}^2 - b_{\delta_c}^2\right) = B_{\delta_c} - C_{\delta_c}$$

$$\frac{1}{5} M_{\delta_c}\left(a_{\delta_c}^2 - c_{\delta_c}^2\right) = C_{\delta_c} - A_{\delta_c} \qquad (2.166)$$

$$\frac{1}{5} M_{\delta_c}\left(b_{\delta_c}^2 - a_{\delta_c}^2\right) = A_{\delta_c} - B_{\delta_c}$$

and

$$\frac{2}{5} M_{\delta_c} a_{\delta_c}^2 = B_{\delta_c}(1 - \gamma_{\delta_{c_2}}) = C_{\delta_c}(1 + \gamma_{\delta_{c_3}})$$

$$\frac{2}{5} M_{\delta_c} b_{\delta_c}^2 = C_{\delta_c}(1 - \gamma_{\delta_{c_3}}) = A_{\delta_c}(1 + \gamma_{\delta_{c_1}}) \qquad (2.167)$$

$$\frac{2}{5} M_{\delta_c} c_{\delta_c}^2 = A_{\delta_c}(1 - \gamma_{\delta_{c_1}}) = B_{\delta_c}(1 + \gamma_{\delta_{c_2}}) \ .$$

These relations imply on the other hand

$$\frac{2}{5} M_{\delta_c} b_{\delta_c} c_{\delta_c} = A_{\delta_c}\sqrt{1 - \gamma_{\delta_{c_1}}^2} \stackrel{\text{def}}{=} \tilde{A}_{\delta_c}$$

$$\frac{2}{5} M_{\delta_c} a_{\delta_c} c_{\delta_c} = B_{\delta_c}\sqrt{1 - \gamma_{\delta_{c_2}}^2} \stackrel{\text{def}}{=} \tilde{B}_{\delta_c} \qquad (2.168)$$

$$\frac{2}{5} M_{\delta_c} a_{\delta_c} b_{\delta_c} = C_{\delta_c}\sqrt{1 - \gamma_{\delta_{c_3}}^2} \stackrel{\text{def}}{=} \tilde{C}_{\delta_c} \ ,$$

where the symbols $\gamma_{...}$ are defined by

$$\gamma_{\delta_{c_1}} = \frac{C_{\delta_c} - B_{\delta_c}}{A_{\delta_c}}\ , \qquad \gamma_{\delta_{c_2}} = \frac{A_{\delta_c} - C_{\delta_c}}{B_{\delta_c}}\ , \qquad \gamma_{\delta_{c_3}} = \frac{B_{\delta_c} - A_{\delta_c}}{C_{\delta_c}}\ . \qquad (2.169)$$

These relations allow it to write the equations (2.164) in the desired form: The first of eqns. (2.164) is multiplied with $\frac{1}{5} M_{\delta_c} b_{\delta_c} c_{\delta_c}$, the second with $\frac{1}{5} M_{\delta_c} a_{\delta_c} c_{\delta_c}$, and the third with $\frac{1}{5} M_{\delta_c} a_{\delta_c} b_{\delta_c}$. Using the above definitions we obtain

$$\tilde{A}_{\delta_c}\dot{\omega}_{\delta \mathcal{F}_1} + A_{\delta_c}\dot{\chi}_{\delta \mathcal{F}_1} + \tilde{B}_{\delta_c}\omega_{\delta \mathcal{F}_2}\chi_{\delta \mathcal{F}_3} - \tilde{C}_{\delta_c}\omega_{\delta \mathcal{F}_3}\chi_{\delta \mathcal{F}_2} - A_{\delta_c}\gamma_{\delta_{c_1}}\chi_{\delta \mathcal{F}_2}\chi_{\delta \mathcal{F}_3} = 0$$

$$\tilde{B}_{\delta_c}\dot{\omega}_{\delta \mathcal{F}_2} + B_{\delta_c}\dot{\chi}_{\delta \mathcal{F}_2} + \tilde{C}_{\delta_c}\omega_{\delta \mathcal{F}_3}\chi_{\delta \mathcal{F}_1} - \tilde{A}_{\delta_c}\omega_{\delta \mathcal{F}_1}\chi_{\delta \mathcal{F}_3} - B_{\delta_c}\gamma_{\delta_{c_2}}\chi_{\delta \mathcal{F}_3}\chi_{\delta \mathcal{F}_1} = 0$$

$$\tilde{C}_{\delta_c}\dot{\omega}_{\delta \mathcal{F}_3} + C_{\delta_c}\dot{\chi}_{\delta \mathcal{F}_3} + \tilde{A}_{\delta_c}\omega_{\delta \mathcal{F}_1}\chi_{\delta \mathcal{F}_2} - \tilde{B}_{\delta_c}\omega_{\delta \mathcal{F}_2}\chi_{\delta \mathcal{F}_1} - C_{\delta_c}\gamma_{\delta_{c_3}}\chi_{\delta \mathcal{F}_1}\chi_{\delta \mathcal{F}_2} = 0 \ .$$

$$(2.170)$$

We have thus established three differential equations for six scalar unknowns (the components of $\boldsymbol{\omega}_\delta$ and $\boldsymbol{\chi}_\delta$). The three "missing" equations are the classical equations relating the change of angular momentum to the torques acting on the body. In the subsequent derivations we first establish these relations separately for core and mantle, then form the sum of the two resulting equations to obtain the missing equations of the Poincaré model.

Exchange of Angular Momentum between Core and Mantle. Having introduced a liquid core and a rigid mantle implies that we have to allow for an exchange of angular momentum between the two components. Equations (I- 3.90) thus have to be generalized as follows (observe that we only need to consider the first of eqns. (I- 3.90) for the Earth and that the index "m" stands for the mantle in this section):

$$
\begin{aligned}
\dot{\boldsymbol{h}}_{\delta m} &= \boldsymbol{\ell}_{\delta m} + \boldsymbol{\ell}_{\delta cm} \\
\dot{\boldsymbol{h}}_{\delta c} &= \boldsymbol{\ell}_{\delta c} - \boldsymbol{\ell}_{\delta cm} \, ,
\end{aligned}
\tag{2.171}
$$

where $\boldsymbol{\ell}_{\delta m}$ and $\boldsymbol{\ell}_{\delta c}$ stand for the *external torques* (caused by Sun, Moon, planets) acting on mantle and core, respectively, whereas $\boldsymbol{\ell}_{\delta cm}$ is the torque caused by the core acting on the mantle. Let us calculate the angular momenta in the PAI-system of the mantle (and the core) at the time considered:

$$
\boldsymbol{h}_{\delta m} = \int_{V_{\delta m}} \rho_{\delta m}(\boldsymbol{r}_{\delta m}) \, \boldsymbol{r}_{\delta m} \times \dot{\boldsymbol{r}}_{\delta m} \, dV_{\delta m} = \mathbf{I}_{\delta m} \cdot \boldsymbol{\omega}_{\delta}
$$

$$
\boldsymbol{h}_{\delta c} = \int_{V_{\delta c}} \rho_{\delta c}(\boldsymbol{r}_{\delta c}) \, \boldsymbol{r}_{\delta c} \times \dot{\boldsymbol{r}}_{\delta c} \, dV_{\delta c}
$$

$$
= \mathbf{I}_{\delta c} \cdot \boldsymbol{\omega}_{\delta} + \int_{V_{\delta c}} \rho_{\delta c}(\boldsymbol{r}_{\delta c})
\begin{pmatrix}
\left(\dfrac{c_{\delta c}}{b_{\delta c}} r^2_{\delta c_{\mathcal{I}_2}} + \dfrac{b_{\delta c}}{c_{\delta c}} r^2_{\delta c_{\mathcal{I}_3}} \right) \chi_{\delta \mathcal{I}_1} \\
\left(\dfrac{a_{\delta c}}{c_{\delta c}} r^2_{\delta c_{\mathcal{I}_3}} + \dfrac{c_{\delta c}}{a_{\delta c}} r^2_{\delta c_{\mathcal{I}_1}} \right) \chi_{\delta \mathcal{I}_2} \\
\left(\dfrac{b_{\delta c}}{a_{\delta c}} r^2_{\delta c_{\mathcal{I}_1}} + \dfrac{a_{\delta c}}{b_{\delta c}} r^2_{\delta c_{\mathcal{I}_2}} \right) \chi_{\delta \mathcal{I}_3}
\end{pmatrix}
dV_{\delta c}
\tag{2.172}
$$

$$
\stackrel{\text{def}}{=}
\begin{pmatrix}
\mathrm{A}_{\delta c} \, \omega_{\delta \mathcal{F}_1} \\
\mathrm{B}_{\delta c} \, \omega_{\delta \mathcal{F}_2} \\
\mathrm{C}_{\delta c} \, \omega_{\delta \mathcal{F}_3}
\end{pmatrix}
+
\begin{pmatrix}
\tilde{A}_{\delta c} \, \chi_{\delta \mathcal{I}_1} \\
\tilde{B}_{\delta c} \, \chi_{\delta \mathcal{I}_2} \\
\tilde{C}_{\delta c} \, \chi_{\delta \mathcal{I}_3}
\end{pmatrix} \, .
$$

Introducing the above expressions into eqns. (2.171) and forming the sum of the two resulting equations leads to the following relations:

$$
\begin{aligned}
\frac{d}{dt} \left(\mathrm{A}_{\delta} \, \omega_{\delta \mathcal{F}_1} + \tilde{A}_{\delta c} \, \chi_{\delta \mathcal{I}_1} \right) &= \ell_{\delta 1} \\
\frac{d}{dt} \left(\mathrm{B}_{\delta} \, \omega_{\delta \mathcal{F}_2} + \tilde{B}_{\delta c} \, \chi_{\delta \mathcal{I}_2} \right) &= \ell_{\delta 2} \\
\frac{d}{dt} \left(\mathrm{C}_{\delta} \, \omega_{\delta \mathcal{F}_3} + \tilde{C}_{\delta c} \, \chi_{\delta \mathcal{I}_3} \right) &= \ell_{\delta 3} \, ,
\end{aligned}
\tag{2.173}
$$

where the index "δ" refers to the entire Earth (core plus mantle), whereas the index "c" refers to the core only. As usual, the equations are transformed into the rotating system leading to the final result:

$$A_\delta \dot{\omega}_{\delta\mathcal{F}_1} + \tilde{A}_{\delta_c} \dot{\chi}_{\delta\mathcal{F}_1} + \gamma_{\delta 1} A_\delta \omega_{\delta\mathcal{F}_2} \omega_{\delta\mathcal{F}_3} + \tilde{C}_{\delta_c} \omega_{\delta\mathcal{F}_2} \chi_{\delta\mathcal{F}_3} - \tilde{B}_{\delta_c} \omega_{\delta\mathcal{F}_3} \chi_{\delta\mathcal{F}_2} = \ell_{\delta 1}$$
$$B_\delta \dot{\omega}_{\delta\mathcal{F}_2} - \tilde{B}_{\delta_c} \dot{\chi}_{\delta\mathcal{F}_2} + \gamma_{\delta 2} B_\delta \omega_{\delta\mathcal{F}_1} \omega_{\delta\mathcal{F}_3} + \tilde{A}_{\delta_c} \omega_{\delta\mathcal{F}_3} \chi_{\delta\mathcal{F}_1} - \tilde{C}_{\delta_c} \omega_{\delta\mathcal{F}_1} \chi_{\delta\mathcal{F}_3} = \ell_{\delta 2}$$
$$C_\delta \dot{\omega}_{\delta\mathcal{F}_3} + \tilde{C}_{\delta_c} \dot{\chi}_{\delta\mathcal{F}_3} + \gamma_{\delta 3} C_\delta \omega_{\delta\mathcal{F}_1} \omega_{\delta\mathcal{F}_2} + \tilde{B}_{\delta_c} \omega_{\delta\mathcal{F}_1} \chi_{\delta\mathcal{F}_2} - \tilde{A}_{\delta_c} \omega_{\delta\mathcal{F}_2} \chi_{\delta\mathcal{F}_1} = \ell_{\delta 3} \; .$$

$$(2.174)$$

In the absence of a liquid core (i.e., for $\tilde{A}_{\delta_c} = \tilde{B}_{\delta_c} = \tilde{C}_{\delta_c} = 0$) the second, forth, and fifth term are zero, and the above equations reduce to to the classical Euler equations of the rigid-body model for the Earth.

Equations of Earth Rotation for the Poincaré Model. Equations (2.174) and (2.170) are the defining equations of the Poincaré Earth model. If Euler's kinematic equations (referring to the mantle-fixed system) are added to these equations, we obtain the complete mathematical description of the Poincaré Earth model:

$$A_\delta \dot{\omega}_{\delta\mathcal{F}_1} + \tilde{A}_{\delta_c} \dot{\chi}_{\delta\mathcal{F}_1} + \gamma_{\delta 1} A_\delta \omega_{\delta\mathcal{F}_2} \omega_{\delta\mathcal{F}_3} + \tilde{C}_{\delta_c} \omega_{\delta\mathcal{F}_2} \chi_{\delta\mathcal{F}_3} - \tilde{B}_{\delta_c} \omega_{\delta\mathcal{F}_3} \chi_{\delta\mathcal{F}_2} \quad = \ell_{\delta 1}$$
$$B_\delta \dot{\omega}_{\delta\mathcal{F}_2} - \tilde{B}_{\delta_c} \dot{\chi}_{\delta\mathcal{F}_2} + \gamma_{\delta 2} B_\delta \omega_{\delta\mathcal{F}_1} \omega_{\delta\mathcal{F}_3} + \tilde{A}_{\delta_c} \omega_{\delta\mathcal{F}_3} \chi_{\delta\mathcal{F}_1} - \tilde{C}_{\delta_c} \omega_{\delta\mathcal{F}_1} \chi_{\delta\mathcal{F}_3} \quad = \ell_{\delta 2}$$
$$C_\delta \dot{\omega}_{\delta\mathcal{F}_3} + \tilde{C}_{\delta_c} \dot{\chi}_{\delta\mathcal{F}_3} + \gamma_{\delta 3} C_\delta \omega_{\delta\mathcal{F}_1} \omega_{\delta\mathcal{F}_2} + \tilde{B}_{\delta_c} \omega_{\delta\mathcal{F}_1} \chi_{\delta\mathcal{F}_2} - \tilde{A}_{\delta_c} \omega_{\delta\mathcal{F}_2} \chi_{\delta\mathcal{F}_1} \quad = \ell_{\delta 3}$$

$$\tilde{A}_{\delta_c} \dot{\omega}_{\delta\mathcal{F}_1} + A_{\delta_c} \dot{\chi}_{\delta\mathcal{F}_1} + \tilde{B}_{\delta_c} \omega_{\delta\mathcal{F}_2} \chi_{\delta\mathcal{F}_3} - \tilde{C}_{\delta_c} \omega_{\delta\mathcal{F}_3} \chi_{\delta\mathcal{F}_2} - A_{\delta_c} \gamma_{\delta c_1} \chi_{\delta\mathcal{F}_2} \chi_{\delta\mathcal{F}_3} = 0$$
$$\tilde{B}_{\delta_c} \dot{\omega}_{\delta\mathcal{F}_2} + B_{\delta_c} \dot{\chi}_{\delta\mathcal{F}_2} + \tilde{C}_{\delta_c} \omega_{\delta\mathcal{F}_3} \chi_{\delta\mathcal{F}_1} - \tilde{A}_{\delta_c} \omega_{\delta\mathcal{F}_1} \chi_{\delta\mathcal{F}_3} - B_{\delta_c} \gamma_{\delta c_2} \chi_{\delta\mathcal{F}_3} \chi_{\delta\mathcal{F}_1} = 0$$
$$\tilde{C}_{\delta_c} \dot{\omega}_{\delta\mathcal{F}_3} + C_{\delta_c} \dot{\chi}_{\delta\mathcal{F}_3} + \tilde{A}_{\delta_c} \omega_{\delta\mathcal{F}_1} \chi_{\delta\mathcal{F}_2} - \tilde{B}_{\delta_c} \omega_{\delta\mathcal{F}_2} \chi_{\delta\mathcal{F}_1} - C_{\delta_c} \gamma_{\delta c_3} \chi_{\delta\mathcal{F}_1} \chi_{\delta\mathcal{F}_2} = 0$$

$$-\sin\varepsilon_\delta \sin\Theta_\delta \dot{\Psi}_\delta + \cos\Theta_\delta \dot{\varepsilon}_\delta = \omega_{\delta\mathcal{F}_1}$$
$$-\sin\varepsilon_\delta \cos\Theta_\delta \dot{\Psi}_\delta - \sin\Theta_\delta \dot{\varepsilon}_\delta = \omega_{\delta\mathcal{F}_2}$$
$$\cos\varepsilon_\delta \dot{\Psi}_\delta + \dot{\Theta}_\delta = \omega_{\delta\mathcal{F}_3} \; .$$

$$(2.175)$$

Mathematically, eqns. (2.175) form a coupled system of nine first order equations for nine scalar unknowns (the components of vectors $\boldsymbol{\omega}_\delta$, $\boldsymbol{\chi}_\delta$ in the mantle-fixed PAI-system and the three Euler angles). These equations are implemented and used in program ERDROT, whenever the option of a liquid core is selected.

For approximate analytical considerations (related to the actual Earth), the system may be reduced considerably by retaining only small quantities up to the first order. If rotational symmetry is assumed, the following equations of motion are obtained:

$$A_{\delta_c}\,\dot{\chi}_{\delta\mathcal{F}_1} + \gamma_\delta\,A_\delta\,\omega_{\delta\mathcal{F}_2}\omega_{\delta\mathcal{F}_3} - A_{\delta_c}\,\omega_{\delta\mathcal{F}_3}\chi_{\delta\mathcal{F}_2} = \ell_{\delta 1}$$
$$A_\delta\,\dot{\omega}_{\delta\mathcal{F}_2} + A_{\delta_c}\,\dot{\chi}_{\delta\mathcal{F}_2} - \gamma_\delta\,A_\delta\,\omega_{\delta\mathcal{F}_1}\omega_{\delta\mathcal{F}_3} + A_{\delta_c}\,\omega_{\delta\mathcal{F}_3}\chi_{\delta\mathcal{F}_1} = \ell_{\delta 2}$$
$$C_\delta\,\dot{\omega}_{\delta\mathcal{F}_3} + C_{\delta_c}\,\dot{\chi}_{\delta\mathcal{F}_3} = 0$$

$$A_{\delta_c}\,\dot{\omega}_{\delta\mathcal{F}_1} + A_{\delta_c}\,\dot{\chi}_{\delta\mathcal{F}_1} - C_{\delta_c}\,\omega_{\delta\mathcal{F}_3}\chi_{\delta\mathcal{F}_2} = 0$$
$$A_{\delta_c}\,\dot{\omega}_{\delta\mathcal{F}_2} + A_{\delta_c}\,\dot{\chi}_{\delta\mathcal{F}_2} + C_{\delta_c}\,\omega_{\delta\mathcal{F}_3}\chi_{\delta\mathcal{F}_1} = 0 \qquad (2.176)$$
$$C_{\delta_c}\,\dot{\omega}_{\delta\mathcal{F}_3} + C_{\delta_c}\,\dot{\chi}_{\delta\mathcal{F}_3} = 0$$

$$- \sin\varepsilon_\delta\,\sin\Theta_\delta\,\dot{\Psi}_\delta + \cos\Theta_\delta\,\dot{\varepsilon}_\delta = \omega_{\delta\mathcal{F}_1}$$
$$- \sin\varepsilon_\delta\,\cos\Theta_\delta\,\dot{\Psi}_\delta - \sin\Theta_\delta\,\dot{\varepsilon}_\delta = \omega_{\delta\mathcal{F}_2}$$
$$\cos\varepsilon_\delta\,\dot{\Psi}_\delta + \dot{\Theta}_\delta = \omega_{\delta\mathcal{F}_3}\,.$$

Free Motion in the Poincaré Model. The key properties of the Poincaré model shall now be illustrated with a few simulations performed with program ERDROT (see Chapter 9 of Part III). In a first series of tests the free motion, i.e., the motion in the absence of the external torques due to Moon and Sun, is studied. Figure 2.57 compares the PM of a rigid Earth with that of a Poincaré-type Earth. The properties (masses, moments of inertia) of the entire Earth body (core plus mantle) were assumed to be identical in both cases. $\chi_{\delta\mathcal{F}}(t_0) = \mathbf{0}$ was assumed at the initial epoch. Figure 2.57 was produced under the assumption of a spherical core-mantle boundary ($\gamma_{\delta_{c_i}} = 0$, $i = 1, 2, 3$). The mass and the moment of inertia of the core were defined by eqn. (2.151). The main difference between the two PMs thus resides in the assumption of a spherical liquid core and one rigid body, respectively. In order to better distinguish the two cases, the initial conditions were chosen to differ slightly (difference of about 10%). Figure 2.57 shows, that the PM is considerably (about 11% (!)) faster when the Poincaré model instead of the rigid Earth model is used.

Figure 2.58 shows the effect of the initial value of vector $\chi(t_0)$ in the case of an ellipsoidal core-mantle boundary in the Poincaré model. The initial conditions used were $\chi(t_0) \stackrel{\text{def}}{=} \mathbf{0}$ (solid line) and $\chi_{\delta_1}(t_0) \stackrel{\text{def}}{=} 10^{-4} \cdot \omega_{\delta_0}$, $\chi_{\delta_2}(t_0) = \chi_{\delta_3}(t_0) = 0$ (dashed line). The integration specifications underlying this figure in essence are those of Figure 2.57. In order to improve the visibility of the effects the radii of PM were chosen to differ by 10%. Also, the integration was only performed over about 90 days. Figure 2.58 reveals one of the key properties of the Poincaré model, namely the existence of a *NDFW (Nearly-Diurnal Free Wobble)* in the PM of the mantle. The amplitude of this

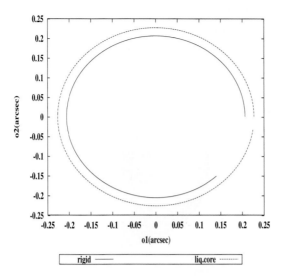

Fig. 2.57. Free motion of pole using a rigid-body and a Poincaré model with spherical core (over first 263 days of year 2000)

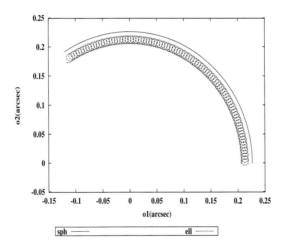

Fig. 2.58. Free motion of pole using Poincaré models with $\chi_{\mathring{\sigma}_1} = (0, 10^{-4}) \cdot \omega_0$ (over first 92 days of year 2000)

wobble obviously depends heavily on the initial conditions for vector $\boldsymbol{\chi}_{\hat{\delta}}$. For $\boldsymbol{\chi}_{\hat{\delta}}(t_0) \stackrel{\text{def}}{=} \mathbf{0}$ no nearly-diurnal free wobble is observed.

Figures 2.59 and 2.60 show what might be called "polar wobble of the core relative to the mantle" for the two initial conditions underlying Figure 2.58.

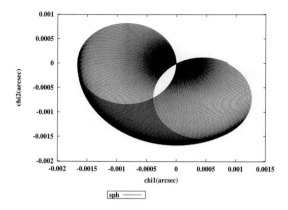

Fig. 2.59. Free motion of the core w.r.t. the mantle for $\chi_{\hat{\delta}_1} = 0$ (over the first 92 days of year 2000)

Whereas the prominent feature of the PM of the mantle is the generalized Euler motion with a period of about 270 days, the prominent feature of Figures 2.59 and 2.60 is the quasi-diurnal motion, the amplitude of which in essence is given the initial condition of vector $\boldsymbol{\chi}_{\hat{\delta}}$.

The main features in Figures 2.57 – 2.60 may be understood by analyzing the equations of motion (2.176) in the absence of external torques: From the third and sixth of these equations one may immediately conclude that

$$
\begin{aligned}
\dot{\omega}_{\hat{\delta}\mathcal{F}_3} &= 0 \;; & \omega_{\hat{\delta}\mathcal{F}_3} &= \omega_{\hat{\delta}_0} \\
\dot{\chi}_{\hat{\delta}\mathcal{F}_3} &= 0 \;; & \chi_{\hat{\delta}\mathcal{F}_3} &= \chi_{\hat{\delta}_0} \;,
\end{aligned}
\tag{2.177}
$$

where $\omega_{\hat{\delta}_0}$ and $\chi_{\hat{\delta}_0}$ are constants.

The constant value $\omega_{\hat{\delta}_0}$ may be interpreted as the constant angular velocity of Earth rotation – the Poincaré model thus does not allow for LOD variations, exactly as the model of a rigid Earth. $\chi_{\hat{\delta}_0}$ approximately characterizes the rotation of the core w.r.t. the mantle. Obviously an arbitrary (but constant) rotation solves the above equations. Following tradition (and common sense) we only consider the case where the rotation rate of the core w.r.t. the inertial space is the same as that of the mantle, i.e.,

$$
\chi_{\hat{\delta}_0} = 0 \;.
\tag{2.178}
$$

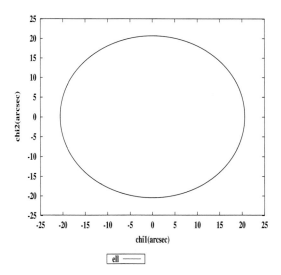

Fig. 2.60. Free motion of the core w.r.t. the mantle for $\chi_{\mathring{\delta}_1} = 1 \cdot 10^{-4} \cdot \omega_{\mathring{\delta}_0}$ (over the first 92 days of year 2000)

As $\chi_{\mathring{\delta}_0}$ enters only in the third and sixth of eqns. (2.176), this assumption does not affect the following considerations concerning PM and the first two components of vector $\chi_{\mathring{\delta}}$. When assuming rotational symmetry a rotation of the core with a constant rate w.r.t. the mantle would thus not affect the rotation of the mantle.

Introducing the result (2.177) into the first, second, forth, and fifth of eqns. (2.176) leads to a coupled, linear, inhomogeneous differential equation system with constant coefficients in the variables $\omega_{\mathring{\delta}\mathcal{F}_i}$, $i = 1, 2$, and $\chi_{\mathring{\delta}\mathcal{F}_i}$, $i = 1, 2$:

$$A_{\mathring{\delta}_c} \dot{\chi}_{\mathring{\delta}\mathcal{F}_1} + \gamma_{\mathring{\delta}} A_{\mathring{\delta}} \omega_{\mathring{\delta}_0}\omega_{\mathring{\delta}\mathcal{F}_2} - A_{\mathring{\delta}_c} \omega_{\mathring{\delta}_0}\chi_{\mathring{\delta}\mathcal{F}_2} = \ell_{\mathring{\delta}\mathcal{F}_1}$$
$$A_{\mathring{\delta}} \dot{\omega}_{\mathring{\delta}\mathcal{F}_2} + A_{\mathring{\delta}_c} \dot{\chi}_{\mathring{\delta}\mathcal{F}_2} - \gamma_{\mathring{\delta}} A_{\mathring{\delta}} \omega_{\mathring{\delta}_0}\omega_{\mathring{\delta}\mathcal{F}_1} + A_{\mathring{\delta}_c} \omega_{\mathring{\delta}_0}\chi_{\mathring{\delta}\mathcal{F}_1} = \ell_{\mathring{\delta}\mathcal{F}_2}$$

$$\text{(2.179)}$$

$$A_{\mathring{\delta}_c} \dot{\omega}_{\mathring{\delta}\mathcal{F}_1} + A_{\mathring{\delta}_c} \dot{\chi}_{\mathring{\delta}\mathcal{F}_1} - C_{\mathring{\delta}_c}\omega_{\mathring{\delta}_0}\chi_{\mathring{\delta}\mathcal{F}_2} = 0$$
$$A_{\mathring{\delta}_c} \dot{\omega}_{\mathring{\delta}\mathcal{F}_2} + A_{\mathring{\delta}_c} \dot{\chi}_{\mathring{\delta}\mathcal{F}_2} + C_{\mathring{\delta}_c}\omega_{\mathring{\delta}_0}\chi_{\mathring{\delta}\mathcal{F}_1} = 0 \ .$$

Using the abbreviation $\xi = A_{\mathring{\delta}_c}/A_{\mathring{\delta}} \approx 0.11$ and dividing the first two of eqns. (2.179) by $A_{\mathring{\delta}}$, the latter two by $A_{\mathring{\delta}_c}$, one obtains

$$\xi \dot{\chi}_{\delta \mathcal{F}_1} + \gamma_\delta \, \omega_{\delta 0} \, \omega_{\delta \mathcal{F}_2} - \xi \, \omega_{\delta 0} \, \chi_{\delta \mathcal{F}_2} = \tilde{\ell}_{\delta \mathcal{F}_1}$$

$$\dot{\omega}_{\delta \mathcal{F}_2} + \xi \dot{\chi}_{\delta \mathcal{F}_2} - \gamma_\delta \, \omega_{\delta 0} \, \omega_{\delta \mathcal{F}_1} + \xi \, \omega_{\delta 0} \, \chi_{\delta \mathcal{F}_1} = \tilde{\ell}_{\delta \mathcal{F}_2}$$

$$\dot{\omega}_{\delta \mathcal{F}_1} + \dot{\chi}_{\delta \mathcal{F}_1} - (1 + \gamma_{\delta c}) \, \omega_{\delta 0} \, \chi_{\delta \mathcal{F}_2} = 0$$

$$\dot{\omega}_{\delta \mathcal{F}_2} + \dot{\chi}_{\delta \mathcal{F}_2} + (1 + \gamma_{\delta c}) \, \omega_{\delta 0} \, \chi_{\delta \mathcal{F}_1} = 0 \;,$$

(2.180)

where

$$\tilde{\ell}_{\delta 1} = \frac{\ell_{\delta 1}}{A_\delta} \;; \quad \tilde{\ell}_{\delta 2} = \frac{\ell_{\delta 2}}{A_\delta} \;.$$

(2.181)

The structure of eqns. (2.180) becomes even simpler, when forming linear combinations of the first and third and the second and forth, respectively. These combinations have the purpose to eliminate *either* the terms $\dot{\chi}_{...}$ or the terms $\dot{\omega}_{...}$:

$$(1 - \xi) \dot{\omega}_{\delta \mathcal{F}_1} + \gamma_\delta \, \omega_{\delta 0} \, \omega_{\delta \mathcal{F}_2} + \xi \gamma_{\delta c} \, \omega_{\delta 0} \, \chi_{\delta \mathcal{F}_2} = \tilde{\ell}_{\delta \mathcal{F}_1}$$

$$(1 - \xi) \dot{\omega}_{\delta \mathcal{F}_2} - \gamma_\delta \, \omega_{\delta 0} \, \omega_{\delta \mathcal{F}_1} - \xi \gamma_{\delta c} \, \omega_{\delta 0} \, \chi_{\delta \mathcal{F}_1} = \tilde{\ell}_{\delta \mathcal{F}_1}$$

$$(1 - \xi) \dot{\chi}_{\delta \mathcal{F}_1} - (1 + \gamma_{\delta c} - \xi) \, \omega_{\delta 0} \, \chi_{\delta \mathcal{F}_2} = -\tilde{\ell}_{\delta \mathcal{F}_1} + \gamma_\delta \, \omega_{\delta 0} \, \omega_{\delta \mathcal{F}_2}$$

$$(1 - \xi) \dot{\chi}_{\delta \mathcal{F}_2} + (1 + \gamma_{\delta c} - \xi) \, \omega_{\delta 0} \, \chi_{\delta \mathcal{F}_1} = -\tilde{\ell}_{\delta \mathcal{F}_1} - \gamma_\delta \, \omega_{\delta 0} \, \omega_{\delta \mathcal{F}_1} \;.$$

(2.182)

In order to explain the above figures, the external torques may be ignored. Moreover, one may take advantage of the fact that the terms $\xi \gamma_{\delta c} \, \omega_{\delta 0} \, \chi_{...}$ on the left-hand sides of the first two of eqns. (2.182) are small compared to the terms $\gamma_\delta \, \omega_{\delta 0} \, \omega_{\delta \mathcal{F}...}$ (because of the factor ξ on one hand and because the quantities $\chi_{...}$ presumably are smaller than the components of the angular velocity vector of the mantle). This structure of eqns. (2.182) allows for an iterative solution of the above differential equation system, by considering the small terms as the inhomogeneous part of the linear system of differential equations, which are computed using the approximate solution vectors of the previous iteration step. Marking the approximated terms by $\tilde{\ }$ one obtains:

$$(1 - \xi) \dot{\omega}_{\delta \mathcal{F}_1} + \gamma_\delta \, \omega_{\delta 0} \, \omega_{\delta \mathcal{F}_2} = -\xi \gamma_{\delta c} \, \omega_{\delta 0} \, \tilde{\chi}_{\delta \mathcal{F}_2}$$

$$(1 - \xi) \dot{\omega}_{\delta \mathcal{F}_2} - \gamma_\delta \, \omega_{\delta 0} \, \omega_{\delta \mathcal{F}_1} = +\xi \gamma_{\delta c} \, \omega_{\delta 0} \, \tilde{\chi}_{\delta \mathcal{F}_1}$$

$$(1 - \xi) \dot{\chi}_{\delta \mathcal{F}_1} - (1 - \xi + \gamma_{\delta c}) \, \omega_{\delta 0} \, \chi_{\delta \mathcal{F}_2} = +\gamma_\delta \, \omega_{\delta 0} \, \tilde{\omega}_{\delta \mathcal{F}_2}$$

$$(1 - \xi) \dot{\chi}_{\delta \mathcal{F}_2} + (1 - \xi + \gamma_{\delta c}) \, \omega_{\delta 0} \, \chi_{\delta \mathcal{F}_1} = -\gamma_\delta \, \omega_{\delta 0} \, \tilde{\omega}_{\delta \mathcal{F}_1} \;.$$

(2.183)

The right-hand sides are set to zero in the first iteration step. This approximation separates the first two from the second two of eqns. (2.183). In order to emphasize the structure of the solution, the amplitudes of the prograde motions are characterized by the symbol "+", those of the retrograde terms with "−". In this approximation the solution of the first of eqns. (2.183) is:

$$
\begin{aligned}
\omega_{\delta\mathcal{F}_1}(t) &= \rho^+ \cos\left((1-\xi)^{-1}\gamma_\delta\,\omega_{\delta 0}t + \alpha^+\right) = \rho^+ \cos(\omega^+ t + \alpha^+) \\
\omega_{\delta\mathcal{F}_2}(t) &= \rho^+ \sin\left((1-\xi)^{-1}\gamma_\delta\,\omega_{\delta 0}t + \alpha^+\right) = \rho^+ \sin(\omega^+ t + \alpha^+) \\
\omega^+ &= (1-\xi)^{-1}\gamma_\delta\,\omega_{\delta 0} \ .
\end{aligned}
\tag{2.184}
$$

In this approximation the rotation pole moves on a circle around the pole of the figure axis with a period of

$$
P_{\text{Poinc}} = \frac{2\pi(1-\xi)}{\gamma_\delta\,\omega_{\delta 0}} = \frac{A_{\delta m}}{A_\delta}\,\frac{2\pi}{\gamma_\delta\,\omega_{\delta 0}} = \frac{A_{\delta m}}{A_\delta}\,P_{\text{Euler}} \approx 270.6 \text{ days} \ .
\tag{2.185}
$$

This formula explains Figure 2.57.

Using the above results in the third and fourth of eqns. (2.183) one may solve the equations for the first two components of vector $\chi_{\delta\mathcal{F}}$. The system is linear and inhomogeneous. As usual, the homogeneous system is solved first:

$$
\begin{aligned}
\chi_{\delta\mathcal{F}_1,\text{hom}} &= +\rho^- \cos\left(\omega^- t + \alpha^-\right) \\
\chi_{\delta\mathcal{F}_2,\text{hom}} &= -\rho^- \sin\left(\omega^- t + \alpha^-\right) \\
\omega^- &= \left(1 + \frac{\gamma_{\delta c}}{1-\xi}\right)\omega_{\delta 0} \ .
\end{aligned}
\tag{2.186}
$$

The period of the free motion of the core thus is

$$
P_{\delta c} = \frac{1}{1 + \frac{\gamma_{\delta c}}{(1-\xi)}}\,P_{\text{sid}} \approx 0.9963 \cdot P_{\text{sid}} \approx 0.9941 \text{ days} \ ,
\tag{2.187}
$$

where P_{sid} is the Earth's rotation period, i.e., one sidereal day.

According to eqn. (2.186) the solution of the *homogeneous equation* governing the motion of the core relative to the mantle is a retrograde, nearly-diurnal circular motion. This is in essence what Figure 2.60 shows.

The solution of the inhomogeneous equation may be obtained with the method of variation of constants. On the other hand one also sees, that a prograde circular motion with constant angular velocity ω^+ (see third of eqns. (2.184)) solves the inhomogeneous equation:

$$
\begin{aligned}
\chi_{\delta\mathcal{F}_1,\text{inh}} &= -\frac{\omega^+}{\omega^+ + \omega^-}\,\rho^+ \cos(\omega^+ t + \tilde\alpha^+) \\
\chi_{\delta\mathcal{F}_2,\text{inh}} &= -\frac{\omega^+}{\omega^+ + \omega^-}\,\rho^+ \sin(\omega^+ t + \tilde\alpha^+) \ .
\end{aligned}
\tag{2.188}
$$

Having established the general solution of the homogeneous and a particular solution of the corresponding inhomogeneous equation for the core, the general solution of the inhomogeneous solution is given by a linear combination of the two mentioned solutions. One easily verifies that the amplitudes of this contribution are about a factor of $\omega^-/\omega^+ \approx 300$ smaller than the amplitude of PM. In first approximation this contribution may be ignored.

The second approximation for PM is thus obtained by introducing the solution (2.186) on the right-hand sides of eqns. (2.183):

$$(1 - \xi)\,\dot{\omega}_{\delta\mathcal{F}_1} + \gamma_\delta\,\omega_{\delta 0}\,\omega_{\delta\mathcal{F}_2} = +\xi\,\gamma_{\delta c}\,\omega_{\delta 0}\,\rho^-\sin\left(\omega^- t + \alpha^-\right)$$
$$(1 - \xi)\,\dot{\omega}_{\delta\mathcal{F}_2} - \gamma_\delta\,\omega_{\delta 0}\,\omega_{\delta\mathcal{F}_1} = +\xi\,\gamma_{\delta c}\,\omega_{\delta 0}\,\rho^-\cos\left(\omega^- t + \alpha^-\right)\ .$$

$$(2.189)$$

After division by the factor $(1 - \xi)$ we obtain the equations:

$$\dot{\omega}_{\delta\mathcal{F}_1} + \omega^+\,\omega_{\delta\mathcal{F}_2} = +\frac{\xi\,\gamma_{\delta c}\,\omega^+}{\gamma_\delta}\,\rho^-\sin\left(\omega^- t + \alpha^-\right)$$
$$\dot{\omega}_{\delta\mathcal{F}_2} + \omega^+\,\omega_{\delta\mathcal{F}_1} = -\frac{\xi\,\gamma_{\delta c}\,\omega^+}{\gamma_\delta}\,\rho^-\cos\left(\omega^- t + \alpha^-\right)\ .$$

$$(2.190)$$

One easily verifies that the following functions solve the equations (2.190):

$$\omega_{\delta\mathcal{F}_1} = -\frac{\xi\,\gamma_{\delta c}\,\omega^+}{\gamma_\delta\,(\omega^- + \omega^+)}\,\rho^-\cos\left(\omega^- t + \alpha^-\right)$$
$$\omega_{\delta\mathcal{F}_2} = +\frac{\xi\,\gamma_{\delta c}\,\omega^+}{\gamma_\delta\,(\omega^- + \omega^+)}\,\rho^-\sin\left(\omega^- t + \alpha^-\right)\ .$$

$$(2.191)$$

The motion of the core relative to the mantle thus causes a retrograde, nearly-diurnal circular motion with period (2.187) which is superimposed to the (generalized) Eulerian motion. The amplitude of this motion is proportional to the amplitude of the motion of the core w.r.t. the mantle. The proportionality factor (see eqn. (2.191)) is, however, rather small:

$$\frac{\xi\,\gamma_{\delta c}\,\omega^+}{\gamma_\delta\,(\omega^- + \omega^+)} \approx \frac{1}{3047}\ .$$

$$(2.192)$$

Having chosen $|\chi_{\delta\mathcal{F}}| \approx 20''$ in the example of Figure 2.58 one expects an amplitude of about 7 mas for the retrograde nearly-diurnal motion – what is confirmed by this figure.

Precession and Nutation in Poincaré's Earth Model. Let us conclude the studies related to the Poincaré model by analyzing the motion of the rotation pole in inertial space, i.e., by considering precession and nutation. Figures 2.61 and 2.62 show the precession plus nutation in longitude and the nutation in obliquity for the year 2000 for a rigid Earth (solid line) and for a Poincaré model of the Earth, which only differ by the initial value of the vector χ_δ. The differences between Earth models are obvious: as opposed

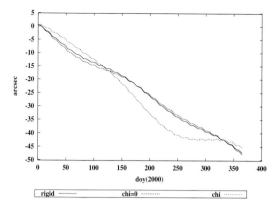

Fig. 2.61. Nutation in longitude for a rigid and a Poincaré Earth model with $\chi_{\mathring{\delta}_1}(t_0) = (0 \text{ or } 10^{-4}) \cdot \omega_{\mathring{\delta}_0}$ for year 2000

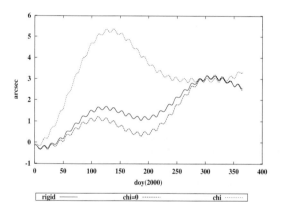

Fig. 2.62. Nutation in obliquity for a rigid Earth and a Poincaré Earth model with $\chi_{\mathring{\delta}_1}(t_0) = (0 \text{ or } 10^{-4}) \cdot \omega_{\mathring{\delta}_0}$ for year 2000

to a purely rigid (or elastic) Earth model a periodic signal of about 350 days with an amplitude greatly depending on the initial state of vector $\chi_{\mathring{\delta}}$ shows up for the Poincaré-type models. This periodic signal is usually called *FCN (Free Core Nutation)*.

Euler's kinematic equations (eqns. $7 - 9$ in the differential equation system (2.175)) show that each spectral line visible in nutation also must show up in the components $\omega_{\mathring{\delta}_i}$, $i = 1, 2$, as retrograde diurnal terms (due to the multipliers $\sin \Theta_{\mathring{\delta}}$ and $\cos \Theta_{\mathring{\delta}}$) in the PM series. Figure 2.63, containing the amplitude spectra of three models (rigid, Poincaré with different initial values for χ_{δ}), underlines this statement. (Observe that the three spectra are slightly offset in the two coordinate directions in order to improve the visibility of the terms). The main difference resides in the spectral line near 0.994 days corresponding to the 350 days free core nutation. Figure 2.63 confirms that the amplitude of this spectral line depends heavily on the initial condition of vector χ. It is worth noting, however, that the amplitude is rather big even for the case $\chi_{\delta_1}(t_0) = 0$.

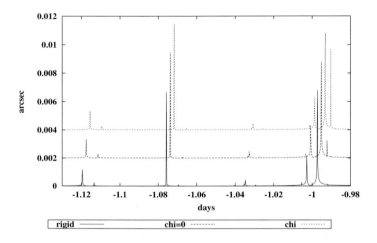

Fig. 2.63. Amplitude spectrum of PM for a rigid Earth and a Poincaré model for $\chi_{\delta_1} = (0 \text{ or } 10^{-4}) \cdot \omega_{\delta_0}$ in the retrograde diurnal band

In order to interpret Figures 2.61, 2.62, and 2.63 we have to introduce our analytical solutions for PM into the last three of the equations (2.175), the so-called kinematic Euler equations. Solving these equations for $\dot{\Psi}_{\delta}$, $\dot{\varepsilon}_{\delta}$, and $\dot{\Theta}_{\delta}$ one obtains:

$$
\begin{aligned}
\dot{\Psi}_{\delta} &= -\left\{ \omega_{\delta\mathcal{F}_1} \sin\Theta_{\delta} + \omega_{\delta\mathcal{F}_2} \cos\Theta_{\delta} \right\} \csc\varepsilon_{\delta} \\
\dot{\varepsilon}_{\delta} &= -\left\{ \omega_{\delta\mathcal{F}_1} \cos\Theta_{\delta} - \omega_{\delta\mathcal{F}_2} \sin\Theta_{\delta} \right\} \\
\dot{\Theta}_{\delta} &= +\left\{ \omega_{\delta\mathcal{F}_1} \sin\Theta_{\delta} + \omega_{\delta\mathcal{F}_2} \cos\Theta_{\delta} \right\} \cot\varepsilon_{\delta} + \omega_{\delta_0} .
\end{aligned}
\tag{2.193}
$$

To give an example we introduce the particular solution (2.191) into the equation for ε_{δ} and obtain

$$\dot{\varepsilon}_{\delta} = - \left\{ \omega_{\delta_{\mathcal{F}_1}} \cos \Theta_{\delta} - \omega_{\delta_{\mathcal{F}_2}} \sin \Theta_{\delta} \right\}$$

$$= + \frac{\xi \gamma_{\delta_c} \omega^+}{\gamma_{\delta} (\omega^- + \omega^+)} \rho^- \cos \left((\omega^- + \omega_{\delta_0}) t + \tilde{\alpha}^- \right) \tag{2.194}$$

$$= + \frac{\xi \gamma_{\delta_c} \omega^+}{\gamma_{\delta} (\omega^- + \omega^+)} \rho^- \cos \left(\frac{\gamma_{\delta_c}}{1 - \xi} \omega_{\delta_0} t + \tilde{\alpha}^- \right) .$$

The equations are solved by

$$\varepsilon_{\delta}(t) = - \xi \frac{\rho^-}{\omega^+ + \omega^-} \sin \left(\frac{\gamma_{\delta_c}}{1 - \xi} \omega_{\delta_0} t + \tilde{\alpha}^- \right)$$

$$= - \xi \frac{\rho^-}{\omega^+ + \omega^-} \sin \left(\frac{A_{\delta}}{A_{\delta_m}} \gamma_{\delta_c} \omega_{\delta_0} t + \tilde{\alpha}^- \right) . \tag{2.195}$$

The motion of the core thus induces in the inertial system a nutation in obliquity with an angular frequency of

$$\omega_{\mathrm{FCN}_{\mathcal{I}}} = \frac{A_{\delta}}{A_{\delta_m}} \gamma_{\delta_c} \omega_{\delta_0} \tag{2.196}$$

corresponding to a period of

$$P_{\mathrm{FCN}_{\mathcal{I}}} = \frac{A_{\delta_m}}{A_{\delta}} \frac{2 \pi}{\gamma_{\delta_c} \omega_{\delta_0}} \approx 354 \text{ days} . \tag{2.197}$$

Furthermore we note that the amplitude of the free core nutation is approximately given by the amplitude of the nearly-diurnal free wobble, reduced, however, by the factor of $\xi \approx 0.11$.

With the advent of free core nutation, the theory of nutation has lost to some extent the "status of complete predictability": Whereas the coefficients of nutation of the rigid or elastic Earth models do not depend on initial conditions (except for the dependence on ω_{δ_0}), free core nutation greatly depends on the initial condition of the vector $\boldsymbol{\chi}_{\delta}$. This may be used as an argument that the motion of the Earth's rotation pole in inertial space should be monitored exactly like PM.

Concluding Remarks Concerning Poincaré-Type Earth Models.
PM and precession plus nutation as resulting from the Poincaré model were compared to the corresponding results using a rigid Earth model. The following aspects are noteworthy:

- The generalized Eulerian PM is faster by a factor of A_{δ}/A_{δ_m} when compared to the rigid body model. Figure 2.57 illustrates the effect. This particular feature of the Poincaré model does not depend on the ellipticity of the core-mantle boundary.

- If the core-mantle boundary is spherical, there is no exchange of angular momentum between core and mantle. Core and mantle rotate about independent axes, where the core axis remains fixed in inertial space. The motion of the core relative to the boundary may not be detected by space geodetic observations in this case.

- Angular momentum *is* exchanged between core and mantle, if the core-mantle boundary is a (rigid) ellipsoid. The prominent feature in PM is a retrograde NDFW as shown in Figure 2.58. The period is given approximately by eqn. (2.187). The corresponding spectral line may be inspected in Figure 2.63.

- The prominent features in nutation are terms with a period of about 350 days in nutation (in longitude and obliquity) as illustrated by Figures 2.61 and 2.62. The effect is called free core nutation (FCN). The period is given by eqn. (2.197) and the amplitude is related to the amplitude of the motion of the core relative to the mantle (reduced by a factor of about $\xi \approx 0.11$).

- Figure 2.63 suggests that free core nutation should be "easily" detectable, e.g., through VLBI monitoring: Even for $\chi_{\delta_1}(t_0) = 0$, an amplitude of the order of more than 10 mas would result – an effect which is more than one order of magnitude above the noise level of today's observation techniques.

- *Resonance:* If the amplitudes of the lunisolar nutation terms as emerging from the rigid and the Poincaré Earth model are compared, one recognizes that the terms with periods close to that of free core nutation have slightly larger amplitudes in the case of the Poincaré model. In the simulations performed here the effect is measurable only for the semiannual term. Figure 2.64, showing the relevant part of the spectrum near $P = 1.003$ days in Figure 2.63, illustrates the effect. For all other lines of the spectrum in Figure 2.63 no effect can be noticed.

  The observed phenomenon is due to a resonance: if a dynamical system governed by a linear differential equation system is exited with a frequency close to one of the Eigenfrequencies of the system (the frequency associated with FCN resp. NDFW are Eigenfrequencies of our system), a resonant magnification of perturbing effect may occur.

  By analyzing the system equations (2.180) one might determine the amplification to be expected for the amplitudes of spectral lines in the nutation spectra as a function of the distance from the resonance frequency.

2.4 Rotation of Earth and Moon: A Summary

In **section 2.1** the essential characteristics (masses, moments of inertia, characteristic rotation and revolution periods) were introduced. Table 2.1 summarizes the essential quantities. The space geodetic observation techniques

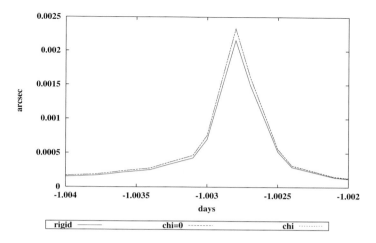

Fig. 2.64. Amplitude spectrum of PM for rigid Earth and Poincaré model for $\chi_{\dot{o}_1} = (0, 10^{-4}) \cdot \omega_{\dot{o}_0}$ near the semiannual nutation term

and their impact on monitoring Earth rotation were also addressed in this introductory section.

In **section 2.2** the N-body-problem Sun, Earth, Moon, and planets was studied under the assumption that Earth and Moon are finite rigid bodies. All the other constituents of the N-body problem were considered as point masses. The orbit of the Moon plays a key role for the understanding of the rotational characteristics of Earth and Moon. This is why the orbital characteristics of the Moon were studied in some detail in section 2.2.1. The orbit of the Moon is in essence governed by the Earth as the central body and the Sun as the main perturbing body. In the absence of the Sun's gravitational attraction, the term "month" would be unique. Table 2.2 lists the different kinds of months, which are due to solar (and planetary) perturbations.

An analysis of the Moon's mean orbital elements shows that there are "only" periodic perturbations in the semi-major axis a and the eccentricity e (the statement would have to be modified if time intervals of more than a few hundred years would be considered). The perturbations in the inclination i w.r.t. the plane of the ecliptic and the longitude of the Moon's ascending node show that the Moon's orbital pole is precessing around the pole of the ecliptic with a period of 18.6 years (see Figures 2.9, 2.10 and 2.11). For Earth rotation it is important that the inclination $\tilde{i}$ of the Moon's orbital plane w.r.t. the equatorial plane changes between the limits $\pm 5.5°$ with the period of 18.6 years induced by the revolution of the lunar node (see Figure 2.12). This characteristic of the Moon's orbit leads to a periodic variation of the mean (monthly, annual) torques exerted by the Moon on the oblate Earth.

It explains the biggest nutation term. Figure 2.13 shows that the Moon's perigee is rotating rather rapidly in the inertial system and also w.r.t. the Moon's ascending node. A Fourier analysis of the Moon's ecliptical longitude (see Figures 2.14 and 2.15) nicely summarizes the classical "anomalies" (as compared to a regular circular motion) of the Moon's orbital motion.

Section 2.2.2 introduces the essential properties of Earth rotation. In a first step the free motion (i.e., the motion in the absence of torques) was analyzed. It was found that in the case of rotational symmetry the Earth's rotational pole would move around the pole of figure (axis of maximum moment of inertia) on a circle with a radius defined by the initial position of the rotation axis. The period, called *Euler period* in the case of a rigid Earth, is a function of the length of the sidereal day and the Earth's dynamical flattening, as represented by eqn. (2.16). In the absence of external torques the Earth's angular momentum axis is fixed in inertial space. The rotation axis consequently shows small periodic diurnal variations in inertial space.

If the external torques caused by Sun and Moon are taken into account, the PM is still governed by the free motion (circular motion around the pole of figure), but a small retrograde nearly-diurnal motion, called Oppolzer motion, shows up (see Figure 2.19). The spectrum of this motion may be inspected in Figure 2.20. By virtue of Euler's kinematic equations (see, e.g., last three equations of the system (2.175)) the Oppolzer part of PM may be interpreted as the first time derivative of the nutation (transformed into the retrograde diurnal band by multiplication with the factors $\sin \Theta_\delta$, $\cos \Theta_\delta$).

The motion of the rotation pole in space is governed by precession and nutation (see Figures 2.24, 2.25, 2.26). The regression of the vernal equinox on the ecliptic with a rate of about $50.4''/y$ and the 18.6-year terms are the dominating effects of the motion of the Earth's axis in space. The motion is of course much more complex. Table 2.4 indicates that the rigid-body model of Earth rotation does account for precession and nutation rather well. Formula (2.39) shows that the term "precession constant" should be used with care: The annual precession rate depends among other on the obliquity of the ecliptic and the semi-major axes of the orbits of Sun and Moon – which, due to perturbations, are not strictly constant.

Whereas precession and nutation are explained quite well by the rigid body model, the same is not true for LOD variations: a rigid Earth with rotational symmetry does not show any LOD variations.

Lunar rotation was studied in section 2.2.3. The principal laws were already stated in the 17th century by Cassini. They were explained or illustrated in this section. The pattern of the analysis was similar to that followed in the case of Earth rotation. The principal difference of lunar rotation in comparison with Earth rotation resides in the fact that the assumption of rotational symmetry does not make sense (for numerical values see Table 2.1). Exactly as in the case of Earth rotation, the free motion (in the absence of the torques exerted by Earth and Sun) was studied first. The result is a generalization of the corresponding result of Earth rotation: The Moon's rotation pole would move on an ellipse with the axes-ratio defined by the two smallest principal moments of inertia. The period of the lunar "Eulerian motion" would be around 150 years. Whereas PM in the presence of torques is still governed by the free motion in the case of the Earth, the same is not true for the Moon: When the torques of Earth and Sun are taken into account, PM becomes rather chaotic (see Figure 2.31), because the equivalent to the Oppolzer motion dominates lunar PM.

The first Cassini law states that the sidereal revolution and rotation periods of the Moon are identical. The law was recognized as a consequence of *gravity stabilization*: The axis of the Moon's smallest moment of inertia always points (approximately) to the Earth. With suitable initial conditions (which were established by tidal friction over millions of years) the strict coupling of the two periods was achieved.

The rotation of the non-rigid Earth was studied in **section 2.3**. Three models were considered:

- an elastic Earth obeying Hooke's law,
- a rigid Earth with atmosphere and oceans, and
- an Earth consisting of a liquid core and a rigid mantle.

We first reviewed the Earth rotation series available from the IERS. We saw that PM is governed by the Chandler motion (period about 435 days) *and* by a term with an annual period and an amplitude of about half of the amplitude of the Chandler term. The essential part of the spectrum may be inspected in Figure 2.38. The spectrum of the measured LOD variations (see Figure 2.40) shows prominent annual and semiannual terms. It is important to note that none of the mentioned characteristics are explained (even approximately) by a rigid Earth model.

The generalizations of the classical Euler equations of motion for a rigid body are based on the concepts already set up in section I-3.3.7. By introducing a *rigid, rotating coordinate system* and by allowing rather arbitrary (but small) deformations of the "real Earth" w.r.t. this system, the problem could be solved with a minimum amount of analytical work: The Liouville-Euler equations (I-3.133) are the generalized equations of motion. The physical content is the same as in the case of the rigid body: The equations simply state that the change of angular momentum is equal to the sum of the applied external torques. The "only" differences w.r.t. the rigid body model reside in the facts that the inertia tensor may no longer be assumed as constant and as diagonal and that a so-called "inner angular momentum" κ_δ may show up. When considering the model of an elastic Earth we may simplify the equations of motion by adopting a Tissérand system with respect to which the inner angular momentum is always zero. The corresponding Liouville-Euler equations are eqns. (I- 3.137).

In order to give the equations for the rotation of an elastic Earth their explicit form, different kinds of deformations were considered in section 2.3.2. We saw that the so-called polar tides are able to explain the transition from the Euler to the Chandler period (see eqn. (2.96)). The lunisolar tidal deformations explain the monthly and bimonthly terms in the LOD variations, but only to a minor extent the annual and semiannual terms.

In section 2.3.3 the equations describing the rotation of a rigid Earth surrounded by the atmosphere and the oceans were developed. It is natural to associate the rigid rotating frame with the rigid Earth and to calculate the angular momentum and the contributions to the inertia tensor from atmosphere and/or oceans from meteorological and oceanographic observation series. Alternatively, one may in space geodetic analyses frequently (e.g., daily, hourly) solve for the PM coordinates and LOD values. From such observation series one may compute the three angular momentum components of the Earth consisting of a rigid body surrounded by the atmosphere and the oceans. The variation of these series may then be compared to atmospheric (or oceanographic) angular momentum series. The results for the third (axial) component of the Earth's angular momentum show a striking agreement: Figures 2.47 and 2.48 suggest that the LOD variations observed in space geodetic analyses of Earth rotation are explained to the greatest extent by interactions between the solid Earth and the atmosphere. The correlation between meteorological and space geodetic measurement series in the other two components of angular momentum is not as striking, but still rather high (see Figures 2.50). It is an established fact, that the features in PM and LOD mentioned initially are to a great extent explained by the Earth model of a rigid Earth surrounded by the atmosphere and the oceans. Subdaily LOD and PM variations could be attributed to the ocean tides.

In the last section 2.3.4 we introduced an Earth model taking into account today's knowledge of the Earth's interior. It should be mentioned that the Poincaré model is in a way a step back: the basic period of PM again is the Euler period (even reduced by about 10%), and tidal deformations do not exist in this model. The Poincaré model does, however, reveal the essential properties of modern Earth models: It does show

- the nearly-diurnal free wobble (NDFW) in PM,
- the free core nutation (FCN) in nutation (both components), and
- increased nutation amplitudes for terms near the period of free core nutation.

More advanced Earth models essentially shift the periods to more realistic values (in particular the Euler to the Chandler period).

3. Artificial Earth Satellites

The motion of artificial satellites is of crucial importance in the wide field of satellite geodesy. In this chapter we focus uniquely on the principal dynamic properties of satellite orbits, whereas, e.g., [106] covers the entire field of satellite geodesy.

3.1 Oblateness Perturbations

3.1.1 A Case Study

Table I- 3.1 in Chapter I- 3.4 tells that the term C_{20}, caused by the Earth's oblateness, is the dominating perturbation term of the Earth's gravitational field. For near Earth satellites the perturbation due to C_{20} is about a factor of 1000 smaller than the main term $\frac{GM}{r}$ and it exceeds all other terms of the Earth's gravitational potential by at least a factor of 200 (in the case of C_{22}), in general by about a factor of 1000.

In an attempt to fix the order of magnitude of the perturbations exprienced by a "normal" satellite, the orbit of a test satellite is integrated over the relatively short time interval of 5 days using the program SATORB, which is documented in Chapter 7 of Part III. The initial osculating orbital elements of the test satellite are given in Table 3.1. The object may be viewed as a typical LEO (Low Earth Orbiter).

The quasi-inertial geocentric reference system used in satellite geodesy is the *mean geocentric equatorial system referring to a particular epoch*. The system $J2000.0$ will be used throughout this Chapter. The system is called *quasi-inertial* (and designated by the symbol $\mathcal{I}$) because its origin is attached to the Earth's center of mass, which revolves on an approximately elliptical curve around the center of mass of the Earth–Moon system; the latter center of mass in turn revolves on an approximately elliptical orbit around the Sun. The origin of the reference system thus follows a rather complicated trajectory in the inertial system. The term quasi-*inertial* is justified, because the system does *not* rotate w.r.t. a truly inertial system.

The osculating orbital elements extracted after the integration are given in Figures 3.1 (as a function of time) over the interval of the first six hours

Table 3.1. Osculating elements of a virtual test satellite at $t_0 = $ January 1, 2001, 0^h UT

Element	Value	Element	Value
a	8000 km	e	0.07
i	35°	Ω	0°
ω	0°	T_0	t_0

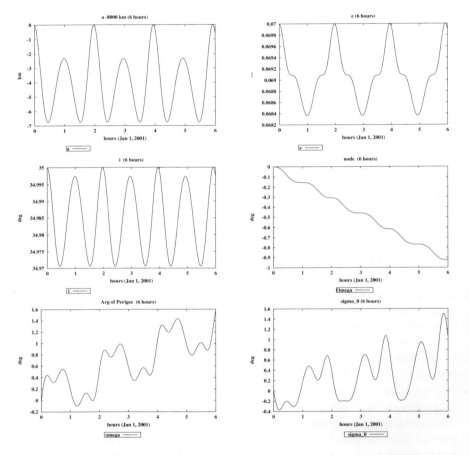

Fig. 3.1. Osculating elements of a satellite orbit in the gravitational field of the oblate Earth (six hours)

(corresponding to about three revolutions), and over the full interval of five days in Figures 3.2. All perturbation terms, except C_{20}, were set to zero in this initial example. In the transformation between the inertial and the Earth-fixed system the polar wobble was ignored and the difference UT1–

UTC was set to zero. The initial epoch is January 1, 2001. The (unperturbed) revolution period is approximately two hours.

Short-periodic perturbations dominate the development of the semi-major axis a, the eccentricity e, and the inclination i in Figure 3.1. The main period of the perturbations in a is $P/2$, half the satellite's revolution period around the Earth. The period of the perturbations in the eccentricity is the satellite's revolution period P.

The amplitudes are (alternatingly) about 3.2 km and 2 km in the case of the semi-major axis a (which, as a matter of fact, is due to a superposition of two periodic signals with periods $P/2$ and P). The amplitudes of the periodic perturbations are about $8 \cdot 10^{-4}$ in the case of the eccentricity, and about $0.015°$ in the inclination i.

Over longer time spans (see Figures 3.2) the right ascension of the ascending node and the argument of perigee are dominated by secular perturbations. Five days seems like a rather short time interval. On the other hand, the degree of difficulty to model an orbit grows with the number of revolutions. For the test-satellite of Table 3.1 the time interval of five days corresponds to 60 revolutions. For a minor planet with a typical revolution period of four years the time interval comparable to five days in satellite geodesy would be approximately 240 years.

Figures 3.2 show that the node rotates in the retrograde sense with an angular velocity of $\dot{\Omega} \approx -3.75°$ per day, the argument of perigee rotates in the prograde sense w.r.t. the ascending node with an angular velocity of about $\dot{\omega} \approx 5.5°$ per day. As the inclination shows mainly short-periodic variations, we may conclude that the orbital pole rotates around the polar axis of the Earth with a period of about 96 days. The perigee performs a prograde rotation w.r.t. the inertial space around the polar axis, where the angular velocity of this rotation is given by the sum $\dot{\Omega} + \dot{\omega} \approx 1.75°$ per day, resulting in a revolution period of about 206 days.

Qualitatively, Figures 3.1 and 3.2 resemble the corresponding figures of three-body perturbations as encountered, e.g., in Chapter 4. This intuitive interpretation is correct, because the equatorial bulge of the Earth may be interpreted (in a very crude approximation) as a third body, with its mass spread out over a circular annulus (orbital curve).

Figures 3.3 and 3.4, showing the development of the mean right ascension Ω of the ascending node and the mean argument of perigee ω for different values of the inclination i, indicate that the angular velocities of the precession of the orbital plane and of the argument of perigee depend heavily on the inclination i of the satellite. The mean elements were computed by averaging the osculating elements over one revolution period of the satellite.

As (perhaps) expected, the regression of the node is maximum for small inclinations $i \approx 0°$ (for $i = 0°$ the angular velocity is even indefinite), for $i = 90°$ the orbital plane does not precess.

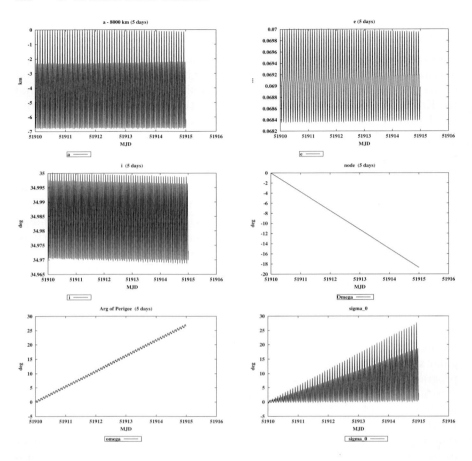

Fig. 3.2. Osculating elements of a satellite orbit in the gravitational field of the oblate Earth (five days)

Using program SATORB one may easily verify that the orbital pole rotates in the prograde sense for inclinations $i > 90°$. The angular velocities are $\dot{\Omega}(180° - i) = -\dot{\Omega}(i)$ for $i > 90°$.

The prograde motion of the argument of perigee w.r.t. the ascending node is maximum for small inclinations $i \approx 0°$, it comes to a standstill for an inclination of $i \approx 63.4°$. The rotation of the perigee w.r.t. the ascending node is retrograde for inclinations $63.4° < i < 116.6°$. For even bigger inclinations the rotation of the perigee becomes prograde, again.

The angle(s) of $i \approx 63.4°$ (and of $i \approx 116.6°$), for which the perigee does not rotate w.r.t. the ascending node, is called the *critical inclination*. The attribute *critical* is misleading – there is nothing critical about this particular inclination (except that denominators in higher-order perturbation theory

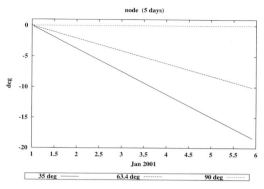

Fig. 3.3. Mean right ascension of node for inclinations of $i = 35°$, $63.4°$ and $90°$ (five days)

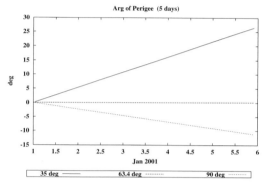

Fig. 3.4. Mean argument of perigee for inclinations of $i = 35°$, $63.4°$ and $90°$ (five days)

might become zero). The effect of a "non-rotating" argument of perigee at $i = 63.4°, 116.6°$ is real and exploited in practice for particular classes of satellites (discussed later on).

3.1.2 Oblateness Perturbations in the Light of First-Order Perturbation Theory

Perturbing Function and Equations of Motion. In section I- 3.4.2 we derived the approximation (I- 3.157) for the Earth's gravitational potential (expressed in the Earth-fixed system), where we only considered the moments up to second order (see eqns. (I- 3.155)) of the Earth's mass distribution. Assuming axial symmetry ($A_\oplus = B_\oplus$), the approximation (I- 3.157) may be further simplified:

$$V(r,\phi) = \frac{GM}{r} + GM\,a_\delta^2\,C_{20}\,\frac{1}{r^3}\left(\frac{3}{2}\sin^2\phi - \frac{1}{2}\right).\tag{3.1}$$

r is the absolute value of the radius vector, ϕ the satellite's latitude in the Earth-fixed equatorial system. In this approximation, the gravitational potential is longitude-independent.

The coordinates in eqn. (3.1) refer to the Earth-fixed system. As the equations of motion assume their simplest form in the quasi-inertial system (no centrifugal or coriolis terms) it would be preferable to express the Earth's potential in this system. In view of the rather complicated transformation between the two systems, this leads to a rather bulky function. For order-of-magnitude considerations it is, however, possible to considerably reduce this transformation. The conventional transformation between the inertial and the Earth-fixed system is given by eqn. (I- 3.56) and it is illustrated by Figure I- 3.4. Equation (I- 3.56) may be given the form

$$\boldsymbol{r}_\mathcal{I} = \mathbf{R}_3(-\Psi_\delta)\,\mathbf{R}_1(\varepsilon_\delta)\,\mathbf{R}_3(-\Theta_\delta)\,\boldsymbol{r}_\mathcal{F} \overset{\text{def}}{=} \mathbf{T}_\delta\,\boldsymbol{r}_\mathcal{F}\,,\tag{3.2}$$

where the satellite coordinates on the left-hand side refer to the inertial ecliptical system, on the right-hand side to the Earth-fixed equatorial system. The inertial reference system in satellite geodesy is the equatorial system referring to a standard epoch, which is why the transformation actually needed is

$$\boldsymbol{r}_\mathcal{I} = \mathbf{R}_1(-\varepsilon_{e0})\,\mathbf{R}_3(-\Psi_\delta)\,\mathbf{R}_1(\varepsilon_\delta)\,\mathbf{R}_3(-\Theta_\delta)\,\boldsymbol{r}_\mathcal{F} \overset{\text{def}}{\approx} \mathbf{R}_3(-\Theta_\delta)\,\boldsymbol{r}_\mathcal{F}\,,\tag{3.3}$$

where ε_{e0} is the mean obliquity referring to the standard epoch of the equatorial coordinate system used. The approximation on the right-hand side is justified by the fact, that we may always select a reference epoch of the inertial coordinate system in such a way that the angle $\Psi_\delta \approx 0$. This is, e.g., achieved by selecting the mean system of the initial epoch of the integration (and not $J2000.0$) as the inertial coordinate system. Using, moreover, the approximation $\varepsilon_\delta \approx \varepsilon_{e0}$ we arrive at the approximative transformation in eqn. (3.3).

Equation (3.3) implies that *approximately* the transformation between the Earth-fixed equatorial and the inertial equatorial systems is given by a rotation about the common third axis of the two equatorial coordinate systems involved.

The Earth's potential (3.1) depends only on the absolute value r of the radius vector of the satellite and on the elevation ϕ above the equatorial plane. Both, r and ϕ, are invariants of the approximate transformation (3.3). Taking into account that in the inertial plane the elevation above the equator is the declination δ, the potential referring to the inertial equatorial system may be simply transcribed from eqn. (3.1) as

$$V(r, \delta) = \frac{GM}{r} + GM \, a_\delta^2 \, C_{20} \, \frac{1}{r^3} \left(\frac{3}{2} \sin^2 \delta - \frac{1}{2} \right) . \tag{3.4}$$

Using

$$\sin \delta = \frac{r_3}{r} \tag{3.5}$$

and introducing for abbreviation the constant

$$\tilde{C}_{20} \overset{\text{def}}{=} GM \, a_\delta^2 \, C_{20} , \tag{3.6}$$

the potential of an oblate Earth may be written as

$$V(\boldsymbol{r}) = \frac{GM}{r} + \tilde{C}_{20} \frac{1}{r^3} \left(\frac{3}{2} \frac{r_3^2}{r^2} - \frac{1}{2} \right) \overset{\text{def}}{=} \frac{GM}{r} + R , \tag{3.7}$$

where R is the perturbation function characterizing the motion of a satellite in the gravitational field of an oblate planet.

The equations of motion of the satellite are obtained by taking the gradient of the above potential

$$\ddot{\boldsymbol{r}} = \boldsymbol{\nabla} \left\{ \frac{GM}{r} + \frac{\tilde{C}_{20}}{r^3} \left(\frac{3}{2} \frac{r_3^2}{r^2} - \frac{1}{2} \right) \right\} \tag{3.8}$$

or, explicitly

$$\ddot{\boldsymbol{r}} = + \frac{1}{r^3} \left\{ - GM \, \boldsymbol{r} + \frac{3}{2} \frac{\tilde{C}_{20}}{r^2} \begin{pmatrix} r_1 \left(1 - 5 \frac{r_3^2}{r^2} \right) \\ r_2 \left(1 - 5 \frac{r_3^2}{r^2} \right) \\ r_3 \left(3 - 5 \frac{r_3^2}{r^2} \right) \end{pmatrix} \right\} . \tag{3.9}$$

The structure of the equations of motion is clearly visible in the representation (3.9): For LEOs one may approximate $r \approx a_\delta$ which results in

$$\frac{3}{2} \frac{\tilde{C}_{20}}{r^2} \approx \frac{3}{2} GM \, C_{20} = \frac{3}{2} GM \, \sqrt{5} \, \bar{C}_{20} = \frac{3}{2} GM \, 1082.6 \cdot 10^{-6} , \tag{3.10}$$

meaning that the second term in the bracket $\{ \dots \}$ of the equations of motion (3.9) is of the order of 0.15% of the first, the two-body term. The numerical value for $\bar{C}_{20}$ was taken from Table I-3.1, and eqn. (I-3.158) was used for the transformation $C_{20} = \sqrt{5} \, \bar{C}_{20}$ between the normalized and un-normalized oblateness terms.

A Particular Solution. We are now in a position to give a closed solution of the equations of motion (3.9) for a special set of initial conditions. For the motion in the equatorial plane the equations of motion are (all terms $\sim \frac{r_3^2}{r^2}$ are zero)

$$\ddot{\boldsymbol{r}} = -GM\left(1 - \frac{3}{2}\frac{a_\oplus^2}{r^2}C_{20}\right)\frac{\boldsymbol{r}}{r^3} . \tag{3.11}$$

If the initial values are defined as

$$\boldsymbol{r}(t_0) = a\begin{pmatrix} \cos u_0 \\ \sin u_0 \\ 0 \end{pmatrix} \quad \text{and} \quad \dot{\boldsymbol{r}}(t_0) = a\,\tilde{n}\begin{pmatrix} -\sin u_0 \\ \cos u_0 \\ 0 \end{pmatrix} , \tag{3.12}$$

where the mean motion is computed as

$$\tilde{n} = \sqrt{\frac{GM\left(1 - \frac{3}{2}\frac{a_\oplus^2}{a^2}C_{20}\right)}{a^3}} , \tag{3.13}$$

one easily verifies that the solution of the initial value problem (3.11), (3.12) actually is a circular motion obeying Kepler's laws with the modified gravitational constant

$$GM' \stackrel{\text{def}}{=} GM\left(1 - \frac{3}{2}\frac{a_\oplus^2}{a^2}C_{20}\right) . \tag{3.14}$$

Observe that $GM' > GM$, because C_{20} has a negative value.

The initial value problem (3.11), (3.12) is not only of academic, but also of practical interest: The orbits of geostationary satellites are solutions of this problem, where the orbit's radius a_{geo} is defined through the revolution period, which must be precisely one sidereal day for this class of satellites. The radius a_{geo} (in SI units) is obtained as the solution of the polynomial equation

$$\left(\frac{2\pi}{86400 \cdot 365.25/366.25}\right)^2 a_{\text{geo}}^3 = GM\left(1 - \frac{3}{2}\frac{a_\oplus^2}{a_{\text{geo}}^2}C_{20}\right) . \tag{3.15}$$

The equation is easily solved iteratively, by using approximate values for a_{geo} on the right-hand side of the above equation ($a_\oplus/a_{\text{geo}} = 0$ is good enough for the first iteration step, and the iteration may be stopped safely after the second step(!)).

Conservation of the Third Component of Angular Momentum and of Energy. The solutions of eqn. (3.9) conserve the third component of the satellites angular momentum and its energy:

1. *Third Component of Angular Momentum:* Using eqn. (3.9) one easily verifies that

$$r_1 \ddot{r}_2 - r_2 \ddot{r}_1 = 0 , \tag{3.16}$$

implying that

$$r_1 \dot{r}_2 - r_2 \dot{r}_1 = h_3 = \text{const.} \tag{3.17}$$

This property is not restricted to the term C_{20}. The third component of the angular momentum is conserved for any rotationally symmetric potential, i.e., for each potential consisting only of zonal terms C_{l0}, $l = 2, 3, \ldots$

2. *Conservation of Energy:* The right-hand side of eqn. (3.9) was computed as the gradient of a potential function:

$$\ddot{\boldsymbol{r}} \cdot \dot{\boldsymbol{r}} = \boldsymbol{\nabla} V \cdot \dot{\boldsymbol{r}} = \dot{V} . \tag{3.18}$$

This implies immediately that the energy is conserved

$$\frac{1}{2} \dot{r}^2 - V = E = \text{const.} \tag{3.19}$$

This property is not restricted to the term C_{20} of the potential, but holds for all zonal terms.

Use of Perturbation Theory: Outline of Methods. Approximate analytical solutions of the equations of motion (3.9) may be produced by using the perturbation theory as developed in Chapter I-6. We may either use the Gaussian version (I-6.88) of the perturbation equations (named after Carl Friedrich Gauss (1777–1855)) or Lagrange's planetary equations (I-6.115) (named after Joseph Louis de Lagrange (1736–1813)) – which is of course an obsolete terminology for our application. The two approaches may be outlined as follow:

Method 1: When using the Gaussian version (I-6.88) of the perturbation equations, we have to transform the perturbing acceleration into the $\mathcal{R}$-system (see Figure I-6.1). The result may be represented by the corresponding term in eqn. (3.9) and a series of matrix multiplications:

$$\begin{pmatrix} R' \\ S' \\ W' \end{pmatrix} = \mathbf{R}_3(u) \, \mathbf{R}_1(i) \, \mathbf{R}_3(\Omega) \, \frac{3 \, \tilde{C}_{20}}{2 \, r^5} \begin{pmatrix} r_1 \left(1 - 5 \frac{r_3^2}{r^2} \right) \\ r_2 \left(1 - 5 \frac{r_3^2}{r^2} \right) \\ r_3 \left(3 - 5 \frac{r_3^2}{r^2} \right) \end{pmatrix} . \tag{3.20}$$

The result, expressed as a function of the orbital elements, is:

$$\begin{pmatrix} R' \\ S' \\ W' \end{pmatrix} = \frac{3 \, \tilde{C}_{20}}{2 \, r^4} \begin{pmatrix} 1 - \frac{3}{2} \sin^2 i + \frac{3}{2} \sin^2 i \, \cos 2u \\ \sin^2 i \, \sin 2u \\ \sin 2i \, \sin u \end{pmatrix} . \tag{3.21}$$

It is an elementary task (and a good exercise) to use the representation (3.21) for the perturbing accelerations in the Gaussian equations (I- 6.88) with the purpose of integrating the terms "analytically", i.e., in closed form. This was, e.g., done in [18].

Method 2: When using the Lagrangian version (I- 6.115) of the perturbation equations, the perturbing function R in eqn. (3.7) has to be represented as a function of the osculating orbital elements and of time t. In rectangular coordinates this perturbing potential was defined by eqn. (3.7). By using the equation $\frac{r_3^2}{r^2} = \sin^2 u \, \sin^2 i$ and the polar equation (I- 4.16) for the absolute value of the radius vector we obtain the expression:

$$
\begin{aligned}
R &= \frac{\tilde{C}_{20}}{r^3} \left(\frac{3}{2} \sin^2 u \, \sin^2 i - \frac{1}{2} \right) \\
&= \frac{\tilde{C}_{20} \, (1 + e \cos v)^3}{a^3 \, (1 - e^2)^3} \left(\frac{3}{2} \sin^2(v + \omega) \sin^2 i - \frac{1}{2} \right) .
\end{aligned}
\tag{3.22}
$$

v stands for the true anomaly, ω is the argument of perigee, and $u = \omega + v$ is the argument of latitude of the satellite.

The second method promises to be more efficient, because one only has to represent one (instead of three) scalar function(s) as a function of the orbital elements.

In order to solve Lagrange's perturbation equations (I- 6.115), one has to take the partial derivatives of function R w.r.t. the orbital elements. This is easily possible, provided the partial derivatives of the true anomaly w.r.t. these elements are known. These derivatives, on the other hand, had to be calculated explicitly when deriving the solution of the variational equations associated with the two-body motion in Chapter I- 5. Extensive use will be made of these results.

A Simple Approximation of the Perturbing Function. Orbits with small eccentricities are an important special case in satellite geodesy. Approximations of the perturbing function up to a certain order in e are therefore important. The simplest approximation up to terms of order 0 in e follows from eqn. (3.22) by approximating the true anomaly v by the mean anomaly $v \overset{\text{def}}{=} \sigma = \sigma(t)$ (see definition in Table I- 4.2):

$$
R = \frac{\tilde{C}_{20}}{a^3} \left(\frac{3}{2} \sin^2(\sigma + \omega) \sin^2 i - \frac{1}{2} \right) + O(e) .
\tag{3.23}
$$

In order to derive formula (3.23) from eqn. (3.22) we simply had to set $v \overset{\text{def}}{=} \sigma$ and $e = 0$, implying that $r = a$. Observe that the perturbation equations containing partial derivatives w.r.t. e (namely those for ω and T_0) become simpler in this approximation.

Perturbations as a Function of the True Anomaly. The differential equation for the argument of latitude (or of the true anomaly) (I- 4.35) was derived in Chapter I- 3:

$$\dot{u} = \dot{v} = \frac{h}{r^2} = \sqrt{\frac{GM}{a^3 (1 - e^2)^3}} \left(1 + e \cos (u - \omega)\right)^2$$

$$= \frac{n}{\sqrt{(1 - e^2)^3}} \left(1 + e \cos v\right)^2 . \tag{3.24}$$

This equation may be used to replace the time t in the perturbation equations by the true anomaly v (or by the argument of latitude u).

If I is one of the orbital elements satisfying the perturbation equation

$$\dot{I} = g(a, e, i, \Omega, \omega, T_0) , \tag{3.25}$$

the transformed equation reads as

$$\frac{dI}{dv} = \frac{dI}{du} = \frac{1}{\dot{v}} g(a, e, i, \Omega, \omega, \sigma_0) = \frac{\sqrt{(1 - e^2)^3}}{n (1 + e \cos v)^2} g(a, e, i, \Omega, \omega, \sigma_0) . \tag{3.26}$$

The transformed equations are particularly well suited to describe short-period perturbations as a function of the (osculating) true anomaly, or to compute mean values of perturbations over one revolution period, i.e., to compute secular perturbations.

The Semi-major Axis a. According to eqns. (I- 6.115) the Lagrangian version of the perturbation equation for the semi-major axis a reads

$$\dot{a} = - \frac{2 a^2}{GM} \frac{\partial R}{\partial T_0} = - \frac{2}{n^2 a} \frac{\partial R}{\partial T_0} . \tag{3.27}$$

Using eqn. (3.22) we obtain

$$\frac{\partial R}{\partial T_0} = \frac{3}{2} \frac{\tilde{C}_{20}}{p^3} \left(1 + e \cos v\right)^2 \frac{\partial v}{\partial T_0} \left\{ - e \sin v \left[3 \sin^2 (v + \omega) \sin^2 i - 1\right] \right.$$

$$\left. + \sin 2 (v + \omega) \sin^2 i \left(1 + e \cos v\right) \right\} . \tag{3.28}$$

Making use of the fact that in the two-body motion

$$\frac{\partial u}{\partial T_0} = - \dot{u} \tag{3.29}$$

and that, according to eqn. (3.6), $\tilde{C}_{20} = n^2 a^3 a_\delta^2 C_{20}$, the following equation for the semi-major axis as a function of the true anomaly is obtained:

$$\frac{da}{du} = C_{20} \frac{3\,a}{(1-e^2)^3} \frac{a_\delta^2}{a^2} (1+e\,\cos v)^2 \Big\{ -e\,\sin v \big[3\,\sin^2 (v+\omega)\,\sin^2 i - 1 \big]$$

$$+ \sin 2\,(v+\omega)\,\sin^2 i\,(1+e\,\cos v) \Big\} \,.$$

$$(3.30)$$

When considering the orbital elements on the right-hand side as constants, i.e., when applying first-order perturbation theory, eqn. (3.30) may be solved in closed form: The brackets $(\dots)^2 \{\dots\}$ are multiplied and the resulting products of sin- and cos-terms are replaced by sin- or cos-functions of linear combinations of the arguments v and ω, using the appropriate relations of trigonometry. This is a slightly tedious, but straight forward task, which may be left as an exercise to the reader. The integration may then be performed term by term.

It is easily verified that in this first-order theory there are no secular terms due to the Earth's oblateness in the semi-major axis a by evaluating the integral of the above perturbation equation over a full revolution period

$$\int_0^{2\pi} \frac{da}{du'}\,du' = 0 \,.$$

$$(3.31)$$

It is interesting to note that in first-order theory there are only terms proportional to e^k, $k = 0, 1, 2, 3$. Considering only the zero-order term we obtain the equation

$$\frac{da}{du} = 3\,C_{20}\,a\,\frac{a_\delta^2}{a^2}\,\sin^2 i\,\sin 2u + O(e) \,,$$

$$(3.32)$$

which may be easily integrated and results in

$$a(u) = -\frac{3}{2}\,a\,\frac{a_\delta^2}{a^2}\,C_{20}\,\sin^2 i\,\cos 2u + C_a + O(e) \,.$$

$$(3.33)$$

The integration constant C_a is defined by the initial condition $a(u_0) = a_0$. The semi-major axis reaches its minimum values for $u = 0°$ and $u = 180°$, its maximum values for $u = 90°$ and $u = 270°$.

The main term of the oblateness perturbation thus consists of a short-period perturbation with period $P/2$, P being the satellite's revolution period. The amplitude is given by

$$A_a = \frac{3}{2}\,a\,\frac{a_\delta^2}{a^2}\,C_{20}\,\sin^2 i \approx 2.717 \text{ km} \,.$$

$$(3.34)$$

The numerical value refers to our example with $i = 35°$ and $a = 8000$ km . This explains the order of magnitude of the perturbations in Figures 3.1 and 3.2.

In order to explain in addition the modulation of the above amplitude, we would have to solve the differential equation (3.30) for the higher orders in e, as well. Observe, that the amplitude A_a of the zero-order term in the eccentricity e is zero for orbits in the equatorial plane $i = 0$.

The Node Ω and the Inclination i. According to eqn. (I- 6.115) the Lagrangian version of the perturbation equations for the inclination and the right ascension of the ascending node are:

$$\frac{di}{dt} = \frac{1}{n\,a^2\sqrt{(1-e^2)}\,\sin i} \left(\cos i\, \frac{\partial R}{\partial \omega} - \frac{\partial R}{\partial \Omega} \right)$$

$$\dot{\Omega} = \frac{1}{n\,a^2\sqrt{(1-e^2)}\,\sin i} \frac{\partial R}{\partial i} \ . \tag{3.35}$$

As the perturbation function due to the Earth's oblateness does *not* depend on the right ascension of the ascending node Ω, the above perturbation equations may be written as

$$\frac{di}{dt} = \frac{\cos i}{n\,a^2\sqrt{(1-e^2)}\,\sin i} \frac{\partial R}{\partial \omega}$$

$$\dot{\Omega} = \frac{1}{n\,a^2\sqrt{(1-e^2)}\,\sin i} \frac{\partial R}{\partial i} \ . \tag{3.36}$$

The partial derivatives required for the equations in i and Ω are obtained from eqn. (3.22):

$$\frac{\partial R}{\partial \omega} = \frac{3}{2} \frac{\tilde{C}_{20}}{r^3} \sin 2u \, \sin^2 i$$

$$\frac{\partial R}{\partial i} = \frac{3}{2} \frac{\tilde{C}_{20}}{r^3} \sin^2(\omega+v) \sin 2i \tag{3.37}$$

$$= \frac{3}{4} \frac{\tilde{C}_{20}}{r^3} \left(1 - \cos 2u\right) \sin 2i \ .$$

With eqns. (3.37) the perturbation equations for i and Ω read as

$$\frac{di}{dt} = \frac{3}{4} C_{20} \frac{n\,a\,a_\delta^2}{r^3\sqrt{1-e^2}} \sin 2i \, \sin 2u$$

$$\dot{\Omega} = \frac{3}{2} C_{20} \frac{n\,a\,a_\delta^2}{r^3\sqrt{1-e^2}} \left(1 - \cos 2u\right) \cos i \ . \tag{3.38}$$

Using eqn. (3.26) we can easily transform the above differential equations in time t into equations in the argument of latitude u. The result is:

$$\frac{di}{du} = \frac{3}{4} C_{20} \frac{a_\oplus^2}{a^2 (1 - e^2)^2} \sin 2i \sin 2u \, (1 + e \, \cos v)$$

$$\frac{d\Omega}{du} = \frac{3}{2} C_{20} \frac{a_\oplus^2}{a^2 (1 - e^2)^2} \, (1 - \cos 2u) \, \cos i \, (1 + e \, \cos v) \ . \tag{3.39}$$

Using the addition theorems of elementary trigonometry, the right-hand side may be written as a sum of sin- and cos-functions of linear combinations of the angles u and ω. In the spirit of perturbation theory of first order, the equations may be easily solved by integrating term-by-term. It is remarkable that this solution contains only terms of order e^0 and e^1. The solution up to terms of order e^0 is:

$$i(u) = -\frac{3}{8} \frac{a_\oplus^2}{a^2 (1 - e^2)^2} C_{20} \sin 2i \cos 2u + C_i + O(e)$$

$$\Omega(u) = \frac{3}{2} \frac{a_\oplus^2}{a^2 (1 - e^2)^2} C_{20} \cos i \left((u - u_0) - \frac{1}{2} \sin 2u \right) + C_\Omega + O(e) \ . \tag{3.40}$$

In this approximation the perturbations in inclination i are short-periodic with period $P/2$. If we include the terms of order e as well, we still obtain only short-period terms. Figure 3.1 shows that the actual solution and the approximation given by first-order theory are in very close agreement. Considering only the main term we may state that the mean values of the inclination i are assumed in the nodes and at maximum and minimum elevations. As C_{20} has a negative value, the maximum values are assumed at $u = 45°, 225°$, the minimum values at $u = 315°, 225°$. In degrees the principal amplitude is given by

$$A_i = \frac{180}{\pi} \frac{3}{8} \frac{a_\oplus^2}{a^2 (1 - e^2)^2} C_{20} \sin 2i \approx 0.0156° \ . \tag{3.41}$$

The numerical value refers to the example specified in Table 3.1. This order of magnitude is confirmed by Figures 3.1 and 3.2.

The equation in Ω contains a secular term. It is easy to verify, that the terms $\sim e$ do not contribute to the secular drift. As $C_{20} \approx -1.082 \cdot 10^{-3}$, the second of equations (3.40) predicts a linear regression of the node which is a function of the argument of latitude. The coefficient is a function of the semi-major axis a, the eccentricity e, and the inclination i.

By evaluating the above formula for $u = u_0$ and $u = u_0 + 2\pi$, it is easy to compute the mean rate of change of Ω as a function of time as

$$\bar{\Omega} = \frac{2\pi}{U} \frac{3}{2} \frac{a_\oplus^2}{a^2 (1 - e^2)^2} C_{20} \cos i = \frac{3}{2} \frac{\sqrt{GM} \, a_\oplus^2}{a^{\frac{7}{2}} (1 - e^2)^2} C_{20} \cos i \ . \tag{3.42}$$

Usually, this drift is expressed in units of degrees per day:

$$\bar{\dot{\Omega}} \, [\,°/\text{day}\,] = \frac{3 \cdot 180 \cdot 86400}{2\,\pi} \sqrt{\frac{GM}{a_\delta^3}} \left(\frac{a_\delta}{a}\right)^{\frac{7}{2}} C_{20} \, \cos i = - \frac{10.0°\,\cos i}{\left(\frac{a}{a_\delta}\right)^{\frac{7}{2}}(1-e^2)^2} .$$

(3.43)

Formula (3.43) is very convenient to use. It tells, e.g., that the regression of the node in the equator of LEO is approximately $\bar{\dot{\Omega}} \, [\,°/\text{day}\,] = 10°\cos i$ and that this regression decreases with $(a/a_\delta)^{-3.5}$. This implies, e.g., that for GPS satellites with $a \approx 26500$ km the regression of the node is reduced to $0.068°$ per day. The corresponding precession period of the orbital plane approximately is 14.5 years.

In the example of Table 3.1 with $a = 8000$ km, $i = 35°$ and $e = 0.07$ we expect a rotation of

$$\bar{\dot{\Omega}} = -3.7 \, [\,°/\text{day}\,] ,$$

(3.44)

which corresponds exactly to the regression of the node in Figure 3.2.

The Argument of Perigee ω. According to Lagrange's perturbation equations (I- 6.115) the argument of perigee obeys the equation

$$\dot{\omega} = \frac{\sqrt{1-e^2}}{e\,n\,a^2} \frac{\partial R}{\partial e} - \frac{\cos i}{n\,a^2\,\sqrt{1-e^2}\,\sin i} \frac{\partial R}{\partial i} = \frac{\sqrt{1-e^2}}{e\,n\,a^2} \frac{\partial R}{\partial e} - \cos i \, \dot{\Omega} . \quad (3.45)$$

The second term is the projection of the regression of the node into the orbital plane (see also Figure 3.5). Replacing the time t as independent argument by the argument of latitude u with eqn. (3.24) we obtain the differential equation

$$\frac{d\omega}{du} = \frac{1}{\dot{u}} \frac{\sqrt{1-e^2}}{e\,n\,a^2} \frac{\partial R}{\partial e} - \cos i \frac{d\Omega}{du} .$$

(3.46)

The partial derivative of R w.r.t. e follows from eqn. (3.22). One easily verifies that the result, when used in the above differential equation, leads to an expression which may be integrated in closed form.

It is also easy to verify that the only non-zero secular term is due to the derivative of the term $(1-e^2)^{-3}$ in the equation (3.22):

$$\frac{\partial R}{\partial e} = 6\,e\,\frac{\tilde{C}_{20}\,(1 + e\,\cos v)^3}{a^3\,(1-e^2)^4} \left(\frac{3}{2}\sin^2(v+\omega)\sin^2 i - \frac{1}{2}\right) + \dots . \quad (3.47)$$

If we are only interested in the secular term, we may further simplify the above equation by replacing $\sin^2 u$ by $\frac{1}{2}(1 - \cos 2u)$ and by retaining only the terms independent of v (and u):

$$\frac{\partial R}{\partial e} = 3\,e\,\frac{\tilde{C}_{20}}{a^3\,(1-e^2)^4} \left(\frac{3}{2}\sin^2 i - 1\right) + \dots . \quad (3.48)$$

Introducing this result into the equation (3.46) for the argument of perigee ω (as a function of u), retaining only the constant term, and making use of the second of eqns (3.39) eventually gives the secular motion of the argument of perigee:

$$\frac{d\omega}{du} = \frac{3}{4} C_{20} \frac{a_\delta^2}{a^2 (1 - e^2)^2} \left(1 - 5 \cos^2 i\right) + \ldots . \tag{3.49}$$

Note, that the terms $+\ldots$ not included in the above equation are of short-periodic nature and do therefore *not* contribute to the secular drift.

The above expression may be used to compute the mean motion of the argument of perigee by evaluating it for $u = u_0$ and $u = u_0 + 2\pi$ and by dividing it by the revolution period:

$$\bar{\dot{\omega}} = \frac{2\pi}{U} \frac{3}{4} C_{20} \frac{a_\delta^2}{a^2 (1 - e^2)^2} \left(1 - 5 \cos^2 i\right) = \frac{3}{4} C_{20} \sqrt{\frac{GM}{a_\delta^3}} \frac{1 - 5 \cos^2 i}{\left(\frac{a}{a_\delta}\right)^{\frac{7}{2}} (1 - e^2)^2} . \tag{3.50}$$

If one furthermore multiplies the result with the length of the day, the motion of the perigee per day is obtained as:

$$\bar{\dot{\omega}} \, [\,°/\text{day}\,] = \frac{3 \cdot 180 \cdot 86400}{4\pi} C_{20} \sqrt{\frac{GM}{a_\delta^3}} \frac{1 - 5 \cos^2 i}{\left(\frac{a}{a_\delta}\right)^{\frac{7}{2}} (1 - e^2)^2}$$

$$= + \frac{5.0° \cdot \left(5 \cos^2 i - 1\right)}{\left(\frac{a}{a_\delta}\right)^{\frac{7}{2}} (1 - e^2)^2} . \tag{3.51}$$

For the example of Table 3.1 with $a = 8000$ km , $i = 35°$, and $e = 0.07$ a positive rotation of the perigee w.r.t. the node of

$$\bar{\dot{\omega}} = +5.4°/\text{day} \tag{3.52}$$

is expected, which corresponds exactly to the drift in Figure 3.2.

The perigee obviously is fixed w.r.t. the node for

$$\cos i = \pm \frac{\sqrt{5}}{5} = \pm 0.44721 , \tag{3.53}$$

which corresponds to

$$i = 63.435° \quad \text{or} \quad i = 116.565° . \tag{3.54}$$

As already mentioned, the above inclinations are called critical inclinations.

For inclinations between $0° < i < 63.435°$ and $116.565° < i < 180°$ the perigee rotates counterclockwise (same sense of rotation as the orbital motion), for inclinations $63.435° < i < 116.565°$ the perigee rotates clockwise w.r.t. the node.

Equation for the Time T_0 of Perigee Passage and the Mean Anomaly $\sigma(t)$. According to eqns. (I-6.115) the differential equation for the time T_0 of perigee passage reads as:

$$\dot{T_0} = \frac{2}{n^2 a} \frac{\partial R}{\partial a} + \frac{1 - e^2}{n^2 a^2 e} \frac{\partial R}{\partial e} \; . \tag{3.55}$$

The problem with the first term becomes obvious already when using the approximation (3.23) for the perturbing function:

$$\frac{\partial R}{\partial a} = -\frac{3 \check{C}_{20}}{a^4} \left(\frac{3}{2} \sin^2(\sigma + \omega) \sin^2 i - \frac{1}{2} \right)$$
$$-\frac{\check{C}_{20} \, 9 \, n \, (t - T_0)}{4 \, a^4} \sin 2(\sigma + \omega) \sin^2 i \; , \tag{3.56}$$

where only the terms of order e^0 were retained. The same structure of the equation for the time T_0 of perigee passage results, if the correct perturbing potential (3.22) is used instead of the approximate (3.23). For analytical solutions the second term, which is proportional to the time argument $t - T_0$, is a nuisance. It is responsible for the short periodic terms with linearly growing amplitude in Figure 3.2 (bottom, right) for σ_0 (using the defining relation $\sigma_0 \stackrel{\text{def}}{=} n(t_0 - T_0)$ one can show that the perturbation equation for σ_0 essentially is of the same structure as that for T_0).

In section I-6.7 the perturbation equation (I-6.137) was derived for the mean anomaly $\sigma(t)$ at time t, which does no longer contain terms proportional to the time t:

$$\dot{\sigma} = n - \frac{2}{n \, a} R_a - \frac{1 - e^2}{n \, a^2 \, e} \frac{\partial R}{\partial e} \; , \tag{3.57}$$

where R_a stands for the partial derivative of the perturbation function $R(a, e, i, \Omega, \omega, v(a, e, T_0))$, where the dependence of the true anomaly v of the semi-major axis a is ignored.

It is interesting to further develop the differential equation for σ for the perturbation function (3.22) associated with the oblateness. As a matter of fact it may be reduced to:

$$\dot{\sigma} = n - \frac{1 - e^2}{n \, a^2 \, e} \left\{ \frac{\partial R}{\partial v} \frac{\partial v}{\partial e} - \frac{3 \, \cos v}{1 + e \, \cos v} R \right\} \; . \tag{3.58}$$

The first term on the right-hand side of eqn. (3.57) is the osculating mean motion as computed from the formulas of the two-body problem. The second term may be written as a linear combination of trigonometric functions of v and u and their multiples. The partial derivative of the true anomaly w.r.t. the eccentricity e was obtained in Chapter I-5 as

$$\frac{\partial v}{\partial e} = \frac{\sin v}{1 - e^2} (2 + e \cos v) \; . \tag{3.59}$$

It is important to note that the term in the brackets $\{\ldots\}$ on the right-hand side of eqn. (3.58) contains a constant term $\sim e$ (a term depending neither on v nor on u). Equation (3.57) may be brought into the form

$$\dot{\sigma} = n \left\{ 1 - \frac{3}{4} C_{20} \frac{a_\oplus^2}{a^2} \frac{3 \cos^2 i - 1}{\sqrt{(1 - e^2)^3}} \right\} + \ldots . \tag{3.60}$$

The terms $+\ldots$ are all of a short-period nature. Equation (3.60) may be viewed as the generalization of Kepler's third law describing the motion of a satellite about an oblate planet. One also recognizes that eqn. (3.13) is a special case of eqn. (3.60).

The constant perturbation term vanishes for

$$\cos i = \pm \frac{\sqrt{3}}{3} , \tag{3.61}$$

i.e., for inclinations

$$i = 54.74° \quad \text{and} \quad i = 125.26° . \tag{3.62}$$

This inclination might be designated as another "critical" inclination. This name was, however, never used – probably because the consequences are less obvious than in the case of the rotation of the perigee.

The angle of $i = 54.74°$ is the angle between the diagonal in the cube and the edge intersecting the diagonal in the same vertex. Observe that $i = 54.74°$ is the nominal inclination of the GPS satellites w.r.t. the Earth's equatorial plane. In the average over the revolution period the mean motion of a GPS satellite in the potential of an oblate Earth therefore is the same as in the potential of a spherical Earth.

Different Kinds of Revolution Periods. There is one unique revolution period in the two-body motion. In the case of satellite motion, the revolution period may, e.g., be defined as the time-interval between subsequent passes of a satellite through perigee, through apogee, or through an arbitrary, but fixed direction in the orbital plane. This is why the distinction has to be made between

- the *anomalistic revolution period* as the time interval between subsequent passes of a satellite through perigee,

- the *draconitic revolution period* as the time interval between subsequent passes of a satellite through the ascending node,

- and the *sidereal revolution period* as the time interval between subsequent passes through a "fixed" inertial direction in the orbital plane.

Equations (3.57) and (3.60) refer to the anomalistic motion. By forming the average of the values defined by eqn. (3.60), we may compare the actual anomalistic revolution period with the mean revolution period as it would be expected according to Kepler's third law:

$$\bar{\dot{\sigma}} = \bar{n} - \frac{3}{4} C_{20} \frac{a_\delta^2}{a^2} \frac{\bar{n}}{\sqrt{(1 - e^2)^3}} (3 \cos^2 i - 1) \stackrel{\text{def}}{=} \bar{n} + \delta n \ . \tag{3.63}$$

Let us define the mean Keplerian revolution period as

$$\bar{U} \stackrel{\text{def}}{=} \frac{2\,\pi}{\bar{n}} \ , \tag{3.64}$$

where $\bar{n}$ has to be formed as the mean value over an anomalistic period of the terms $\sqrt{GM/a^3}$, a being the osculating semi-major axis. At this point it is appropriate to provide the (approximate) difference between the mean semi-major axis $\bar{a}$ and the osculating semi-major axis a as assigned to a latitude u (the formula follows directly from eqn. (3.33)):

$$a - \bar{a} \approx -\frac{3}{2} a \frac{a_\delta^2}{a^2} C_{20} \sin^2 i \cos 2u \ , \tag{3.65}$$

allowing it to compute the mean semi-major axis from the osculating semi-major axis. The formula holds for orbits with small eccentricities in the gravitational field of an oblate Earth.

The anomalistic revolution period is then computed as

$$U_\omega \stackrel{\text{def}}{=} \frac{2\,\pi}{\bar{\dot{\sigma}}} = \frac{2\,\pi}{\bar{n}\left(1 + \frac{\delta n}{\bar{n}}\right)} = \bar{U}\left\{1 + \frac{3}{4} C_{20} \frac{a_\delta^2}{a^2} \frac{3 \cos^2 i - 1}{\sqrt{(1 - e^2)^3}}\right\} \ , \tag{3.66}$$

where terms of the order $\left(\frac{\delta n}{\bar{n}}\right)^2$ were neglected.

The difference $U_\omega - U_\Omega$ of the anomalistic and the draconitic period is obtained by multiplying the rate (3.50) of perigee precession by the revolution period and dividing it through the mean motion of the satellite:

$$U_\omega - U_\Omega = \frac{3}{4} C_{20} \frac{a_\delta^2}{a^2 (1 - e^2)^2} (1 - 5 \cos^2 i)\,\bar{U} \ . \tag{3.67}$$

By similar arguments we may also compute the difference between the sidereal and the draconitic period. Figure 3.5 shows that per revolution the angle $\Delta\gamma = -\dot{\Omega} U \cos i$ has to be covered by the satellite in excess to one draconitic revolution in order to arrive at "the same" inertial position within the orbital plane: Based on these considerations the difference between the sidereal and the draconitic period may be computed as

$$U_s - U_\Omega = \frac{\Delta\gamma}{n} = -\frac{3}{2} \frac{a_\delta^2}{a^2 (1 - e^2)^2} C_{20} \cos^2 i \ \bar{U} \ . \tag{3.68}$$

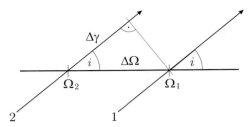

Fig. 3.5. Draconitic and sidereal revolution period

For inclinations $i < 90°$ the draconitic revolution period is always smaller than the sidereal revolution period.

Note, that the sidereal revolution period might also be defined in a different way as the time period between subsequent passes of the satellite through the same right ascension (such a definition might make sense for special applications). The resulting sidereal period would depend on the actual choice of the right ascension. With the help of Figure 3.5 one easily verifies that the sidereal revolution period, when using the right ascension of the ascending node of the first pass through the node (marked with "1" in the figure) as reference, would be computed as

$$U_s(\Omega) - U_\Omega = \frac{\Delta\gamma}{n\cos^2 i} = -\frac{3}{2}\frac{a_\oplus^2}{a^2(1-e^2)^2}C_{20}\bar{U}\,, \qquad (3.69)$$

whereas the above formula would have to be multiplied by $\cos i$, if the right ascension of the point of maximum elevation (corresponding to $u = 90°$) would be selected as reference.

3.1.3 Exploitation of the Oblateness Perturbation Characteristics

The Critical Inclination. It was known well before the beginning of the space era that certain types of orbits would be particularly interesting. Geostationary satellites are, e.g., well suited for communication or for surveying geographical areas in special longitude slots. Geostationary orbits have disadvantages, as well: the reception of signals is not optimum for high latitudes. It is therefore not amazing that Russia developed and deployed satellites of the Molnija-type. These satellites are in orbits with eccentricities $e \approx 0.72$, with revolution periods of half a sidereal day, corresponding to semi-major axes of approximately $a \approx 26'550$ km, and with inclinations of $i \approx 63.4°$. The arguments of perigee are initially set to $\omega \approx 270°$, which means that the apogee (at a height above the Earth's surface of about $h_{\text{apo}} \approx 39360$ km) resides over the (maximum) Northern latitude of $i \approx 63.4°$ achievable with this type of orbit. Thanks to the fact that a satellite with the inclination

$i = 63.4°$ does not suffer from perturbations in the argument of perigee ω due to the Earth's oblateness, it does not take much energy (fuel) to keep the perigee in place.

Thanks to Kepler's law of areas such a satellite spends about $8 - 10$ hours per revolution over the Northern hemisphere, corresponding to about 66-83% of the revolution period.

The same principle might be used for communication purpose over the Southern hemisphere by using the same orbit characteristics, except that the argument of perigee would have to be set to $\omega = 90°$.

According to Flury [41] the Russian Tundra satellites are communication satellites exploiting the critical inclination, as well. The revolution period is one sidereal day in this case and the eccentricity is $e \approx 0.27$. With only three satellites of this class it is possible to cover the vast Russian high-latitude territory.

Sun-Synchronous Satellites. For Earth observing satellites it may be useful or even a requirement to have always the same (or at least similar) illumination conditions of the Earth's surface below the satellite. In order to achieve maximum contrast it may be advisable to follow the terminator, the light-shadow boundary on the Earth's surface. For Sun-observing satellites this particular orbit may be suitable as well. Orbits of this type result, if the angle between the orbital pole direction and the vector pointing from the Earth to the Sun does not change in time. This condition may be met approximately if the angle between the geocentric unit vector to the Sun and the geocentric direction to the ascending node of the satellite orbit does not change. Such a condition can only be met in the average over one year because the Sun revolves around the Earth (formulation for satellite geodesists) with variable angular velocity (Kepler's second law) in the ecliptic (and not in the equator).

In practice mean Sun-synchronous orbits may be achieved if the ascending node is rotating in the prograde sense with an angular velocity of one revolution per year, corresponding to $\dot{\Omega} \approx +360°/365.2422 \approx 0.9856°/\text{day}$ in order to compensate for the annual geocentric motion of the Sun.

In the previous section we gave the first-order solution for the angular velocity of the node in formula (3.43). From this formula we see that a prograde rotation of the node is only achievable for inclinations $i > 90°$. The same formula shows furthermore that the rotation rate is in addition a function of the semi-major axis and the eccentricity. Table 3.2 gives an impression of the inclinations to be selected as a function of the height h above a spherical Earth of $a_\delta = 6378$ km for circular orbits ($e \stackrel{\text{def}}{=} 0$). Table 3.2 shows that Sun-synchronous LEOs are in retrograde motion and have inclinations close to $90°$. This undoubtedly is an advantage for Earth-observing satellites because they essentially fly over the entire globe.

Table 3.2. Sun-synchronous orbits in a height h above a spherical Earth with a_δ =6378 km

Height h [km]	Inclination i [°]	Revolution Period U [min]
400	97.03	92.6
600	97.79	96.7
800	98.61	100.9
1000	99.48	105.1
1200	100.42	109.4

More information related to the topic of Sun-synchronous orbits, in particular for orbits with exact ground-track repeatability, may be found in [41].

3.1.4 Higher-Order Oblateness Perturbations

First-order perturbation theory due to the term C_{20} of the Earth's gravitational field explains short-period perturbations with periods equal to the revolution period of the satellite and fractions thereof, and secular perturbations of the right ascension of the ascending node Ω and of the argument of perigee ω. The secular motions of the node and of the perigee induce long periods, which are expected to play an important role in the actual motion of the satellite.

For our initial example with the elements defined by Table 3.1 the precession period of the node was found to be about 97 days (see Figure 3.2 and formula (3.43)) and the rotation period of the perigee w.r.t. the node about 67 days (see Figure 3.2 and formula (3.51)).

Figure 3.6 shows the development of the mean orbital elements a, e, i, Ω, and ω over the time interval of two years with the initial epoch January 1, 2001. The force field uniquely consisted of the main term and the term C_{20} of the Earth's gravitational field. The average period was taken to be 12 anomalistic periods (about one day). Two years correspond to about 8700 revolutions of the test satellite of Table 3.1. The transformation between the inertial and the Earth-fixed system was handled as explained previously, i.e., the polar wobble and the difference UT1–UTC were neglected.

Figure 3.6 shows that long-period perturbations exist in the semi-major axis and in the eccentricity. The amplitudes are very small, however: The observed amplitude in the semi-major axis is about $\Delta a \approx 5$ m (the difference $a - 7995$ km is given in units of meters), in the eccentricity it is about $\Delta e \approx 6 \cdot 10^{-6}$. The amplitudes are about a factor of 1000 smaller than the amplitudes of the first-order perturbations in these elements. The period is approximately 33.5 days for both, the perturbations in the semi-major axis

a and the eccentricity e. This corresponds to half the rotation period of the perigee.

The inclination i and the right ascension Ω of the ascending node show an interesting characteristic: We observe periodic perturbations with a linearly growing amplitude. The period is about 97 days, i.e., the period of the rotation of the node. In order to see the interesting part in the perturbation of the node, the secular regression of about 3.7° per day was removed. The linear growth of the amplitude has nothing whatsoever to do with the oblateness perturbations: it is caused by the fact that the orbital elements are referred to the inertial system corresponding to mean equator and equinox of the initial epoch. The linear growth is caused by the precession of the Earth's pole of figure around the pole of the ecliptic. The linear growth would "disappear",

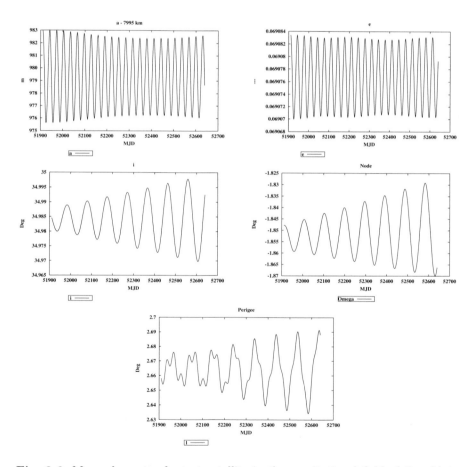

Fig. 3.6. Mean elements of a test satellite in the gravitational field of the oblate Earth over two years (MJD (Modified Julian Date))

if either the precession would be neglected, or if the precession of the orbital plane would be referred to a moving, non-inertial equatorial coordinate system.

The linear growth of the perigee of about 5.4° per day was removed, as well, in Figure 3.6. Obviously, long period perturbations also exist in the perigee. The period and the explanation are the same as in the case of the other two Eulerian angles.

Figure 3.6 proves that long-period perturbations due to the Earth's oblateness exist, but that the amplitudes are very small. Therefore, our initial remark, that "long-period perturbations due to the Earth's oblateness are very important", seems hardly justified. We will see in the next section, however, that these periods *are* most significant for the perturbations due to the higher-order terms of the Earth's potential.

3.2 Higher-Order Terms of the Earth Potential

Figure 3.7 compares the mean oblateness perturbations (already shown in Figure 3.6) with the mean perturbations resulting from an integration with identical initial conditions (those of Table 3.1), but using the Earth's gravitational potential up to terms of degree and order $n = m = 4$ (see Table I- 3.1).

Figure 3.7 shows that terms other than C_{20} cause substantial perturbations. The offset of about 15 m in the perturbations of the semi-major axis proves that the mean motion is significantly influenced by the higher-order terms.

A common trend was removed from the development of the node and also of the perigee. We see that higher-order terms (actually the zonal terms C_{l0}) give rise to secular perturbations, even in the same sense of rotation. The rates are, however, much smaller: Instead of a few degrees per day we observe a few degrees per year.

The perturbations of the mean eccentricity and of the mean argument of perigee are quite interesting: The amplitudes of the long-period perturbations caused by the higher-order terms are much bigger than the second-order effects due to the term C_{20}. The period of 67 days is that of the rotation of the perigee due the Earth's oblateness. Obviously the period of the higher-order terms is taken over from the period of the rotation of the perigee (mainly) due to the term C_{20}.

Figure 3.8 gives more insight into the role of the term C_{20} for the higher-order terms C_{lm} of the Earth's potential: It compares the perturbations in eccentricity e and inclination i due to the full Earth potential up to degree

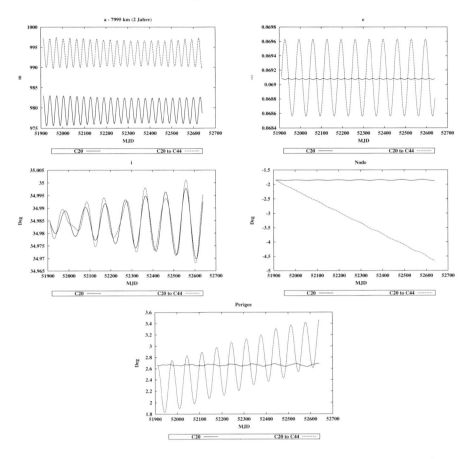

Fig. 3.7. Mean elements over two years using only the oblateness term C_{20} (solid lines) and using all terms up to degree and order 4 (dotted line)

and order four (these are the perturbations already shown in Figure 3.7, dotted lines) with the perturbations which would result, if the oblateness term $C_{20} = 0$ would be zero (and all other terms identical).

Figure 3.8 demonstrates that the oblateness attenuates the higher-order perturbations and reduces their periods. Due to the changing orbital geometry (perigee and orbital plane) the higher-order perturbations are of much shorter period (the period of the rotation of the argument of perigee) and the amplitudes are much smaller than they would be without the term C_{20}. The perturbations with $C_{20} = 0$ almost look like secular ones in Figure 3.8. Actually, the period would be the rotation period of the perigee due to the higher than second order zonal terms. Figure 3.8 indicates that the period would be of the order of decades.

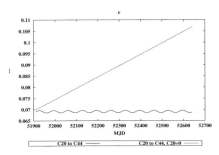 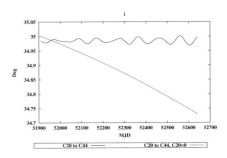

Fig. 3.8. Mean eccentricity and inclination over two years for an Earth with (solid line) and without (dotted line) term C_{20}

Cum grano salis we may state that the Earth's oblateness has a stabilizing influence on the orbits of artificial satellites (at least where higher-order perturbations are concerned). Let us mention, that the example with $C_{20} = 0$ is unreal.

Due to the limited space available, only few examples could be presented here. The reader is encouraged to use program SATORB (see Chapter 7 of Part III) for more detailed studies. It would be instructive to study the differences of the perturbations due to the zonal, tesseral, and sectorial terms. We confine ourselves to the study of particular tesseral and zonal terms, namely those terms giving rise to resonance.

3.3 Resonance with Earth Rotation

The revolution period of the satellite is said to be *commensurable* with the sidereal rotation period of the Earth, if a small integer number k_2 of sidereal days is equal to a small number k_1 of revolution periods of the satellite:

$$\frac{P_{\Earth}}{P_s} = \frac{k_1}{k_2} , \quad k_1 \text{ and } k_2 \text{ integers} . \tag{3.70}$$

In analogy to commensurabilities in the planetary system, the type of commensurability (3.70) may give rise to resonant perturbations. If condition (3.70) holds, the satellite experiences (almost) the same perturbations due to the Earth's potential after k_1 revolutions (which correspond to k_2 rotations of the Earth w.r.t. inertial space).

Equation (3.70) also tells that after one revolution the satellite experiences the same perturbations due to the terms C_{lk_1} of order k_1 (or multiples thereof) because after this time period the Earth has rotated by an angle of $\frac{k_2}{k_1} \cdot 360°$ and the terms of order k_1 have a period of $\frac{1}{k_1}$ days.

Commensurability of the satellite's revolution period with the sidereal day is a necessary, but not yet a sufficient condition for resonant perturbations to occur. Resonance is only encountered, if at least one of the orbital elements shows a net non-zero accumulated perturbation due to the term considered over one revolution period. Only tesseral and sectorial terms, i.e., the longitude-dependent terms of the Earth's potential may give rise to resonance.

Subsequently, resonances associated with geostationary satellites and with GPS satellites will be studied in some detail. For both cases $k_2 = 1$, which allows us to write condition (3.70) as

$$\frac{P_{\text{sid}}}{P_{\text{sat}}} = k , \qquad (3.71)$$

where $k = 1$ for geostationary and $k = 2$ for GPS satellites.

3.3.1 Geostationary Satellites

Case Studies. Figure 3.9 shows the mean semi-major axes (over one anomalistic revolution) of two hypothetical geostationary satellites over a time interval of ten years. Program SATORB (see Chapter 7 of Part III) was used to perform the integration. The initial epoch was selected as January 1, 2001. Only perturbations of the Earth's gravitational field up to terms of degree $n = 2$ and order $m = 2$ were taken into account. The satellites' geocentric longitudes initially were $\lambda_1 = 0°$ in one case (solid line) and $\lambda_2 = 45°$ (dotted line) in the other case. The attempt was made to produce a geostationary satellite in the Greenwich meridian and in a meridian over the Persian gulf.

The perturbations in a are of a very long period: Almost four years in the first case, still about two and a half years in the second case. The amplitudes $\Delta a_1 \approx 32$ km and $\Delta a_2 \approx 17.5$ km are orders of magnitude bigger than other short- or long-period perturbations in this element.

Figure 3.10 shows the development of the geocentric longitudes of the same two test satellites over the years $2001 - 2010$. The Figure proves that in general it is not possible to deploy a truly geostationary satellite – neither over the Greenwich meridian nor in a longitude of $\lambda_2 = 45°$. In the first case we observe an oscillation in longitude with an amplitude of about $\Delta\lambda_1 = 75°$ centered at a longitude of about $\lambda_{22} = 75°$ with a period of about 1450 days, in the second case an oscillation with an amplitude of about $\Delta\lambda_2 = 30°$ with a period of about 870 days, centered at the same longitude as in the first case, tentatively called $\lambda_{22} = 75°$. Below this "natural" or "zero" longitude will be identified with the reference longitude of the term J_{22} of the Earth's potential (representation (I- 3.163)). The periods of the oscillations in longitude λ are identical with the periods observed in the semi-major axis in Figure 3.9.

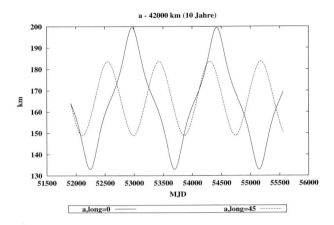

Fig. 3.9. Semi-major axes of two geostationary satellites deployed at longitudes $\lambda_1 = 0°$, $\lambda_2 = 45°$

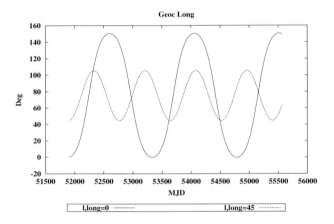

Fig. 3.10. Geocentric longitudes of two geostationary satellites deployed at longitudes $\lambda_1 = 0°$, $\lambda_2 = 45°$

It is easily possible to verify that only in the longitudes $\lambda_{22} \approx 75°$ (meridian over the Indian ocean) and $\lambda_{22} \approx 255°$ (meridian over Galapagos) stable geostationary satellites may be deployed. Deployment of a 24-hours satellite at an arbitrary longitude λ will result in oscillations in longitude with an amplitude of $\Delta\lambda = |\lambda - \lambda_{22}|$ or $\Delta\lambda = |\lambda - \lambda_{22} - 180°|$ – whatever is smaller.

Geostationary Satellites Viewed by Perturbation Theory. In order to explain the perturbations experienced by a geostationary satellite, we first have to represent the relevant part of the perturbation function as a function of the satellite's orbital elements. It is easily verified that the perturbations in Figures 3.9 and 3.10 are due to the terms C_{22} and S_{22} of the Earth's potential. Using the representation (I-3.163) the perturbation function may be written as:

$$R(r, \phi, \lambda) = -\frac{3\,GM\,a_\oplus^2}{r^3}\,J_{22}\,\cos^2\phi\,\cos 2(\lambda - \lambda_{22})\,, \qquad (3.72)$$

where the associated Legendre function $P_2^2(\sin\phi)$ was set to $P_2^2(\sin\phi) = 3\cos^2\phi$ and where, according to Table I-3.1 and eqns. (I-3.162), the numerical values for J_{22} and λ_{22} are:

$$\begin{aligned} J_{22} &= 0.181553 \cdot 10^{-5} \\ \lambda_{22} &= 75.0709° \,. \end{aligned} \qquad (3.73)$$

Subsequently, we will confine ourselves to the discussion of circular orbits in the equatorial plane. In this approximation the perturbation function (3.72) may be represented as a function of the orbital elements:

$$R(a, \tilde{\omega}, T_0) = -\frac{3\,GM\,a_\oplus^2\,J_{22}}{a^3}\,\cos 2\,(\tilde{\omega} + \sigma - \Theta - \lambda_{22})\,, \qquad (3.74)$$

where $\Theta(t) \approx \dot\Theta(t - t_0) + \Theta_0$ is Greenwich sidereal time, $\tilde{\omega}$ the right ascension of the perigee, and $\sigma = n(t - t_0) + \sigma_0$ the mean anomaly. The satellites longitude at t_0, $\lambda_0 \overset{\text{def}}{=} \lambda(t_0)$, thus may be written as $\lambda_0 \overset{\text{def}}{=} \tilde{\omega} + \sigma_0 - \Theta_0$. The simplified transformation equation (3.3) allows it therefore to write the longitude $\lambda(t)$ as

$$\lambda(t) = \tilde{\omega} + \sigma - \Theta \approx \tilde{\omega} + \sigma_0 - \Theta_0 + (n - \dot\Theta)(t - t_0) = \lambda_0 + (n - \dot\Theta)(t - t_0)\,. \qquad (3.75)$$

According to eqn. (I-6.115) the differential equation for the semi-major axis reads as

$$\begin{aligned} \dot{a} = -\frac{2}{n^2 a}\frac{\partial R}{\partial T_0} &= \frac{12\,GM\,a_\oplus^2\,J_{22}}{n\,a^4}\,\sin 2\,(\tilde{\omega} + \sigma - \Theta - \lambda_{22}) \\ &\approx \frac{12\,GM\,a_\oplus^2\,J_{22}}{n\,a^4}\,\sin 2\,(\lambda_0 - \lambda_{22})\,, \end{aligned} \qquad (3.76)$$

where the latter transformation made use of the fact that in our application $n(t - t_0) \approx \dot\Theta(t - t_0)$. This equation holds exactly when applying first-order perturbation theory. It shows that first-order perturbation theory breaks down when resonance perturbations are considered. The first-order solution of the above differential equation simply is a linear function of time:

$$a(t) = a_0 + (t - t_0)\,\frac{12\,GM\,a_\oplus^2\,J_{22}}{n\,a^4}\,\sin 2\,(\lambda_0 - \lambda_{22})\,. \qquad (3.77)$$

The solution is, by the way, not bad in the vicinity of the epoch t_0: It is the approximation by the Taylor series truncated after the terms of the first order in time t. Figure 3.9 shows that the solution is acceptable for time intervals up to a few months. The solution is also correct for a geostationary satellite deployed at $\lambda = \lambda_{22} = 75.0709°$ (where the solution simply is $a(t) = a_0$).

But solutions based on first-order perturbation theory will never be able to reveal the period of a resonant motion. Even an approximate solution of the above differential equation for the semi-major axis requires to take the time-dependence of the mean motion in the sin-argument in the differential equation (3.76) into account. It is not acceptable to use the approximation in the second line of eqn. (3.76).

So, in some sense, we failed to explain the resonant motion of a geostationary satellite using first-order perturbation theory. On the other hand, it is intuitively clear that the motion in longitude λ of a geostationary satellite is closely related to the motion of a pendulum with its rest-position at $\lambda = \lambda_{22}$. Let us further pursue this aspect.

For that purpose we take the first time derivative of the above equation (3.75) for the geocentric longitude of a geostationary satellite and obtain

$$\dot{\lambda} = \dot{\omega} + \dot{\sigma} - \dot{\Theta} \approx n - \dot{\Theta} . \tag{3.78}$$

The latter equation holds because we only consider circular orbits in the equatorial plane for which there are no out-of-plane perturbation components. By taking the second time derivative of the longitude λ we eventually obtain

$$\ddot{\lambda} = \ddot{\sigma} = \dot{n} = -\frac{3\,n}{2\,a}\,\dot{a} , \tag{3.79}$$

where the derivative of the angular velocity of Earth rotation and the second derivatives of ω and Ω were assumed to be zero – an assumption which is amply justified for our order-of-magnitude considerations.

Replacing $\dot{a}$ in the above equation by eqn. (3.76) gives a simple differential equation for the longitude of the satellite:

$$\ddot{\lambda} = -\frac{18\,GM\,a_\delta^2\,J_{22}}{a^5}\,\sin 2\,(\lambda - \lambda_{22}) . \tag{3.80}$$

Introducing the auxiliary variable

$$x \stackrel{\text{def}}{=} 2\,(\lambda - \lambda_{22}) \tag{3.81}$$

into the above differential equation reduces this equation to the standard equation for the mathematical pendulum:

$$\ddot{x} = -\left(\frac{6\,n\,a_\delta\,\sqrt{J_{22}}}{a}\right)^2\,\sin x \stackrel{\text{def}}{=} -\nu_0^2\,\sin x . \tag{3.82}$$

Using the numerical values

$$
\begin{aligned}
n &= 6.300 \ [\text{ rad/day }] \\
\frac{a_{\text{o}}}{a} &= 0.15 \\
J_{22} &= 1.8155 \cdot 10^{-6} \ ,
\end{aligned}
\tag{3.83}
$$

the angular velocity in eqn. (3.82) is

$$
\nu_0 = \frac{6 \, n \, a_{\text{o}} \, \sqrt{J_{22}}}{a} = 7.640 \cdot 10^{-3} \ [\text{ rad/day }] \ ,
\tag{3.84}
$$

and the corresponding period is

$$
P = \frac{2 \, \pi}{\nu_0} = 823 \ [\text{ days }] \ .
\tag{3.85}
$$

For small amplitudes we may use the approximation $\sin x \approx x$. The differential equation then assumes the form

$$
\ddot{x} \overset{\text{def}}{=} -\nu_0^2 \, x \ ,
\tag{3.86}
$$

which is solved by a cos-function with angular frequency ν_0 and period $P = 823$ days. The correct solution of eqn. (3.82) involves elliptical integrals (see, e.g., [1], pp. 589ff). For arbitrary amplitudes the period is approximated by the following series:

$$
P = \frac{2 \, \pi}{\nu_0} K(\alpha) = \frac{2 \, \pi}{\nu_0} \left\{ 1 + \left(\frac{1}{2}\right)\alpha + \left(\frac{1 \cdot 3}{2 \cdot 4}\right)^2 \alpha^2 + \left(\frac{1 \cdot 3 \cdot 5}{2 \cdot 4 \cdot 6}\right)^2 \alpha^3 + \cdots \right\} \ ,
\tag{3.87}
$$

where

$$
\alpha = \sin^2 \Delta\lambda \ .
\tag{3.88}
$$

$\Delta\lambda$ is the amplitude of the solution.

Long-Term Development of the Orbital Poles of Geostationary Satellites. Figure 3.11 shows the projection on the equatorial plane of the mean orbital pole of a geostationary satellite moving in the gravitational field of the Earth (potential complete up to terms of degree and order $n = m = 4$), the Moon, and the Sun over a time interval of 60 years. The projection of the components of the unit vector normal to the orbital plane were multiplied by $\frac{180°}{\pi}$ in order to visualize (approximately) the angle between the orbital pole and the equatorial pole.

The orbital pole precesses counterclockwise around an axis which lies between the pole of the Earth and the pole of the ecliptic, about 7.4° away from the pole of the Earth. The angle between the orbital pole and the axis

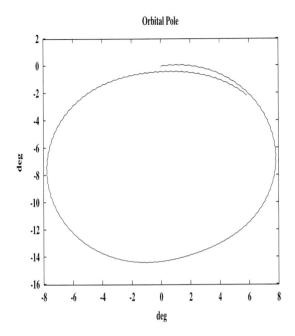

Fig. 3.11. Projection of the orbital pole of a geostationary satellite moving in the gravitational field of Earth (complete up to terms of degree and order $n = m = 4$), Moon, and Sun over 60 years

mentioned is (always) about 7.4°. The period is about 52 years. Very similar results are obtained for other examples of geostationary satellites (for all initial conditions leading to geostationary orbits).

The result may be understood as follows: If the Earth were spherically symmetric, and if both, Sun and Moon would have their orbits in the ecliptic, the orbital pole of a geostationary satellite would have to precess around the pole of the ecliptic. The angle between the orbital pole and the pole of the ecliptic would be the obliquity of the ecliptic $\varepsilon \approx 23.5°$. This is exactly the effect observed initially: The orbital plane is moving rapidly away from the equatorial plane. If we would "switch off" the gravitational attractions by Sun and Moon and "switch on" the term J_{22} after a while, the satellite's orbital plane would have to precess around the polar axis of the Earth, the angle being given by the angle i between the two poles at the time of switching off the gravitational attractions by Sun and Moon.

In reality we observe a superposition of three precessional motions, namely (1) precession around the Earth's rotation axis, mainly caused by the oblateness, (2) precession around the pole of the ecliptic due to the gravitational attraction of the Sun, and (3) precession around the pole of the Moon's orbital plane due to the gravitational attraction exerted by the Moon. It is remarkable that in a fair approximation the resulting motion may be understood as the motion around a single axis with an almost constant angular velocity.

The plane normal to the pole of the resulting precessional motion is the *generalized Laplace plane* (named in honour of Pierre Simon de Laplace (1749–1827)) of the particular three-body problem oblate Earth, Sun, Moon. It is interesting to note that Laplace treated a similar problem in order to describe the orbital motion of Japetus in the gravitational field of the oblate Saturn and the Sun.

The general problem of the precession of orbital planes for artificial Earth satellites is extensively treated in [3]. It is remarkable that this general treatment appeared only few years after the launch of the first artificial satellites.

Figure 3.12 shows that the theoretical expectations are confirmed by observation: The figure contains the orbital poles of all known active and passive (space debris) geostationary satellites in 1996 as collected by the ESA. That the right-hand side of the precession cone is more densely populated than the left-hand side is due to the fact that the first geostationary satellites were only deployed towards mid 1960s and that they could leave the equatorial plane only after the discontinuation of orbital manoeuvres. Unfortunately it is not difficult to predict that after about the year 2050 the entire precession cone associated with geostationary satellites will be fully populated. For more information the reader is referred to [57].

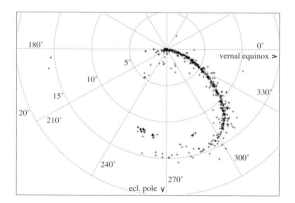

Fig. 3.12. Orbital poles of (formerly) geostationary satellites and debris in 1996 (according to ESA)

3.3.2 GPS Satellites

Heuristic Analysis. GPS satellites have revolution periods of half a sidereal day. They are in deep (2:1)-resonance with Earth rotation. Potentially, every term J_{ik} of order $k = 2, 4, 6, \ldots,\ i \leq k$ may give rise to resonance. GPS satellites are in orbital planes with inclinations of $i \approx 55°$. The six orbital planes of the constellation are separated by $60°$ on the equator. Currently (in summer 2001) there are 28 active GPS satellites available.

Not all the terms mentioned actually give rise to strong resonance perturbations. There is an easy way, due to Urs Hugentobler [57], to gain insight into the manifestation of resonant perturbation terms for the orbits of GPS satellites. We follow this approach to introduce the problem.

Figures 3.13 should be viewed as a Mercator-projection of the Earth (the horizontal axis reflecting the longitude over an interval of $360°$, the vertical axis the latitude between $\pm 90°$).

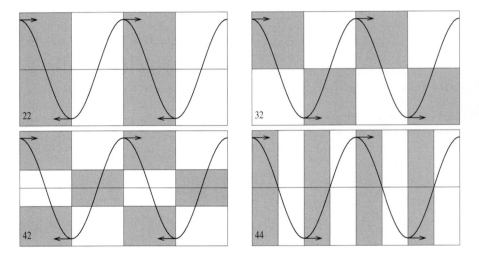

Fig. 3.13. Perturbation of GPS orbits due to the terms J_{22}, J_{32}, J_{42}, and J_{44}

Instead of continents and oceans the four maps show those regions of the Earth, where the perturbation functions related to the terms J_{22}, J_{32}, J_{42}, and J_{44} ((top, left), (top, right), (bottom, left), (bottom, right)) are positive (white areas) and negative (shaded areas). The maps are arranged in such a way that their left boundaries coincide with boundaries of white and shaded areas.

Each map shows in addition the sub-satellite-track of a GPS satellite over one entire day (the "Earth-fixed revolution period" of GPS satellites is one sidereal day). Each track starts at the North-West corner of the map (and returns to that point one day later). These special sub-satellite tracks promise maximum perturbation effects.

Let us now assume that the orbits are circular (a condition closely met by the orbits of GPS satellites). Under this assumption the maps corresponding to the terms J_{22} and J_{42} in Figure 3.13 let us expect that the accumulated effect in one of the orbital elements, in particular in the semi-major axis a, is zero after one (sidereal) revolution: For each sub-satellite point in the Northern hemisphere we find a corresponding point in the Southern hemisphere, where the potential term has the same absolute value, but opposite sign. This must give rise to perturbing accelerations of the same size, but pointing into opposite directions in a coordinate system following the trajectory (e.g., first axis along-track, second cross-track in the orbital plane, third out-of-plane). Two examples for corresponding points, related to the epochs $(t_0, t_0 + 6^h)$ and $(t_0 + 12^h, t_0 + 18^h)$ are contained in each of the maps. The maps corresponding to the terms J_{32} and J_{44} let us expect, on the other hand, that there is a accumulated effect over one entire revolution: Related to each point in the Northern hemisphere we find a point in the Southern hemisphere generating identical perturbing accelerations in the coordinate system accompanying the satellite.

Figure 3.13 also tells that the accumulated perturbation effect significantly depends on the initial longitude of the sub-satellite track. If we shift, e.g., the orbit in the figure corresponding to the term J_{32} by 90° to the East (corresponding to half a sidereal revolution) w.r.t. the map, the accumulated effect over one revolution is again non-zero, but of opposite sign. If the track is shifted only by 45° (corresponding to a quarter of a sidereal revolution) the accumulated effect is zero already after half a revolution (thus also over one entire revolution). This longitude-dependence has an important consequence: GPS satellites situated in one and the same orbital plane, but equally spaced in the argument of latitude will produce significantly different resonance effects. This behavior is in particular a nuisance where the semi-major axis a and (as a consequence thereof) the argument of latitude is concerned. Frequent orbit manoeuvres (about one per year and satellite) are required to maintain a reasonable spacing between the satellites in the same orbital planes.

The maps of Figure 3.13 let us expect that the term J_{32} is the dominating resonance term, that J_{44} is the next-important term (attenuated by the factor $\frac{a_\oplus}{a} \approx 0.24$ w.r.t. the term J_{32}), and that the other two terms will "only" generate resonances proportional to the eccentricity e (or higher powers thereof).

Case Studies. In order to gain insight into the order of magnitude of the perturbations due to resonance (and mainly due to the term J_{32}) we integrate the orbits of seven GPS-like satellites situated in one and the same orbital plane. The geocentric longitude of the ascending node was set to

$\lambda_{32} = 72.8117°$, the reference longitude of the term J_{32}. The orbital elements of these satellites are given in Table 3.3. The terms up to degree and

Table 3.3. Osculating elements at t_0 = January 1, 2001, 0^h for seven GPS-like satellites

Element	Value	Element	Value
P	0.5^d (sid)	e	0.001
i	$55°$	$\Omega + \Theta_0$	72.8117°
ω	$k \cdot 30°$	$\sigma(t_0)$	0°
	$k = 0, 1, \ldots, 6$		

order $n = m = 4$ were taken into account for the integration. With the initial elements in Table 3.3 we make sure that the seven initial orbits (essentially) only differ in the initial argument of latitude, which varies from 0° to 180°. According to our heuristic treatment, the range from 180° to 360° would generate perturbations with the same absolute value, but of different sign as the corresponding orbits in the range between 0° to 180°.

The mean perturbations of the semi-major axes a for the seven test objects may be found in Figure 3.14, where the averaging period was two sidereal revolutions. Figure 3.14 must be interpreted with some care, because not only the resonance terms J_{22}, J_{32}, J_{42}, J_{44} contribute to the perturbations.

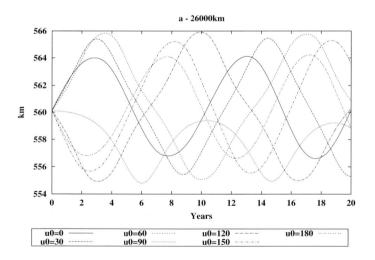

Fig. 3.14. Mean semi-major axes of seven GPS satellites

A more detailed investigation based on program SATORB (see Chapter 7 of Part III) would, however, show, that in essence the perturbations in Figure 3.14 are due to the resonance terms, and mainly to the term J_{32}.

Figure 3.14 shows long-period perturbations with periods ranging between about eight years to fifteen years and amplitudes ranging between 1–5 km . Long periods are associated with big amplitudes, a typical characteristic of resonance phenomena. Observe that all test objects share one and the same orbital plane, and that the remarkably different perturbation characteristics are only due to one initial element, namely the argument of perigee ω (which, for orbits with small eccentricities, is equivalent to different initial arguments of latitude).

The long-period changes in the semi-major axes cause long-period changes in the mean motion of the satellites via Kepler's third law. The order of magnitude of the effect may be established as follows: A change of the element a induces the following change in the mean motion:

$$\Delta n = -\frac{3}{2} \frac{n}{a} \Delta a \ .$$

Using degrees (°), days, and kilometers as units we obtain (the mean motion per day of GPS satellites in these units is $n \approx 720°/\text{day}$):

$$\Delta n \, [°/\text{day}] \approx -\frac{1.5 \cdot 720}{26560} \Delta a \ ,$$

where Δa has to be supplied in km. Multiplying this value by 365.25 gives the offset in the mean motion in degrees per year:

$$\Delta n \, [°/\text{year}] \approx -14.85 \, \Delta a \ ,$$

where Δa has to be provided in km.

The resulting effect in the mean anomaly is

$$\Delta l \, [°] = \int \Delta n(t') \, dt' \ .$$

In order to assess the order of magnitude we approximate $n(t)$ by a pure sin-function:

$$\Delta n = \Delta n_0 \, \sin\left(\frac{2\pi}{P} t\right) = -14.85 \, \Delta a_0 \, \sin\left(\frac{2\pi}{P} t\right) \ ,$$

where P is the period of the variations in $a(t)$ and $n(t)$ and Δa_0 is the amplitude associated with Δn_0.

This allows us to derive the following formula for the oscillations in the mean anomaly:

$$\Delta l \, [^\circ] = \int \Delta n(t') \, dt' = \frac{\Delta n_0 \, P}{2 \, \pi} \, \cos\left(\frac{2 \, \pi}{P} t\right) \stackrel{\text{def}}{=} \Delta l_0 \, \cos\left(\frac{2 \, \pi}{P} t\right) \, .$$

For GPS satellites we obtain:

$$\Delta l_0 \, [^\circ] = \frac{14.85}{2 \, \pi} \, P \, \Delta a = 2.36 \cdot P \, \Delta a \, . \tag{3.89}$$

Using the typical values $P = 8$ years and $\Delta a = 5$ km we obtain $\Delta l_0 \, [^\circ] \approx 97^\circ$. It is thus clear that the maintenance of a reasonable distribution of GPS satellites in one and the same orbital plane requires frequent manoeuvring of satellites. This remark is confirmed by Figure 3.15 where the mean axis of one particular GPS satellite (PRN14) is shown over the time interval $1992 - 1995$. The discontinuities of the order of about 2.7 km have nothing to do with the natural development of the semi-major axis of this satellite orbit, but with manoeuvres (an along-track thrust of a few seconds duration). The discontinuity thus might be seen as a consequence of "Celestial Mechanics of the second kind".

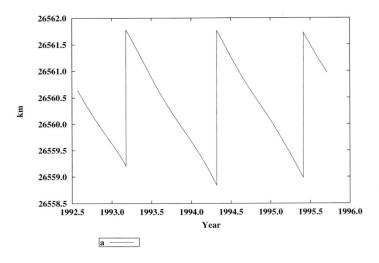

Fig. 3.15. Mean semi-major axis of GPS satellite PRN14

Resonance perturbations do not only occur in the semi-major axis. Figure 3.16 shows the eccentricities of the seven integrations associated with the initial elements in Table 3.3.

All eccentricities were originally very close to zero. Due to resonant perturbations the satellites start developing considerably larger eccentricities than the original mean eccentricity of about $e \approx 0.001$. It would be fascinating to study the development of eccentricities over hundreds of years with the

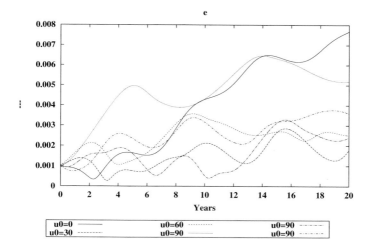

Fig. 3.16. Mean eccentricities of seven GPS satellites in one orbital plane

goal of deciding whether or not there is a chaotic component in the motion of resonant satellites.

Figure 3.17 shows the development of the inclinations i. The common part of the perturbations in the inclinations has nothing to do with resonance. It is a consequence of the continued use of one and the same inertial reference system ($J2000.0$) to describe the orbital elements. This common trend would disappear if the elements were referred to the mean (or true) systems of date. Apart from this common effect there are significant differences between the seven inclinations which are due to resonance. When analyzing the mean elements of the GPS constellation, one can very well see that the constellation slightly degenerates as a function of time: The distribution of the eccentricities and the inclinations is much broader today than it was shortly after the deployment.

Figure 3.18 documents that resonance is a function of the revolution period, therefore of the semi-major axis: The perturbation in the semi-major axis corresponding to $\omega = 0°$ of Table 3.3 is compared to the perturbation of a hypothetical satellite with a revolution period of half a synodic day ($0.5 \cdot \frac{366.25}{365.25}$ sidereal days).

The semi-major axis is only about 47 km bigger than the semi-major axis of a GPS satellite (this value was subtracted in Figure 3.18). 47 km correspond to a change of about 0.2% in the semi-major axis. This small change dramatically reduces the length of the period and of the amplitude of the resonant perturbation.

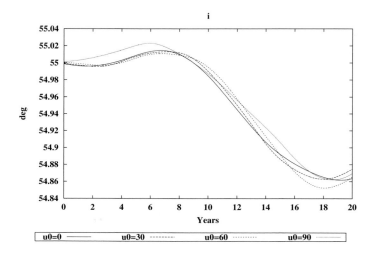

Fig. 3.17. Mean inclinations of seven GPS satellites in one orbital plane

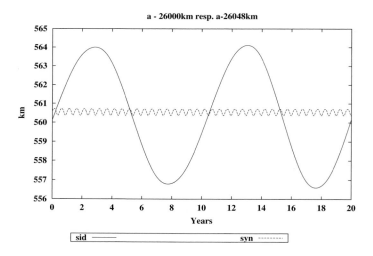

Fig. 3.18. Mean semi-major axes in deep (solid line) and shallow 2:1 (dotted line) resonance with Earth rotation

In summary we may state that the deep 2:1 resonance of the GPS with Earth rotation poses a few unnecessary problems for the maintenance of the system. This aspect is much better taken care of in the case of the GLONASS (Global Navigation Satellite System), the Russian counterpart of the GPS. The revolution period is $11^{\mathrm{h}}15.8^{\mathrm{m}}$, which is far away from the 2:1 resonance.

Resonances Viewed by Perturbation Theory. We will now apply first-order perturbation theory to the deep (2:1)-resonance with Earth rotation. We consider in particular the perturbations due to the terms J_{22} and J_{32}. We will base our developments on circular orbits. In this approximation the perturbation functions associated with the two terms may be written as follows (see eqn. (I- 3.163) and/or (3.72)):

$$R_{22} = -\frac{3\,GM\,a_\oplus^2\,J_{22}}{a^3}\ \cos^2\phi\ \cos 2(\lambda - \lambda_{22})$$
$$R_{32} = -\frac{15\,GM\,a_\oplus^3\,J_{32}}{a^4}\ \sin\phi\ \cos^2\phi\ \cos 2(\lambda - \lambda_{32})\ . \tag{3.90}$$

The simplified transformation between the inertial and the Earth-fixed system (neglecting polar wobble, precession, and nutation) allows it to replace the spherical coordinates λ and ϕ in the Earth-fixed system by the corresponding quantities α and δ (right ascension and declination) in the inertial system:

$$\lambda = \alpha - \Theta$$
$$\phi = \delta\ , \tag{3.91}$$

where Θ is Greenwich sidereal time. The perturbation functions may now be related to the inertial system:

$$R_{22} = \tilde{J}_{22}\ \cos^2\delta\ \cos 2(\alpha - \tilde{\lambda}_{22})$$
$$R_{32} = \tilde{J}_{32}\ \sin\delta\ \cos^2\delta\ \cos 2(\alpha - \tilde{\lambda}_{32})\ , \tag{3.92}$$

where:

$$\tilde{\lambda}_{ik} = \lambda_{ik} + \Theta$$
$$\tilde{J}_{22} = -\frac{3\,GM\,a_\oplus^2\,J_{22}}{a^3}$$
$$\tilde{J}_{32} = -\frac{15\,GM\,a_\oplus^3\,J_{32}}{a^4}\ . \tag{3.93}$$

It is remarkable that in this approximation the third spherical coordinate, the length r of the satellite's position vector, does not show up in eqns. (3.92) (it is considered as a constant $r = a$).

The transformation between the orbital system and the inertial system gives the relationship of the orbital elements and the spherical coordinates α and δ (see eqns. (I- 4.66)):

$$\begin{pmatrix} \cos\alpha\,\cos\delta \\ \sin\alpha\,\cos\delta \\ \sin\delta \end{pmatrix} = \mathbf{R}_3(-\Omega)\,\mathbf{R}_1(-i)\,\mathbf{R}_3(-\omega) \begin{pmatrix} \cos v \\ \sin v \\ 0 \end{pmatrix}$$

$$= \begin{pmatrix} \cos\Omega\,\cos u - \cos i\,\sin\Omega\,\sin u \\ \sin\Omega\,\cos u + \cos i\,\cos\Omega\,\sin u \\ \sin i\,\sin u \end{pmatrix} \stackrel{def}{=} \begin{pmatrix} A \\ B \\ C \end{pmatrix}. \tag{3.94}$$

It is now a straight forward process to express the perturbation functions (3.92) by the orbital elements. Let us define the auxiliary function

$$\begin{aligned} \chi_{ik} &\stackrel{def}{=} \cos^2\delta\,\cos 2(\alpha - \tilde{\lambda}_{ik}) \\ &= \cos^2\delta\,\Big\{ \cos 2\alpha\,\cos 2\tilde{\lambda}_{ik} + \sin 2\alpha\,\sin 2\tilde{\lambda}_{ik} \Big\} \\ &= \cos^2\delta\,\Big\{ (\cos^2\alpha - \sin^2\alpha)\cos 2\tilde{\lambda}_{ik} + 2\sin\alpha\,\cos\alpha\,\sin 2\tilde{\lambda}_{ik} \Big\} \\ &= (A^2 - B^2)\cos 2\tilde{\lambda}_{ik} + 2\,A\,B\,\sin 2\tilde{\lambda}_{ik}\ , \end{aligned} \tag{3.95}$$

where the products $\cos\delta\,\cos\alpha$ and $\cos\delta\,\sin\alpha$ could be replaced by the first two lines of the transformation equations (3.94), i.e., the expressions A and B.

Using the basic theorems of trigonometry, the function may be written explicitly as a function of the orbital elements:

$$\begin{aligned} \chi_{ik} &\stackrel{def}{=} \cos^2\delta\,\cos 2(\alpha - \tilde{\lambda}_{ik}) \\ &= \Big\{ \tfrac{1}{2}\sin^2 i\,\cos 2\Omega + \big(1 - \tfrac{1}{2}\sin^2 i\big)\cos 2\Omega\,\cos 2u \\ &\quad - \cos i\,\sin 2\Omega\,\sin 2u \Big\}\cos 2\tilde{\lambda}_{ik} \\ &\quad + \Big\{ \tfrac{1}{2}\sin^2 i\,\sin 2\Omega + \big(1 - \tfrac{1}{2}\sin^2 i\big)\sin 2\Omega\,\cos 2u \\ &\quad + \cos i\,\cos 2\Omega\,\sin 2u \Big\}\sin 2\tilde{\lambda}_{ik} \\ &= \tfrac{1}{2}\sin^2 i\,\cos 2(\tilde{\lambda}_{ik} - \Omega) + \big(1 - \tfrac{1}{2}\sin^2 i\big)\cos 2(\tilde{\lambda}_{ik} - \Omega)\cos 2u \\ &\quad + \cos i\,\sin 2(\tilde{\lambda}_{ik} - \Omega)\sin 2u\ . \end{aligned} \tag{3.96}$$

According to Lagrange's perturbation equations (I- 6.115) the equations for the semi-major axis due to the two terms of interest read as:

$$\dot{a}_{22} = -\frac{2}{n^2 a} \frac{\partial R_{22}}{\partial T_0} = -\frac{2 \tilde{J}_{22}}{n^2 a} \frac{\partial \chi_{22}}{\partial T_0}$$

$$\dot{a}_{32} = -\frac{2}{n^2 a} \frac{\partial R_{32}}{\partial T_0} = -\frac{2 \tilde{J}_{32} \sin i}{n^2 a} \frac{\partial (\sin u \, \chi_{32})}{\partial T_0} .$$

(3.97)

In view of the structure of the auxiliary function χ_{22} provided by eqn. (3.96) we see immediately that there are only short-period terms in the differential equation for a_{22}. This result implies that there are *no* resonant terms of the order e^0 in the semi-major axis. This confirms our heuristic considerations.

In order to deal with the perturbations due to J_{32} we first have to compute the function $\sin u \, \chi_{32}$. From eqn. (3.96) we obtain:

$$\begin{aligned}
\sin u \, \chi_{32} &= \tfrac{1}{2} \sin^2 i \, \cos 2(\tilde{\lambda}_{32} - \Omega) \sin u \\
&\quad + \left\{1 - \tfrac{1}{2} \sin^2 i\right\} \cos 2(\tilde{\lambda}_{32} - \Omega) \sin u \cos 2u \\
&\quad + \cos i \, \sin 2(\tilde{\lambda}_{32} - \Omega) \sin u \sin 2u \\
&= \tfrac{1}{4} \sin^2 i \left\{\sin \left(u - 2(\tilde{\lambda}_{32} - \Omega)\right) + \ldots\right\} \\
&\quad - \tfrac{1}{4} \left\{1 - \tfrac{1}{2} \sin^2 i\right\} \left\{\sin \left(u - 2(\tilde{\lambda}_{32} - \Omega)\right) + \ldots\right\} \\
&\quad - \tfrac{1}{4} \cos i \left\{\sin \left(u - 2(\tilde{\lambda}_{32} - \Omega)\right) + \ldots\right\} \\
&= \tfrac{1}{4} \left\{\tfrac{3}{2} \sin^2 i - \cos i - 1\right\} \sin \left(u - 2(\tilde{\lambda}_{32} - \Omega)\right) + \ldots .
\end{aligned}$$

(3.98)

Because for satellites in deep (2:1)-resonance with Earth rotation

$$n \approx 2 \dot{\Theta} ,$$

(3.99)

the terms $+ \ldots$ in eqn. (3.98) *all are* short-period terms in the sense of first-order perturbation theory. The term retained in this equation is, on the other hand, time-independent in first-order perturbation theory. In order to obtain the differential equation for a_{32} we have to take the partial derivative of $\sin u \, \chi_{32}$ w.r.t. T_0. Using the development (3.98) we eventually obtain the following differential equation for the semi-major axis due to the term J_{32}:

$$\begin{aligned}
\dot{a}_{32} &= \frac{15 \, n \, a \, J_{32}}{2} \left(\frac{a_\oplus}{a}\right)^3 \sin i \left\{\frac{3}{2} \sin^2 i - \cos i - 1\right\} \\
&\quad \cdot \cos \left(\omega + \sigma - 2 \Theta - 2(\lambda_{32} - \Omega)\right) + \ldots ,
\end{aligned}$$

(3.100)

where we used the definition (3.93) to represent $\tilde{\lambda}_{32}$ and replaced the argument of latitude by $u = \omega + \sigma$ (circular orbit). The neglected terms are of a short-period nature. This equation is identical with the equation given by

Hugentobler [57] where, to the best our knowledge, the resonance behavior of GPS-like satellites was dealt with for the first time in depth.

The resonance argument for the term J_{32} is defined as

$$\Psi_{32} \stackrel{\text{def}}{=} \omega + \sigma - 2\Theta - 2\left(\lambda_{32} - \Omega\right) . \tag{3.101}$$

By observing that its second derivative is approximately defined by (neglecting the accelerations in ω, Ω, and Θ)

$$\ddot{\Psi}_{32} \approx \dot{n} = -\frac{3\,n}{2\,a}\,\dot{a} , \tag{3.102}$$

we obtain the following differential equation for this angular argument:

$$\ddot{\Psi}_{32} = -\frac{45}{4}\,n^2\left(\frac{a_{\check{\delta}}}{a}\right)^3 J_{32}\,\sin i\left\{\frac{3}{2}\,\sin^2 i - \cos i - 1\right\}\cos\Psi_{32} + \dots . \tag{3.103}$$

This is again (note the analogy with the case of a geostationary satellite) the equation of a mathematical pendulum where the stable point is at $\Psi_{32} = 90°$. For oscillations with small amplitudes the angular frequency ν_0 is defined by

$$\nu_0^2 = \frac{45}{4}\,n^2\left(\frac{a_{\check{\delta}}}{a}\right)^3 J_{32}\,\sin i\left|\left\{\frac{3}{2}\,\sin^2 i - \cos i - 1\right\}\right| , \tag{3.104}$$

resulting in a period of

$$P_0 = \frac{2\,\pi}{\nu_0} . \tag{3.105}$$

The general solution of eqn. (3.103) involves the evaluation of elliptic integrals. The resulting period is formally identical with eqn. (3.87) already derived for geostationary satellites (of course with a different definition of the parameter)

$$P = \frac{2\,\pi}{\nu_0}\,K(\alpha) = \frac{2\,\pi}{\nu_0}\left\{1 + \left(\frac{1}{2}\right)\alpha + \left(\frac{1\cdot 3}{2\cdot 4}\right)^2\alpha^2 + \left(\frac{1\cdot 3\cdot 5}{2\cdot 4\cdot 6}\right)^2\alpha^3 + \dots\right\} , \tag{3.106}$$

where

$$\alpha = \sin^2 \Delta\Psi_{32,\text{max}} . \tag{3.107}$$

$\Delta\Psi_{32,\text{max}}$ is the amplitude of the solution.

Using the numerical values

$$J_{32} = 3.7441 \cdot 10^{-7}$$
$$n \;\; = 12.601 \; [\, \text{rad/day} \,]$$
$$\frac{a_{\leftmoon}}{a} = 0.2401 \qquad\qquad (3.108)$$
$$i \;\;\; = 55° \,,$$

we obtain

$$\nu_0 = 2.074 \cdot 10^{-3} \; [\, \text{rad/day} \,]$$
$$P_0 = \frac{2\,\pi}{\nu_o \, 365.25} = 8.30 \; [\, \text{years} \,] \,. \qquad\qquad (3.109)$$

The comparison of these theoretical predictions with the results of the numerical integrations in Figure 3.14 is in general quite good. The shortest periods were predicted to be of the order of 8.3 years, which is confirmed by Figure 3.14. The heuristic treatment would let us expect the shortest periods and the largest amplitudes associated with $u = 0°$. This behavior is observed in Figure 3.14 as well. We would expect, however, the period associated with $u = 90°$ (corresponding to $w = 90°$ in Table 3.3) to be substantially longer.

This reduction of the period (and the amplitude) is due to the attenuating influence of the oblateness perturbations. If we repeat the numerical experiments documented in Figure 3.14 but take into account only the term J_{32}, we obtain the perturbations shown in Figure 3.19. Figure 3.19 is in full agreement with our theoretic expectations: The shortest perturbation periods are of the order of $8 - 9$ years and very long periods (of the order of well over twenty years) do occur.

The similarity of Figures 3.14 and 3.19 (with the exception of the orbit with $w = 90°$) underlines that the perturbations actually are mainly due to the term J_{32} (remember that the complete potential up to terms of degree and order $n = m = 4$ are used in Figure 3.14, whereas only the main term and the term J_{32} were used in Figure 3.19).

There are substantial differences in the results associated with $w = 90°$, where the largest amplitudes and the longest periods are expected: The period of about 15 years in Figure 3.14 corresponds to a period substantially longer than twenty years in Figure 3.19. The difference once more must be explained by the attenuating influence of the obliquity: The perturbations that would be of very long period and of large amplitudes are greatly reduced in both, period and amplitude, thanks to the presence of the obliquity perturbation.

Let us mention that it is not trivial to find the semi-major axis precisely corresponding to "the deepest" (2:1)-resonance. In the program SATORB the perturbations due to C_{20} are approximately taken into account to find

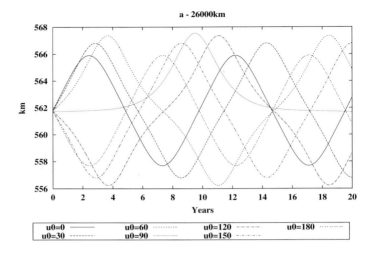

Fig. 3.19. Mean semi-major axes of seven GPS satellites ($C_{22} = 0$, terms up to degree and order 4)

the resonant semi-major axis. But the approximations are based on first-order perturbation theory (and actually only the terms proportional to e^0 are taken into account). In order to get a full picture of the characteristics of 2:1 for GPS satellites one should scan through this resonance by systematically varying the revolution period – very much in the same way as it is done to investigate the resonant behavior of minor planets (see Chapter 4).

3.4 Perturbations due to the Earth's Stationary Gravitational Field in Review

In the previous three sections we discussed the orbital perturbations of an artificial satellite due to the Earth's stationary gravitational field. In general, the oblateness-term C_{20} gives rise to orbital perturbations exceeding those of the other terms C_{nm}, S_{nm} by about three orders of magnitude. This is (of course) a consequence of the hierarchy of the terms of the Earth's gravitational field (see Table I- 3.1).

The hierarchy is not necessarily seen when studying orbits with revolution periods commensurable with the sidereal day. Using the example of geostationary satellites (1:1-commensurability) and of GPS satellites (2:1-commensurability) we encountered perturbations (due to C_{22}, S_{22} and others for geostationary, C_{32}, S_{32} and others for GPS satellites) exceeding the long-period perturbations due to C_{20} by orders of magnitude.

The oblateness term C_{20} was also seen to attenuate the perturbations due to the other terms C_{nm}, S_{nm} thanks to a relatively rapid change of the orbital geometry (elements Ω and ω).

General perturbation methods based on first-order theory proved to give correct qualitative explanations of the orbital perturbations of artificial satellites. First-order perturbation theory revealed "only" short-period terms (with the revolution period P of the satellite or entire fractions thereof as periods) and secular terms in the right ascension of the ascending node Ω, the argument of perigee ω, and in the mean anomaly σ.

Long-period perturbations with the rotation period of these three angles show up only in a higher-order treatment of the perturbation equations. Very small long-period perturbations could actually be observed when integrating a test satellite in the gravitational field of the main term and of C_{20}. The resulting orbital perturbations proved to be very small (by a factor of about 1000 smaller than the first-order perturbations).

Long-period perturbations with periods due to C_{20} play a very important role, however, when considering the perturbations due to terms C_{nm}, S_{nm} (other than C_{20}): The drifts in Ω, ω, and σ due to C_{20} must be taken into account when applying perturbation theory to the terms C_{nm}, S_{nm}, otherwise the amplitudes and periods of the perturbations due to these terms are grossly overestimated. In this sense the term C_{20} has a stabilizing influence on the orbits of artificial Earth satellites.

First-order perturbation theory in all cases provides accurate and reliable values for the first derivative of the orbital elements. Under resonance conditions (see eqn. (3.70)), i.e., if an entire number of revolution periods k_1 is contained in an entire number k_2 of days, first-order theory breaks down (at least for some of the elements) because the time dependence may be eliminated (in essence) due to the occurrence of arguments of type $k_1 n - k_2 \dot{\Theta} = 0$ in some of the trigonometric functions on the right-hand side of the perturbation equations. First-order theory still gives the correct instantaneous time derivative of these elements, but no longer the correct perturbation of the elements over longer time intervals.

Special treatments based on the correct perturbation functions may be used to overcome this difficulty. We outlined the solution method for two important cases, namely the motion of a geostationary satellite in the equatorial plane due to the resonant terms C_{22}, S_{22} and the motion of GPS satellites under the influence of the terms C_{32}, S_{32}. In both cases, the motion of a resonant argument can be described by the differential equation of the mathematical pendulum. The resonant perturbations are of a very long period (up to a few decades) with amplitudes comparable to or even greater than the amplitudes of the short-period perturbations due to the oblateness.

Resonance is a central aspect of satellite motion. As opposed to the motion of minor planets in the planetary system, the long-term development of artificial satellites is not yet well established. Numerical experiments comparable to those conducted in the planetary system should reveal whether or not chaotic motion plays a significant role in the development of satellite orbits over time periods of thousands of years.

In the framework of first-order theory we found that, by replacing time t by the argument of latitude u as an independent argument, the transformed perturbation equations for the oblateness term C_{20} could be solved in closed form in terms of elementary integrable functions. Use was made of this property when developing the formulas for the rotation rates of the perigee, of the right ascension of the node, and of the mean anomaly. The question, whether this interesting property is common to other perturbation terms C_{nm}, S_{nm}, will be addressed below.

The preceding sections were related to a few specific terms of the stationary gravitational field. In the following two paragraphs some of the findings will be generalized. In paragraph 3.4.1 the perturbation function R_{lm} for an arbitrary term of the Earth's gravitational field will be developed and some consequences of this representation will be discussed. In paragraph 3.4.2 we will deal with the perturbation equations for the general term R_{lm} when using the argument of latitude u as a general argument.

3.4.1 First-Order General Perturbation Solutions

Using the simplified transformation between the Earth-fixed and the inertial system ($\phi = \delta$ and $\lambda = \alpha - \Theta$), the general term of the development (I- 3.163) may be written as:

$$R_{lm}(r, \lambda, \phi) = -\frac{GM\, a_\oplus^l}{r^{l+1}}\, P_l^m(\sin\phi)\, J_{lm}\, \cos m(\lambda - \lambda_{lm})$$

$$R_{lm}(r, \alpha, \delta) = -\frac{GM\, a_\oplus^l\, J_{lm}}{r^{l+1}}\, P_l^m(\sin\delta)\, \cos m(\alpha - \Theta - \lambda_{lm})$$

$$= -\frac{GM\, a_\oplus^l\, J_{lm}}{a^{l+1}\,(1 - e^2)^{l+1}}\, (1 + e\cos v)^{l+1}\, \tilde{P}_l^m(\sin i\,\sin u) \qquad (3.110)$$
$$\cdot \cos^m \delta\, \cos m(\alpha - \Theta - \lambda_{lm})$$

$$= -\frac{GM\, a_\oplus^l\, J_{lm}}{a^{l+1}\,(1 - e^2)^{l+1}}\, (1 + e\cos v)^{l+1}\, \tilde{P}_l^m(\sin i\,\sin u)$$
$$\cdot \cos^m \delta\, \big\{ \cos m\alpha\, \cos m(\Theta - \lambda_{lm}) + \sin m\alpha\, \sin m(\Theta - \lambda_{lm}) \big\}\,,$$

where $P_l^m(\sin i\,\sin u) = \tilde{P}_l^m(\sin i\,\sin u)\cos^m \delta$ is the associated Legendre function of degree l and order m as defined by eqn. (I- 3.151). According to this definition $\tilde{P}_l^m(\sin i\,\sin u)$ is a polynomial of degree $l - m$ in its argument, which either has only terms of even or of odd powers (depending on whether $l - m$ is even or odd).

Well-known theorems of trigonometry (see, e.g., [25]) allow it to establish the relations

$$\cos m\alpha = \binom{m}{0} \cos^m \alpha - \binom{m}{2} \cos^{m-2} \alpha \sin^2 \alpha + \binom{m}{4} \cos^{m-4} \alpha \sin^4 \alpha - \ldots$$

$$\sin m\alpha = \binom{m}{1} \cos^{m-1} \alpha \sin \alpha - \binom{m}{3} \cos^{m-3} \alpha \sin^3 \alpha$$
$$+ \binom{m}{5} \cos^{m-5} \alpha \sin^5 \alpha - \ldots .$$

$$(3.111)$$

Using these relations in eqn. (3.110) implies that its last line may be written as a sum of terms

$$(\cos \alpha \, \cos \delta)^i \quad (\sin \alpha \, \cos \delta)^k \ .$$

The transformation equations (3.94) then allow it to write each of the terms (...) as a function of the orbital elements and of the argument of latitude u. This implies in turn, that eventually the entire perturbation function R_{lm} of eqn. (3.110) may be written as a finite linear combination of sin- or cos-functions of linear combinations of the angles u, Ω, ω, v, and Θ.

The representation still used today in satellite geodesy is due to Kaula. It is based on a theoretical development given by Tissérand in 1888 [121] (!). The form cited here is that used by Hugentobler [57]:

$$R_{lm} = -\frac{GM \, J_{lm} \, a_\oplus^l \, (1 + e \cos v)^{l+1}}{a^{l+1} \, (1 - e^2)^{l+1}} \sum_{p=0}^{l} F_{lmp}(i) \begin{cases} \cos \Psi_{lmp} \, , & l - m \ \text{even} \\ \sin \Psi_{lmp} \, , & l - m \ \text{odd} \end{cases} ,$$

$$(3.112)$$

where

$$\Psi_{lmp} = (l - 2\,p)(\omega + v) + m\,(\Omega - \Theta - \lambda_{lm}) \ . \qquad (3.113)$$

The functions $F_{lmp}(i)$ are called *inclination functions*. Up to degree and order $l = m = 4$ they are, e.g., provided in [62].

The development (3.112) may be considered as the final representation of the perturbation function when considering the perturbations of a circular orbit. When studying the perturbations as a function of the argument u the development (3.112) is the appropriate representation, as well (see section 3.4.2).

In order to obtain integrable functions in first-order theory when using the time t as the independent argument we have to replace the true anomaly v in the above equation by the mean anomaly σ. This leads to the representation of the type (see, e.g., Kaula [62]):

$$R_{lm} = -\frac{GM \, a_\oplus^l \, J_{lm}}{a^{l+1}} \sum_{p=0}^{l} \sum_{q=-\infty}^{+\infty} F_{lmp}(i) \, G_{lmq}(e) \begin{cases} \cos \Psi_{lmpq} \, , & l - m \ \text{even} \\ \sin \Psi_{lmpq} \, , & l - m \quad \text{odd} \end{cases} ,$$

$$(3.114)$$

where

$$\Psi_{lmpq} = (l - 2\,p)\,\omega + (l - 2\,p + q)\,\sigma + m\,(\Omega - \Theta - \lambda_{lm})\,. \qquad (3.115)$$

$G_{lmq}(e)$ is the so-called *eccentricity function* proportional to $e^{|q|}$.

The first-order perturbations due to the oblateness term C_{20} only generate secular and short-period perturbations in first-order theory. First-order theory, applied to an arbitrary term J_{lm} in equations (3.114) and (3.115), tells that perturbations of the following kind may result:

- Short-period terms with sin- or cos-functions containing the argument u or multiples thereof.
- Secular terms growing linearly with time.
- Long-period terms with sin- or cos-functions in the argument $(l - 2\,p)\,\omega$ (and no dependence on either u or Θ).
- m-daily terms due to sin- or cos-functions in the argument $m\,\Theta$ (and no dependence on u).

In addition to these four types of perturbations quasi-secular terms may result in the case of resonances, when $(l - 2\,p + q)\,n - m\,\dot{\Theta} \approx 0$ for some index combinations, because eqn. (3.115) becomes (almost) time-independent in this case.

The development (3.114) is perfectly suited for formal integration in first-order theory. It contains only sin- or cos-terms of linear functions of time. Its derivation is somewhat laborious. Tissérand [121] gave this derivation in 1888 which is still referred to today.

It is of course much easier to develop the above representation today (to any order desired) by making use of readily available algebraic computer packages: One "merely" has to represent u as a series in σ (in time) and make use of the appropriate trigonometric relations.

3.4.2 Perturbation Equations in the Argument of Latitude u

If the argument of latitude u serves as independent argument instead of the time t, the representation (3.112) for the perturbation function R_{lm} may be used instead of the development (3.114), a circumstance promising a significantly simpler treatment. We may now ask, whether the integration may be performed in an elementary way in all cases – as it was the case for the term C_{20}.

In order to perorm the transformation of the independent argument we have to replace the time-derivative by the derivative w.r.t. the argument of latitude u and we have to multiply the right-hand sides of the same equations with $\dfrac{\sqrt{(1-e^2)^3}}{n\,(1+e\cos v)^2}$ (see eqn. (3.26)).

Consequently, the resulting equations will have a pre-factor proportional to $(1 + e \cos v)^k$, $k \geq 0$: The perturbation function (3.112) has a pre-factor proportional to $(1 + e \cos v)^k$, $k \geq 3$ for the terms of interest. Therefore, the exponent of this pre-factor is positive or zero in the resulting perturbation equation in u, because the exponent of the perturbation function is in the worst case reduced by 1 when taking the required partial derivatives (see eqns. (I-6.115)). This allows us to state that *all perturbation equations associated with the zonal terms* R_{l0}, $l = 2, 3, \ldots$ *may be integrated term-by-term in an elementary way with u as independent argument.*

Consequently, it is possible to give for every term R_{l0}, $l \geq 2$ the "true" first-order expressions for the rotation rates of the three angles Ω, ω and σ, or to calculate the true short-period perturbations as a function of the argument u.

At first sight one would assume that the perturbation equations for the tesseral and sectorial terms have the same property. Unfortunately this is not the case, because Θ needs to be known as a function of the argument u. Furthermore the resulting functions of u have to be brought into a form allowing a formal integration. As we may write

$$\Theta = \Theta_0 + \dot{\Theta}\,(t - t_0) = \Theta_0 + \frac{\dot{\Theta}}{n}\,(u - u_0) + O(e)\ ,$$

it is easily possible to solve the perturbation equations with u as independent argument related to any term R_{lm}, $m \neq 0$ to the order e^0.

A higher-order approximation in the eccentricity e requires the inclusion of the higher-order terms in the eccentricity in the above equation for Θ. But then the equations in u loose their elegance and simplicity and one may equally well solve the equations with the time t as independent argument.

3.5 Non-Gravitational Forces

Non-gravitational accelerations are essential for an adequate description of the orbits of artificial Earth satellites. The area-to-mass ratio A/m (where m is the mass of the satellite and A the area of the projection of the satellite onto a plane relevant for the perturbation considered) plays the role of a proportionality factor when calculating the accelerations due to non-gravitational forces.

Atmospheric drag usually is the dominating non-gravitational acceleration for LEOs, *solar radiation pressure* plays that role for satellites in orbits higher than about 2000 km above the surface of the Earth. A is the area of the projection of the satellite onto the plane normal to the orbital velocity vector $\dot{r}$ in the case of drag, it is the area of the projection normal to the direction Sun $\to$ satellite in the case of radiation pressure.

Geodetic satellites (e.g., for the LAGEOS satellites) should have a minimal area-to-mass ratio. Table 3.4 lists the A/m–ratios for four satellites including the only natural Earth satellite.

Table 3.4. Area-to-mass ratios A/m for five Earth satellites

Satellite	A/m [m^2/kg]
Lageos 1 and 2	0.0007
Starlette	0.001
GPS(Block II)	0.02
Moon	$1.3 \cdot 10^{-10}$

The area-to-mass ratio of the oldest geodetic satellite, namely the Moon, is by many orders of magnitude smaller (thus better suited for geodetic purposes) than that of any artificial satellite – despite the fact that the attempt was made to minimize the ratios A/m for the first three satellite types in Table 3.4. The small A/m–ratio for the Moon explains why the Moon's orbit, when observed and determined with cm-accuracy, is so well suited for tests of general relativity, for the determination of UT1–UTC, etc.

The first three satellites in Table 3.4 were designed as geodetic satellites. Lageos 2 (launched in 1992) is, e.g., a spherical satellite with a diameter of 0.6 m, a weight of 405 kg, and 426 corner cubes inlaid in its surface (see Figure I-2.4). Lageos 2 is a close relative of Lageos 1 (launched in 1976). The two Lageos satellites are in almost circular orbits about 6000 km above the surface of the Earth, which is why atmospheric drag is virtually non-existent and why the acceleration due to radiation pressure is small compared to most of the gravitational accelerations (see also Tables 3.7, 3.8 at the end of this chapter). Starlette, with a diameter of about 27 cm, is similar in construction to the two Lageos satellites, but it is in a much lower orbit.

GPS satellites have a rather big area-to-mass ratio. Values of this order of magnitude are typical for navigation satellites (and other bulky satellites) equipped with big solar panels of a few m^2. The mass of a GPS satellite is about 1000 kg . As these satellites orbit the Earth in a height of more than 20000 km , drag does not need to be considered, but radiation pressure is the central issue for precise orbit determination (to be discussed below).

Future generations of geodetic satellites will eliminate (better: greatly reduce) the influence of non-gravitational accelerations by using onboard *accelerometers* to measure non-graviational effects, which may then be compensated by manoeuvring the spacecraft. Accelerometers are shielded against the non-gravitational forces by the surface of the satellite. They measure the accelerations (at best) in three orthogonal directions. Naturally, these accelerometers do not only measure the accelerations affecting the orbit (like, e.g., drag and radiation pressure) but also the effects due to the non-inertiality of the satellite-fixed reference frame (i.e., centrifugal and coriolis accelerations).

When studying the effect of non-gravitational accelerations on the orbits of satellites, one has to make the distinction between *a priori models* for these effects and the *a posteriori determination* or *improvement* of model parameters. A priori models are based on the knowledge of the properties of the satellite (surface, mass, reflectivity, attitude, etc.) and of the properties of the perturbing force (e.g., the density of the upper atmosphere as a function of height). The a posteriori determination or improvement of these accelerations often involves the determination of scaling parameters of the a priori models and/or purely empirical parameters. We will rather focus on a priori models in this Chapter whereas the parameter estimation aspect will be considered in Chapter I-8. There are many more non-gravitational effects than just radiation pressure and drag. A complete list must include

- drag due to the electrically neutral atmosphere,
- drag due to charged particles on the satellite's surface,
- direct radiation pressure (models of varying complexity),
- albedo radiation pressure due to the sunlight reflected or re-emitted by the surface of the Earth,
- thermic emission of radiation by the satellite, and
- effects induced by the Earth's magnetic field.

In the following two sections 3.6 and 3.7 we will, however, focus on atmospheric drag and on radiation pressure. For a concise overview of the non-gravitational effects listed above (and others) we refer to [74].

3.6 Atmospheric Drag

Above a height of about 50 km the density of the neutral atmosphere is sufficiently low to assume laminar air currents. Assuming furthermore that the atmosphere is co-rotating with the Earth (an assumption in essence ignoring winds) and neglecting the thermal motion of the molecules, it is relatively easy to calculate the transfer of linear momentum from the atmosphere to the satellite: During a short time interval Δt the velocity $\dot{\boldsymbol{r}}'$ of the satellite relative to the particles may be assumed constant. Assuming furthermore that all molecules and atoms encountered by the satellite's cross section normal to $\dot{\boldsymbol{r}}'$ are absorbed by the satellite, the linear momentum lost by the satellite equals the product of the volume ($|\dot{\boldsymbol{r}}'|\,\Delta t\,A$) of the cylinder with height $\dot{\boldsymbol{r}}'\,\Delta t$ and ground surface A with the density $\rho(\boldsymbol{r})$ and the velocity $-\dot{\boldsymbol{r}}'$ of the molecules relative to the satellite at the current position of the satellite. The loss of linear momentum by the satellite may thus be computed as

$$\Delta p = -\rho(r)\, A\, \dot{r}'^2\, \Delta t\, \frac{\dot{r}'}{|\dot{r}'|}\ . \tag{3.116}$$

The velocity change experienced during the time interval Δt is obtained by dividing this expression through the time interval Δt and the mass of the satellite m. The acceleration $\boldsymbol{a}_d$ of the satellite due to drag is obtained by the limit-process $\Delta t \to 0$ and reads as:

$$\boldsymbol{a}_d = -\rho(r)\, \frac{A}{m}\, \dot{r}'^2\, \frac{\dot{r}'}{|\dot{r}'|}\ . \tag{3.117}$$

The acceleration due to drag is therefore anti-parallel to the velocity of the satellite in the Earth-fixed system, proportional to the square of the velocity in this system and to the area-to-mass ratio A/m.

The assumptions underlying eqn. (3.117) are not fully met in reality: Only a fraction of the particles is actually absorbed by the satellite, the rest is reflected by the satellite's surface. This circumstance complicates matters considerably, because the linear momentum transferred by a reflected molecule to the satellite is twice the projection of its linear momentum on the normal to the surface element. In order to take such effects into account we have to know the shape and the attitude of the satellite, except for satellites with rotational symmetry w.r.t. the velocity vector in the Earth-fixed system. Prominent examples of this latter class of satellites are cannonball satellites (e.g., the Lageos and Starlette satellites).

Subsequently we will only consider accelerations in the direction of the satellite velocity vector and use the following model to account for drag:

$$\boldsymbol{a}_d = -\frac{C}{2}\, \rho(r)\, \frac{A}{m}\, \dot{r}'^2\, \frac{\dot{r}'}{|\dot{r}'|}\ , \tag{3.118}$$

where the coefficient is $C = 2$ for spherical satellites (independent of the fraction of absorbed and reflected particles) and for satellites absorbing all molecules hitting the satellite. In general one may assume that

$$2 \leq C \leq 2.5\ . \tag{3.119}$$

For the sake of completeness we mention that the velocity of the satellite relative to the particles in the upper atmosphere at rest in the Earth-fixed system is calculated (in non-relativistic approximation) by

$$\dot{r}' = \dot{r} - \boldsymbol{\omega}_\oplus \times r\ , \tag{3.120}$$

where $\boldsymbol{\omega}_\oplus$ is the momentary angular velocity vector of the Earth.

This leaves us with the discussion of the density $\rho(r)$ of the upper atmosphere.

3.6.1 Density of the Upper Atmosphere

Only with the advent of the space era it was possible to gather reliable information about the density and composition of the Earth's upper atmosphere above 50 km. The information stems from analyzing the orbits of artificial satellites and from satellite missions carrying mass spectrometers and scatterometers.

First reliable information was made available in the *US Standard Atmosphere (1976)*. In parallel, several versions of the CIRA (COSPAR International Reference Atmosphere) were developed and made available in the years 1972, 1986, and 1992. One of the more elaborate models openly available is called MSIS (Mass Spectrometer and Incoherent Scatter), identifying the principal source of the data underlying the model. The version *MSISe-90* (where e stands for extended) was developed by A. E. Hedin et al. from Nasa Goddard Space Flight Center. It is available over the internet (http://nssdc.gsfc.nasa.gov) and documented in [50] and [51]. It may be used as an a priori model for the density and temperature of the (constituents of the) upper atmosphere.

MSISe-90 gives the density of the atmosphere as a function of the height above the Earth's surface, of the DoY (Day of Year), of the geocentric latitude and longitude, of the time of day, and of true (local) solar time. The solar flux $F(10.7\,\text{cm})$ at 10.7 cm (corresponding to 2800 MHz) and the magnetic index A_p are scaling factors of the model. When using realistic values for these parameters, the variations due to the 11 year solar cycle are therefore implicitly taken into account.

Figures 3.20 and 3.21 show that the solar flux $F(10.7\,\text{cm})$ and the magnetic index A_p are subject to significant variations of several frequencies within one year. The two figures documenting the year 1999 stem from the series *Solar-Geophysical Data, prompt reports* [28]. The MSISe-90 computer programs allow it to specify mean values of these indicators over longer time periods or to use recent values. We do not address such details here and confine ourselves to present some prominent features of the upper atmosphere.

Figures 3.22-3.25 were produced using MSISe-90. They give an impression of the many variations that should be taken into account when implementing an a priori model for atmospheric drag. The most significant density variation is due to the height above the surface. Locally (i.e., over a few tens of km), this variation may be accounted for by the barometric formula

$$\rho(h) = \rho_0 \; e^{-\frac{h-h_0}{H_0}} \; , \tag{3.121}$$

where ρ_0 is the density at the reference height h_0 and H_0 is the scaling height at h_0.

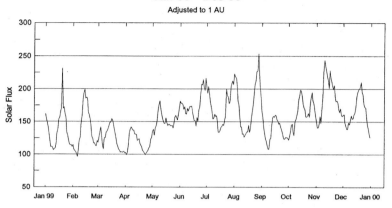

Fig. 3.20. Solar flux $F(10.7\,\text{cm})$ (Penticton) for 1999 in solar flux units

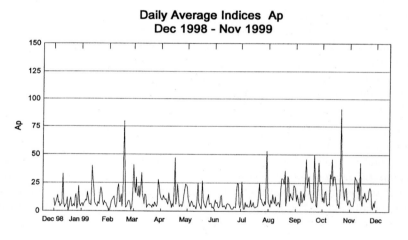

Fig. 3.21. Magnetic A_p-index for 1999

Figure 3.22 shows the logarithm (referring to base 10) of the density of the atmosphere (in units of kg/m^3) for March 24, 10$^\text{h}$ UT for geographical longitude 7.5° and latitude 45.5° (corresponding to a location in Switzerland). A solar flux of $F(10.7\,\text{cm}) = 150$ (in solar flux units) and an index of $A_p = 4$ were assumed. Figure 3.22 reveals, that *MSISe-90* is a composed model. The *homosphere*, consisting of *troposphere* $(0 - 15$ km), *stratosphere* $(15 - 50$ km)

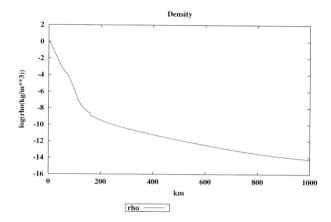

Fig. 3.22. Density of the atmosphere profile according to MSISe-90

and *mesosphere* $(50 - 90$ km), the *thermosphere* $(90 - 400$ km) and the *exosphere* $(400 - \infty$ km) obviously were dealt with separately.

Figure 3.22 shows that the density drops within the first 50 km by about a factor of 1000, at a height of 100 km the density is only of the order of 10^{-6} of the value at sea level. Between 100 km and 200 km and between 200 km and 1000 km the density is reduced by about a factor of ten thousand (in each case).

Figure 3.22 shows that even for orbits with small eccentricities the density at perigee is orders of magnitude higher than in apogee. For a satellite with an apogee height $h_{apo} \approx 200$ km and a perigee height $h_{per} \approx 126$ km , corresponding to $a \approx 6541$ km and $e \approx 0.006$, Figure 3.22 lets us expect $\frac{\rho_{per}}{\rho_{apo}} \approx 40$. For close Earth satellites the velocity w.r.t. the rotating frame is of the order of $6 - 7$ km/s . Using formula (3.118) one must expect significant accelerations due to drag up to heights of about 1000 km.

Figures 3.23 - 3.25 show the variations of the density in a height of 100 km as a function of latitude, of the time of the day, and of the day of the year. The longitudinal variations are not very pronounced and therefore not documented here. Four curves (corresponding to spring, summer, fall, and winter) are given in Figures 3.23 and 3.24. Figure 3.23 shows a pronounced latitudinal variation corresponding to a factor of about 3 in density. In spring and fall the density distribution is symmetric w.r.t the equator whereas strong asymmetries are observed in summer and winter (the seasons in Figures 3.23 - 3.25 refer to the Northern hemisphere).

Figure 3.24 documents, that daily variations of the density of the atmosphere are significant and must be taken into account. Figure 3.25 shows the variation of the density as a function of the day of the year (for "Switzerland").

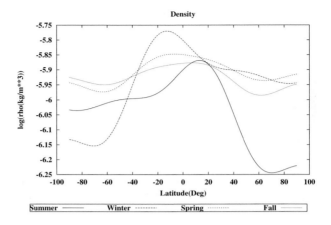

Fig. 3.23. Density of the atmosphere in a height of 100 km as a function of latitude

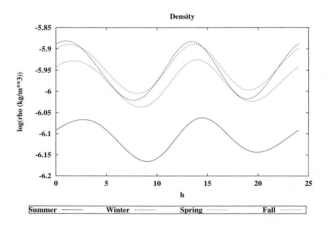

Fig. 3.24. Daily variations of the density of the atmosphere in a height of 100 km at mid-latitude

At a height of 100 km the density is about a factor of 2 lower in summer than in winter. (Experience tells that, fortunately, effects of the same order of magnitude are *not* observed at "normal" altitudes in Switzerland!).

3.6.2 Effect of Drag on Satellite Orbits

Equation (3.118) and the density model MSISe-90 are used in program SATORB (see Chapter 7 of Part III) to take into account the atmospheric drag. Parameters C and A/m of eqn. (3.118) as well as the solar flux

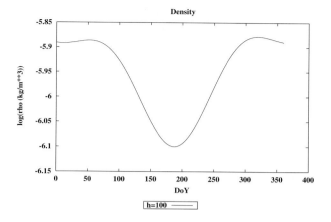

Fig. 3.25. Annual variation of density in a height of 100 km above Switzerland

$F(10.7\,\text{cm})$ and the magnetic index A_p may be defined by the program user. In this section we confine ourselves to present and discuss a few examples using the program SATORB.

In order to fix the ideas we model the effect of atmospheric drag for the research satellite GPS/MET, documented by Melbourne et al. [73], which was designed as a test satellite to use spaceborne GPS receivers for *atmospheric limb sounding*. The proof-of-concept mission, deployed in 1995, was a full success, and the concept is being used (and is going to be used) in many LEO missions. Here, we are uniquely interested in the perturbations of the orbit of GPS/MET by atmospheric drag. The relevant orbit and satellite characteristics of the spacecraft are contained in Table 3.5. Nominally the eccentricity of GPS/MET is $e = 0$. Due to short-period perturbations the eccentricity varies within the limits $0 < e < 0.002$.

Table 3.5. Characteristics of GPS/MET in 1995

Satellite/Orbit Property	Numerical Value
a	7100 km
e	0.0
i	70°
A/m	0.02

Figure 3.26 shows the effect of drag on the orbit of GPS/MET over one day, assuming in the first case $F(10.7\,\text{cm}) = 150$ and $A_p = 4$ and in the second $F(10.7\,\text{cm}) = 200$ and $A_p = 8$. The first case corresponds to calm conditions as they are typically encountered during the years of minimum solar activ-

ity, the other would rather characterize an active phase (but by no means extreme conditions) within the 11 years solar cycle. The simulation, where all perturbation terms due to the Earth's gravitational field were neglected, is uniquely meant to illustrate the effect of drag. In reality the short-period perturbations with amplitudes of few kilometers (mainly) due to the Earth's oblateness would be superimposed to the curve in Figure 3.26.

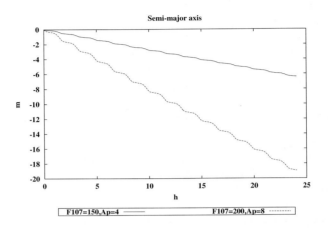

Fig. 3.26. Semi-major axis of GPS/MET over one day

The semi-major axis a decreases in both cases (by about 6 m per day in the first, by about 18 m per day in the second case). Obviously, the actual decrease of the semi-major axis is heavily dependent on the solar activity. This confirms the well-known fact that many more LEOs "decay" in the Earth's atmosphere during the years of maximum solar activity than during the years of low activity. The solar cycle thus has the nice side-effect of "cleaning" the atmosphere below heights of about 1000 km. The "pollution" of these lower parts of the atmosphere by satellites and in particular space debris is not as dangerous (seen from the long-term perspective) as that of regions like the geostationary belt.

The behavior is not as linear as one might expect. The periodic effects are caused by the variations of the density discussed in the previous paragraph and by the fact that the surface of the Earth is (in good approximation) an ellipsoid, implying that the height above surface of a circular orbital curve is growing with latitude (and has a minimum over the equator). The characteristics in Figure 3.26 are only typical for circular orbits. As soon as the eccentricity significantly differs from zero, the deceleration due to drag is much more pronounced in perigee than in the other parts of the orbit. Usually a good mathematical approximation is achieved by feigning an instantaneous velocity change (decrease) in perigee (see discussion in next paragraph).

Figures 3.27 - 3.29 illustrate the mechanism. The initial conditions in Table 3.5 were modified by assuming an elliptical orbit with $e = 0.05$ instead of a circular orbit. The consequences of this subtle change are rather dramatic: The perigee is at a geocentric distance of $r_{per} = a(1 - e) = 6745$ km corresponding to a height of $h_{per} = 6745\,\text{km} - 6378\,\text{km} = 367\,\text{km}$ (assuming a spherical Earth with radius $a_\delta \approx 6378$ km), the apogee, however, is at a height of $h_{apo} = a(1 + e) - a_\delta = 1077$ km . As we see from Figure 3.22, the density at apogee height is negligible w.r.t. the density at perigee height. Figure 3.27, which was generated with $F(10.7\,\text{cm}) = 150$ and $A_p = 4$, should be compared to the solid curve in Figure 3.26. The reduction of the semi-major axis takes place almost uniquely during a short time interval centered at the time of perigee (the integration was started in perigee). Due to the fact that the perigee is much lower than in Figure 3.26 the mean decrease of the semi-major axis over one day is about 275 m in Figure 3.27 instead of about 6 m in Figure 3.26.

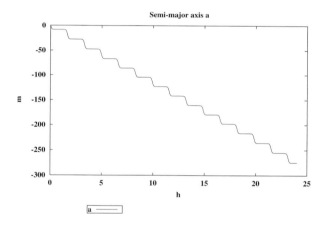

Fig. 3.27. Semi-major axis of GPS/MET with $e = 0.05$ over one day

Figure 3.28 illustrates that the eccentricity, as the semi-major axis, decreases as a function of time. Figures 3.27 and 3.28 are highly correlated in the sense that the elements decrease "only" while the satellite is close to the perigee.

Figure 3.29 shows the development of the perigee $a(1 - e)$ and of the apogee $a(1 + e)$. One can see that the perigee decreases only slowly (about 20 m per day), whereas the apogee decreases at a rate of more than 500 m per day. Obviously, atmospheric drag has the tendency to bring down the apogee height to the perigee height.

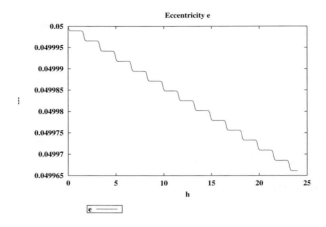

Fig. 3.28. Eccentricity of GPS/MET with $e_0 = 0.05$ over one day

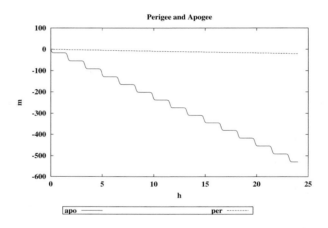

Fig. 3.29. Perigee and apogee of GPS/MET with $e_0 = 0.05$ over one day

3.6.3 Theoretical Interpretation of Drag Perturbations

We will now apply the theory of perturbations of first order as developed in Chapter I-6 in order to describe the perturbations encountered in Figures 3.26 - 3.29. We have to use the Gaussian (and not the Lagrangian) perturbation equations because atmospheric drag is a non-conservative force. In the approximation of eqn. (3.118) atmospheric drag is anti-parallel to the velocity vector, implying that the N'- and W'-components are zero (see also Figure I-6.1). The T' component may be simply taken over from eqn. (3.118):

$$T' = -\frac{C}{2}\frac{A}{m}\dot{\boldsymbol{r}}'^2\,\rho(\boldsymbol{r})\;. \tag{3.122}$$

As $N' = W' = 0$, the Gaussian perturbation equations (I-6.91) are reduced to

$$\dot{a} = \frac{2}{n^2 a} |\dot{\boldsymbol{r}}| \, T' = -\frac{1}{n^2 a} C \frac{A}{m} |\dot{\boldsymbol{r}}|^3 \, \rho(\boldsymbol{r})$$
$$\dot{e} = \frac{2(\cos v + e)}{|\dot{\boldsymbol{r}}|} T' = -(\cos v + e) C \frac{A}{m} |\dot{\boldsymbol{r}}| \, \rho(\boldsymbol{r}) \, . \tag{3.123}$$

Note, that we approximated the velocity $\dot{\boldsymbol{r}}'$ w.r.t. the rotating atmosphere by the velocity $\dot{\boldsymbol{r}}$ of the satellite in a non-rotating Earth system for the purpose of the subsequent discussion. For LEOs with a circular velocity in the inertial system of about 7 km per second, this approximation introduces an error below 10%. Unnecessary to say that the correct version of equation (3.118) is used in program SATORB.

For orbits with small eccentricities one may furthermore simplify these equations by neglecting the terms of first order in e on the right-hand side of eqns. (3.123):

$$\dot{a} = -C \frac{A}{m} n \, a^2 \, \rho(\boldsymbol{r}) + O(e)$$
$$\dot{e} = -C \frac{A}{m} n \, a \, \cos \sigma \, \rho(\boldsymbol{r}) + O(e) \, , \tag{3.124}$$

where σ is the mean anomaly. Equation (3.124) may be used to compute the secular change of the elements a and e for circular orbits, where it is assumed that the density $\rho(\boldsymbol{r}) \overset{\text{def}}{=} \rho(a) \overset{\text{def}}{=} \rho_0$ is constant.

The eccentricity e is quickly dealt with: The perturbation is purely periodic and the net effect over the revolution is zero. The short-period variations would imply negative eccentricities, a somewhat nonsensical result revealing the limitations of approximate solutions.

The net decrease of the semi-major axis per day is obtained by multiplying the above constant drift with the number of seconds per day:

$$\Delta a/\text{day} = -86400 \, n \, C \frac{A}{m} a^2 \, \rho_0 \, . \tag{3.125}$$

The comparison of the result (3.125) with the results achieved by simulation is not entirely trivial, because in the simulation underlying Figure 3.26 the Earth's surface was assumed to be an ellipsoid and the correct ellipsoidal height was used to compute the height $h(t)$ of the satellite above the surface. As a crude guess we compute the mean height of the satellite as $h_{\text{mean}} \approx \left(a - \frac{1}{2}\left(a_{\delta\text{equ}} - a_{\delta\text{pol}}\right)\right) \approx 732.5$. The mean density at this height is $\rho_0 \approx 6 \cdot 10^{-14}$ kg/m³. The decrease per day thus is expected to be

$$\Delta a/\text{day} = -86400 \cdot 2 \cdot 0.02 \, a^2 \, \rho_0 \approx 11 \text{ m} \, . \tag{3.126}$$

This overestimates the actual drift by almost a factor of 2. This is explained by the complexity of the MSISe-90 model: The mean value of $\rho(r)$ over one revolution was rather $\rho \approx 3.5 \cdot 10^{-14}$ and not $\rho_0 \approx 6 \cdot 10^{-14}$ kg/m^3 as assumed above. The resulting decrease of a per day is $\Delta a/\text{day} = 6.4$ m which is in excellent agreement with the result actually obtained in Figure 3.26.

The case $e \neq 0$ is somewhat more complicated. We treat it approximately for orbits with small eccentricities only. We will still use formulas (3.124) (i.e., we retain only the terms $\sim e^0$), but we will use the correct density corresponding to the vicinity of the perigee.

Let us approximate the density of the atmosphere in the vicinity of the perigee with the barometric formula (3.121), where $h_0 = a\,(1-e) - a_{\delta}$. The argument $h - h_0$ may then be approximated as:

$$h - h_0 = \big(a\,(1 - e\,\cos\sigma) - a_{\delta}\big) - \big(a\,(1-e) - a_{\delta}\big) = a\,e\,(1 - \cos\sigma) \approx \tfrac{1}{2}\,a\,e\,\sigma^2 , \tag{3.127}$$

where $\sigma \overset{\text{def}}{=} n\,(t - T_0)$ is the mean anomaly. In the vicinity of the perigee the density ρ is approximated by:

$$\rho = \rho_0\,e^{-\frac{a\,e}{2\,H_0}\sigma^2} \overset{\text{def}}{=} \rho_0\,e^{-\frac{\sigma^2}{2\,m_0^2}} , \tag{3.128}$$

where

$$m_0 \overset{\text{def}}{=} \sqrt{\frac{H_0}{a\,e}} . \tag{3.129}$$

Using the above approximations in the equation for the semi-major axis in eqns. (3.124) gives

$$\dot{a} \approx n\,\frac{da}{d\sigma} = -\,C\,\frac{A}{m}\,n\,a^2\,\rho_0\,e^{-\frac{\sigma^2}{2\,m_0^2}} . \tag{3.130}$$

Cum grano salis we may now approximate the change of the semi-major axis per revolution by:

$$\Delta a/\text{rev} = -\,C\,\frac{A}{m}\,a^2\,\rho_0 \int\limits_{-\infty}^{+\infty} e^{-\frac{\sigma^2}{2\,m_0^2}}\,d\sigma = -\,C\,\frac{A}{m}\,a^2\,\rho_0\,\sqrt{2\,\pi}\,m_0 . \tag{3.131}$$

The mean drift per day is then obtained from the above expression by multiplication with the number of revolutions per day:

$$\Delta a/\text{day} = -\,n_{\text{rev/day}}\,\sqrt{\frac{2\,\pi\,H_0}{a\,e}}\,C\,\frac{A}{m}\,a^2\,\rho_0 . \tag{3.132}$$

With a semi-major axis of $a = 7100$ km and an eccentricity of $e = 0.05$ we obtain a perigee height (over the equator) of about 367 km and a value of $a\,e = 355$ km. The air density at perigee is $\rho_0 \approx 9 \cdot 10^{-12}$ kg/m^3. The scaling height H_0 was taken from Figure 3.22 as $H_0 \approx 54$ km in the height of the satellite's perigee. This gives a value of $m_0 = \sqrt{\frac{H_0}{a\,e}} \approx 0.39$. With $n_{\text{rev/day}} \approx 14.5$ we obtain eventually the following estimation for the drift in Figure 3.27 using eqn. (3.132):

$$\Delta a/\text{day} = -257 \text{ m} , \tag{3.133}$$

which is pretty much what Figure 3.27 shows.

Let us quickly assess the order of magnitude for the development of the eccentricity (second of eqns. (3.124)). Assuming again that the main contribution to the perturbations occurs in the vicinity of the perigee (what is by the way confirmed by Figure 3.28), we may in a very crude approximation set $\cos v \approx 1$. In this approximation we obtain the simple result (compare the two equations (3.124)):

$$\dot{e} = \frac{\dot{a}}{a} . \tag{3.134}$$

But this implies

$$\Delta e(t) \approx \frac{\Delta a(t)}{a} . \tag{3.135}$$

For the above example we expect therefore a drift in the eccentricity of $\Delta e/\text{day} = -\frac{257}{7'100'000} = -3.67 \cdot 10^{-5}$. This is a remarkably good approximation of the minute computation shown in Figure 3.28.

In this approximation it is rather easy to predict the development of perigee and apogee:

$$\frac{d\{a\,(1 \pm e)\}}{dt} = \dot{a}\,(1 \pm e) \pm a\,\dot{e} = \begin{cases} \approx +2\,\dot{a} & \text{apogee} \\ \approx 0 & \text{perigee} \end{cases} . \tag{3.136}$$

These results agree very well to those of the stricter computations in Figure 3.29.

The above order-of-magnitude estimates are but crude approximations. If one uses eqns. (3.123) instead of eqns. (3.124), much better approximations involving Bessel functions may be provided. The reader is referred to [41] for details.

Formulas for $\Delta a(t)$ and $\Delta e(t)$ of the kind derived in this section may be used to estimate the lifetime of LEOs. The equation for the eccentricity may, e.g., be used for a crude approximation to calculate the time it takes for an elliptical orbit to become circular. Also, experience tells that an orbit is no longer stable if the revolution period drops below $U_0 \approx 87$ minutes (see [41]).

It is rather difficult to predict the precise time and the geographical location of a satellite decay. It is, however, easy to predict that the geographical location must lie approximately in the orbital plane (at the time of the decay).

3.7 Radiation Pressure

3.7.1 Solar Radiation and Radiation Pressure

Quantum mechanics says that each photon of frequency ν and wavelength $\lambda = c/\nu$ (where c is the speed of light in vacuum) carries the energy

$$E = h\nu \qquad (3.137)$$

and the linear momentum

$$\boldsymbol{p} = \frac{h\nu}{c}\,\boldsymbol{e}\;, \qquad (3.138)$$

where

$h = 6.62\cdot 10^{-34}$ Js is Planck's constant (named after Max Karl Ernst Ludwig Planck (1858–1947)) and

$\boldsymbol{e}$ is the unit vector of the propagation of the photon.

The momentum transferred per time unit onto a unit surface in a radiation field is called *radiation pressure*. Radiation pressure therefore is a vectorial quantity.

In a general radiation field it may be quite an elaborate task to calculate the radiation pressure acting on a surface element. The underlying principle is simple, however: We have to calculate the change $\Delta\boldsymbol{p} = \boldsymbol{p}_a - \boldsymbol{p}_b$ of linear momentum $\boldsymbol{p}$ for each of the photons hitting the surface element S (the index b characterizes the momentum before, a that after hitting the surface). The momentum transferred to the surface element must be $\boldsymbol{p}_s \stackrel{\text{def}}{=} -\Delta\boldsymbol{p} = \boldsymbol{p}_b - \boldsymbol{p}_a$. After having computed the momentum transferred by each of the photons we simply have to compute the vectorial sum of the momenta $\boldsymbol{p}_s$.

The mechanism of momentum transfer to the surface by a single photon is illustrated by Figure 3.30.

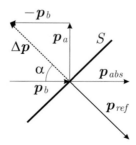

Fig. 3.30. Absorption and reflection of radiation by a surface element S

According to the law of conservation of linear momentum $-\Delta p$ is the momentum gained by the surface element. If the photon is absorbed by the surface, $p_a = 0$ and the surface gains the momentum $p_{abs} = p_b$. If the photon is reflected by the surface, the photon is reflected according to Snellius' law of reflection. The surface gains linear momentum normal to its surface with absolute value $2\,|\,p_b\,|\,\cos\alpha = |\,p_{ref}\,|$.

The primary radiation source to be considered in satellite geodesy is the Sun. The radiation pressure due to the direct solar radiation is also referred to as *direct radiation pressure*. Other sources of radiation are the Earth, which reflects and/or re-emits the radiation received by the Sun, or to a much lesser extent, the Moon, reflecting the solar radiation. We will mainly focus on the direct radiation pressure, but will also briefly address the radiation pressure due to the sunlight reflected by the Earth towards the end of this chapter.

The radiation field due to the direct solar radiation may be considered as parallel to the direction Sun $\rightarrow$ Satellite. Consequently, the momentum transferred per second to a unit surface normal to this direction may be expressed by the solar constant S, giving the energy flowing through this surface per time unit at the distance of 1 AU. According to [68] the value of the solar constant (which we remember (hopefully) from school as twenty Calories (kcal) per minute and m^2) is

$$S = 1368 \text{ Watt/m}^2 \ . \tag{3.139}$$

If 100% of the radiation is absorbed, the linear momentum

$$p = \frac{S}{c}\, e \tag{3.140}$$

is gained in one second by a surface element of 1 m^2 normal to the direction Sun $\rightarrow$ surface element at 1 AU. The absolute value of the vector is

$$|p| = \frac{S}{c} = 4.56316 \cdot 10^{-6} \text{ N/m}^2 \ . \tag{3.141}$$

Obviously, the surface of the cross section of the satellite normal to the direction Sun $\rightarrow$ satellite and the mass m of the satellite play the same roles as the corresponding quantities when dealing with drag. We will use the same symbol A as in the case of drag, but point out that the interpretation is different from the preceding section.

Assuming rotational symmetry of the satellite w.r.t. the axis Sun $\rightarrow$ satellite, the acceleration due to the direct solar radiation may be written as:

$$a_{rad} = \frac{\tilde{C}}{2} \frac{A_\delta^2}{|r - r_\odot|^2} \frac{S}{c} \frac{A}{m} \frac{r - r_\odot}{|r - r_\odot|} \ , \tag{3.142}$$

where

$\tilde{C}$ depends on the reflective properties of the satellite surface,

   r is the geocentric position vector of the satellite,

$r_\odot$ is the geocentric position vector of the Sun,

$A_\delta = 149{,}597{,}870{,}610$ m (numerical value from [68]) is the Astronomical Unit,

   m is the mass of the satellite, and

   A is the cross section of the satellite normal to the direction Sun $\rightarrow$ satellite.

Rotational symmetry w.r.t. the axis Sun $\rightarrow$ satellite naturally may not always be assumed. Nevertheless, formula (3.142) is in practice an excellent first-order approximation to take radiation pressure into account. If the radiation is mainly absorbed (as it should be the case, e.g., for solar panels), formula (3.142) is correct for satellites of arbitrary shape. Subsequently we will use formula (3.142) to describe direct radiation pressure. More accurate radiation pressure models will be considered in section 3.7.5.

It is remarkable that the direct radiation pressure as represented by eqn. (3.142) practically does not depend on the height of the satellite above the Earth's surface, whereas atmospheric drag is exponentially decreasing with height. This is one of the reasons why radiation pressure is the dominating non-gravitational perturbation above heights of about 600 km .

Radiation pressure obviously is "turned off" if the radiation is blocked by an "obstacle" between the Sun and the satellite. The Moon or the Earth may "serve" as obstacles. The shadowing of the sunlight by the Moon is often not considered. For precise analyses (like the ones performed by the IGS Analysis Centers) this effect must be taken into account. Subsequently, we will, however, only deal with the shadowing of the sunlight by the Earth.

Figure 3.31 illustrates the shadow geometry for the Earth as "obstacle". We assume that (a) the Earth is spherical with radius a_δ, and that (b) solar radiation is parallel in the Earth's environment. In this case the sunlight-shadow boundary in space is a right cylinder with radius a_δ and with its axis on the continuation of the line Sun $\rightarrow$ Earth (lying in the plane of the figure).

Figure 3.31 may be used to compute the maximum duration of the eclipse phase in relation to the revolution period for orbits with small eccentricities. The longest eclipse period for a given semi-major axis a results, if the Sun lies in the orbital plane. We may therefore assume that the plane of drawing is the orbital plane. Consider now satellite S_1 which just crosses the light-shadow boundary. At shadow entry its geocentric position vector forms an angle of γ with the axis of the shadow cone. At shadow exit (not shown in Figure 3.31) its geocentric position vector will again form an angle of γ with the axis of the shadow cylinder, and the radius vector will have covered the angle of 2γ in the shadow. Figure 3.31 shows that the angle γ has to be computed as:

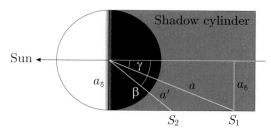

Fig. 3.31. Shadow geometry

$$\gamma = \arcsin \frac{a_{\oplus}}{a} \ . \tag{3.143}$$

The fraction of the revolution spent in the eclipse thus is

$$\frac{\Delta t_{\text{ecl}}}{P} = \frac{2\,\gamma}{2\,\pi} \ , \tag{3.144}$$

where P is the satellite's revolution period.

For GPS/MET, with $a = 7100$ km , the maximum duration of the eclipse phase per revolution is $\Delta t_{\text{ecl}}/P \approx 0.355$ (corresponding to 35 minutes), for a GPS satellite with $a \approx 26'500$ km it is $\Delta t_{\text{ecl}}/P \approx 0.077$ (corresponding to about 56 minutes), and for a geostationary satellite with $a \approx 42'164$ km it is $\Delta t_{\text{ecl}}/P \approx 0.048$ (corresponding to 70 minutes).

Figure 3.31 and formulas (3.143) and (3.144) therefore tell that the eclipsed parts of the orbital curve may be significant (up to almost half a revolution period) and that they have to be taken into account.

Figure 3.31 also may be used to find, for a given semi-major axis a', the maximum angle β_0 between the geocentric unit vector $e_{\odot}$ of the Sun and the orbital plane for which eclipses still occur. We are free to choose the orbital plane orthogonal to the drawing plane of Figure 3.31. Satellite S_2 is a satellite in such a plane just touching the shadow cylinder. Obviously the angle between the orbital plane and the unit vector to the Sun observes the equation

$$\sin \beta_0 = \frac{a_{\oplus}}{a'} \tag{3.145}$$

for a satellite touching the shadow cylinder.

The satellite actually enters the Earth's shadow cylinder if the angle β between the orbital plane and the geocentric unit vector of the Sun is

$$\beta < \beta_0 = \arcsin\left(\frac{a_{\oplus}}{a'}\right) \ . \tag{3.146}$$

For GPS/MET the maximum angle is $\beta_0 = 63.9°$, for a GPS satellite $\beta_0 = 13.9°$, and for a "geostationary" satellite $\beta_0 = 8.7°$.

Whether or not eclipses occur for each revolution or only during certain "eclipse seasons", depends on the inclination $\tilde{i}$ of the orbital plane w.r.t. the plane of the ecliptic. This inclination angle varies between the two limits:

$$|i - \varepsilon| \leq \tilde{i} \leq |i + \varepsilon| \,, \tag{3.147}$$

where $\varepsilon \approx 23.5°$ is the obliquity of the equatior w.r.t. the ecliptic. As the angle β between the geocentric unit vector of the Sun $e_\odot$ and the orbital plane for a given constellation is

$$|\beta| \leq \tilde{i} \,, \tag{3.148}$$

we see, e.g., that eclipses always occur (independently of the actual inclination angle $\tilde{i}$) for satellites with low inclinations i w.r.t. the equator, if

$$i + \varepsilon < \beta_0 \,. \tag{3.149}$$

For a satellite like GPS/MET this would be the case for inclinations

$$i + \varepsilon < 63.9° \text{ i.e. } i < 40.4° \,.$$

One should not conclude that a satellite for which (3.149) does *not* hold will never be eclipsed: If the Sun is near the (ecliptical) node of the orbital plane, the angle β is zero and deep eclipses will occur at least twice per year (if the precession rate of the orbital plane is small). The frequency of these "eclipse seasons" may be higher for LEOs, where the precession period of the node may be a few months only.

Our discussion of the shadowing effects was based on Figure 3.31, which in turn was based on three simplifying assumptions:

1. The Earth was assumed to be spherical, whereas in reality it is an ellipsoid of rotational symmetry, with the two axes differing by about 21.4 km .

2. The boundary between the sunlit and the eclipsed part of space was assumed to be a right cylinder. Due to the angular diameter of the Sun of about 0.5° the shadow cylinder should be replaced by two shadow cones, one for the *umbra* and one for the *penumbra*. Approximately, the boundaries of the two cones are symmetric relative to the cylindrical boundary in Figure 3.31, and inclined to the cylinder boundary by 0.25° (the angular radius of the Sun).

3. The Earth's atmosphere was neglected. Refraction effects occur and should be taken into account.

The first approximation easily may be dealt with when numerical procedures are used to generate the orbits. The impact of the second approximation on the orbital accuracy is often overestimated. One should be aware of the symmetry inherent in the problem: when adopting a cylindrical shadow boundary (where the radiation pressure is instantaneously "switching on/off") instead of taking into account the two conic boundaries (with a continuous light ↔ shadow transition between the penumbra and umbra boundaries), the neglected accelerations before and after crossing the cylinder almost compensate each other. The cylindrical approximation for the shadow boundary, where radiation pressure is switched on or off instantaneously, is a sufficient approximation for the true process in practice. The reader who, nevertheless, wishes to be more familiar with penumbra effects, is referred to the refined treatment in [75].

When taking into account radiation pressure in a procedure for numerical integration (and there is no real alternative to that) one must take care *not* to introduce discontinuities at the epochs of shadow entry or exit. Advanced numerical integration tools are based on the assumption of analytical functions, where discontinuities (in the accelerations) must be avoided. Collocation algorithms may "easily" cope with this problem if the following procedure is followed:

1. From the initial position vector at the epoch t_i of the current integration step it is checked whether or not the satellite is in sunlight at time t_i.

2. If the satellite is in sunlight at t_i, the radiation pressure acceleration (3.142) is switched on. Otherwise it is switched off.

3. These settings are not altered when performing the integration step in the interval $[t_i, t_{i+1}]$, where t_{i+1} is the right interval boundary as defined by an automatic stepsize procedure.

4. The numerical solution of the initial value problem pertaining to the interval $[t_i, t_{i+1}]$ is now used to check whether or not the shadow cylinder was crossed during this interval.

5. If a crossing of the cylinder took place, let us say at the epoch t^*, the current numerical solution is used to compute new initial conditions at time t^*, and a new initial value problem is invoked.

6. The algorithm proceeds with step (1).

The procedure outlined above allows it to treat the radiation pressure acceleration without introducing any errors (apart from the rounding and approximation errors associated with every numerical integration procedure). It is, by the way, more difficult to treat the problem correctly when using multistep methods. First-order corrections of the scheme of differences do not eliminate higher-order terms. From our point of view only a procedure of the kind outlined above is capable of coping with the problem of radiation pressure in a satisfactory way. The above procedure is implemented in program SATORB.

3.7.2 Simulations

In order to gain insight into the orbit perturbations due to radiation pressure we integrate the orbits of a satellite with GPS characteristics (see, e.g., Figure I- 2.6) and of the US/French altimetry satellite TOPEX/Poseidon (see Figure I- 2.8). The orbital and satellite characteristics (meant to be typical for two kinds of spacecrafts) are contained in Table 3.6. In reality the area-to-mass ratio A/m is a function of the attitude (and thus of time) for both, the GPS and the TOPEX/Poseidon satellites. For our order-of-magnitude assessment it is sufficient to use the values provided in Table 3.6. Figure 3.32 shows the development of the orbital elements of a GPS-like satellite under the influence of radiation pressure. On the left-hand side we find the osculating elements over the time interval of one day, i.e., over two revolutions, on the right-hand side the mean elements over one year (the averaging period being one revolution). The characteristics of Table 3.6 were used, with $e = 0.02$. The simulation documented by Figure 3.32 is rather unrealistic because the Earth's gravitational field was assumed to be spherically symmetric. The implications of this assumption will be dealt with later on.

Table 3.6. Characteristics of GPS satellites and of TOPEX/Poseidon

Characteristics	GPS satellite	TOPEX/Poseidon
a	26550 km	7714 km
e	$0.001 - 0.020$	0.001
i	$55°$	$66°$
A/m	$0.02 \ \mathrm{m^2/kg}$	$0.008 \ \mathrm{m^2/kg}$

The osculating semi-major axis a shows a periodic signal of an amplitude of about 13 m in Figure 3.32. This amplitude is very small compared to the amplitude of the short-period oblateness perturbation or to the amplitude of the gravitational perturbations caused by Sun and Moon. When looking at the mean semi-major axis over one year, we see that "usually" there is no net effect over one revolution. The two eclipse seasons of about seven weeks duration are an exception. But even then the effect is small (of the order of a few meters). Note, that the net effect in the semi-major axis after two eclipse seasons is very close to zero, whereas the mean effect after one eclipse season differs significantly from zero. This behavior is typical for the orbits of all GPS satellites. The actual shape of the perturbation in a is highly dependent on the date (time within the year) of the eclipse periods and on the eccentricity and the argument of perigee. The net effect of one eclipse season is almost zero for circular orbits ($e = 0$). If the initial conditions are chosen in such a way that the dates of the eclipse seasons are symmetric relative to the time of the solstices, the net effect of one eclipse season is to change the semi-major

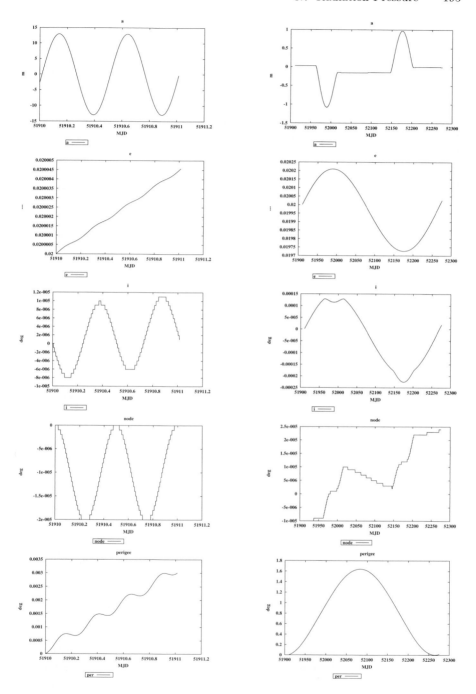

Fig. 3.32. Radiation pressure effects on the orbit of a GPS satellite over one day (left) and over one year (right)

axis by about four meters, an effect which will be reversed by the following eclipse season (compare [18]). In summary we may state that the net effect of radiation pressure on the semi-major axis is not significant.

The perturbations in the eccentricity e show a different pattern in Figure 3.32. The osculating eccentricity grows almost linearly with time over the time interval of one day (two revolutions). The mean eccentricity has a clear annual signal with an amplitude of about $\Delta e \approx 2 \cdot 10^{-4}$. This does not look like a big effect, but it implies that perigee- and apogee-heights of a typical GPS satellite will vary by about $\pm a \, \Delta e \approx \pm 5$ km over one year – an effect which is notable.

The perturbations of both, inclination and right ascension of the ascending node, are very small. Note that the step-function in the osculating inclinations and nodes are artifacts due to the limited number of digits in the data files. The perigee, on the other hand, shows a strong annual signal of an amplitude of about $\Delta \omega \approx 0.8°$.

Figure 3.33 demonstrates that radiation pressure is an important characteristic of GPS orbits: The mean development of the eccentricity over one year in a rather realistic force field (Earth's potential complete up to degree and order $n = m = 8$, gravitational attractions by Sun and Moon taken into account) is clearly different for the orbits including the radiation pressure acceleration (solid curve) and those ignoring it (dashed curve). Note, however, that the net effect due to radiation pressure is more or less averaged out after one year.

Figure 3.34 documents the development of the mean orbital elements for a TOPEX-like satellite over one year due to radiation pressure. The results in this figure should be compared to Figure 3.32: Only the main term of the gravitational field and radiation pressure define the force field. The general picture is similar for both types of satellites: Eccentricity and perigee are much more affected by radiation pressure than the other elements.

It is, however, interesting to note that the perturbations of the orbital plane (elements i and Ω) are slightly more significant and less symmetric than in the case of GPS satellites. This is explained by the long shadow periods. There actually was no eclipse-free period in this experiment: The right ascension of the ascending node was defined as $\Omega = 0°$. Due to absence of the oblateness term C_{20} the node did not precess, and the inclination of the orbital plane w.r.t. the ecliptic was constant at the value of $(\tilde{i} = 66° - \varepsilon \approx 42.5°)$. Using the arguments developed in the previous paragraph, we conclude that there actually could not be a shadow-free period under these conditions. If we would "switch on" the oblateness perturbations, we would observe shadow-free periods and, as a consequence, an averaging effect of the perturbations in i and Ω – very much like in the case of GPS satellites – would take place.

When comparing the perturbations in Figures 3.32 and 3.34 we have to take into account that, according to Table 3.6 we have $A/m \approx 0.02$ for GPS-

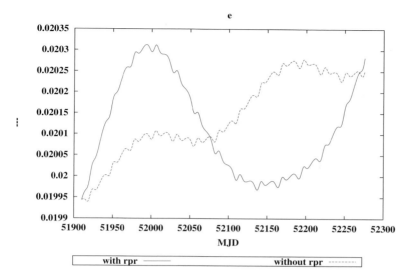

Fig. 3.33. Mean eccentricity in a realistic force field with and without radiation pressure

like satellites, but only $A/m \approx 0.008$ for TOPEX/Poseidon. Therefore, the amplitudes in Figure 3.34 have to be multiplied by a factor of 2.5, in order to compare the effect of the same perturbing acceleration on a high and a low orbiting satellite. When comparing the perturbations in e in the two figures we note that the ratio is more like a factor of six. This ratio and other effects will be explained in the next section.

3.7.3 Theoretical Considerations Concerning Radiation Pressure

Following tradition we consider radiation pressure as a non-conservative force and therefore use the Gaussian perturbation equations for explaining its effects on the orbital elements (we refer to paragraph 3.7.4 for additional remarks). It turns out that version (I- 6.88), based on the decomposition into the components R', S', and W' (compare also Figure I- 6.1), is best suited for our purpose.

Let us slightly simplify the acceleration (3.142) by replacing the direction Sun $\rightarrow$ satellite by the unit vector pointing from the Sun to the center of the Earth. Moreover we neglect the variations of the distance between Sun and Earth (in view of the eccentricity of $e \approx 0.016$ of the Earth's orbit a scale factor varying between 1 ± 0.032):

$$\boldsymbol{a}_{\mathrm{rad}} = -\frac{\tilde{C}}{2}\frac{S}{c}\frac{A}{m}\,\boldsymbol{e}_{\odot}\ . \qquad (3.150)$$

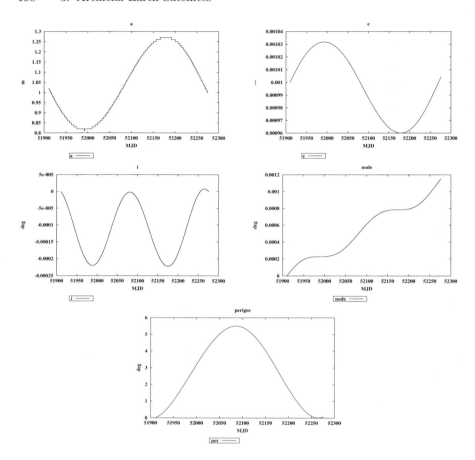

Fig. 3.34. Perturbation of the mean orbital elements of TOPEX/Poseidon due to radiation pressure

Furthermore, we will replace the varying position of the Sun over one revolution by its fixed position at the midpoint of the time interval considered. Let us assume that the Sun's spherical coordinates in the orbital system (reference plane = orbital plane, first axis ascending node) are the argument of latitude $u_\odot$ and the elevation $\beta_\odot$ above the orbital plane (see Figure 3.35).

Figure 3.35 shows that the perturbing acceleration reads as:

$$
\begin{pmatrix} R' \\ S' \\ W' \end{pmatrix} = -\frac{\tilde{C}}{2}\frac{S}{c}\frac{A}{m} \begin{pmatrix} \cos\beta_\odot \ \cos(u - u_\odot) \\ -\cos\beta_\odot \ \sin(u - u_\odot) \\ \sin\beta_\odot \end{pmatrix} . \tag{3.151}
$$

With these assumptions the W'-component is a constant. A closer look at eqns. (I- 6.88) shows that in first-order there are no secular terms in the two

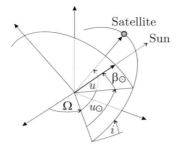

Fig. 3.35. Position of the Sun w.r.t. the orbital plane

elements i and Ω (replace the independent argument t by u and integrate over one revolution). A similar analysis shows that the same is true for the perturbations of the semi-major axis a. This leaves us with the other three elements e, ω, and σ as the elements influenced by radiation pressure. As an example, we focus now on the long-term behavior of the eccentricity e. Retaining only the terms $\sim e^0$ we obtain the equation:

$$
\begin{aligned}
\dot{e} &\approx \frac{1}{n\,a}\{\sin v\,R' + 2\,\cos v\,S'\} \\
&= -\frac{\tilde{C}}{4}\,\frac{S}{c}\,\frac{A}{m}\,\frac{\cos\beta_\odot}{n\,a}\{3\,\sin(u_\odot - \omega) + \sin(u_\odot - 2\,v - \omega)\}\,.
\end{aligned}
\tag{3.152}
$$

Equation (3.152) may be used to calculate the mean drift of the eccentricity over one revolution as

$$
\bar{\dot{e}} = -\frac{3\,\tilde{C}}{4}\,\frac{S}{c}\,\frac{A}{m}\,\frac{\cos\beta_\odot}{na}\,\sin(u_\odot - \omega)\,,
\tag{3.153}
$$

which obviously is different from zero. The change of the eccentricity per day is obtained by multiplying the above expression with the number of seconds contained in a day:

$$
\Delta e/\text{day} = -\frac{3\,\tilde{C}}{4}\,\frac{S}{c}\,\frac{A}{m}\,\frac{\sin(u_\odot - \omega)\,\cos\beta_\odot}{n\,a}\cdot 86400\,.
\tag{3.154}
$$

Using the relation $(n\,a)^{-1} = \sqrt{a/\mu}$ emerging from eqn. (I-4.41) one expects that the drift in the eccentricity due to radiation pressure for two satellites with the same values $u_\odot - \omega$ and $\beta_\odot$ is

$$
\frac{\bar{\dot{e}}_1}{\bar{\dot{e}}_2} = \sqrt{\frac{a_1}{a_2}}\,\frac{A_1}{m_1}\,\frac{m_2}{A_2}\,,
\tag{3.155}
$$

i.e., the satellite with the bigger semi-major axis is expected to show the larger perturbation due to radiation pressure. For the examples of the GPS satellite and the TOPEX/Poseidon spacecraft we would expect a ratio

$$\frac{\bar{\dot{e}}_{\mathrm{GPS}}}{\bar{\dot{e}}_{\mathrm{TOPEX}}} = \sqrt{\frac{a_{\mathrm{GPS}}}{a_{\mathrm{TOPEX}}}} \left(\frac{A_1}{m_1}\right)_{\mathrm{GPS}} \left(\frac{m_2}{A_2}\right)_{\mathrm{TOPEX}} = \sqrt{\frac{26550}{7714}} \frac{0.02}{0.008} \approx 4.65 \,.$$

(3.156)

A comparison of Figures 3.32 and 3.34 show a good agreement of theory and experiment.

3.7.4 Radiation Pressure as a Dissipative Force

When speaking of non-gravitational forces one usually also means *dissipative* forces, i.e., forces leading to a loss of energy and angular momentum of the satellite. As the energy is in essence represented by the satellite's semi-major axis a, a dissipative force is expected to reduce the satellite's semi-major axis. Atmospheric drag is a typical example of a dissipative form.

Apparently direct radiation pressure, as represented by eqn. (3.142) or its "strap-down" version (3.150), does not give rise to a secular reduction of the semi-major axis. This is not amazing, because one easily verifies that the radiation pressure acceleration (3.142) formally may be expressed as the gradient of the following scalar perturbation function:

$$R_{\mathrm{rad}} = \frac{\tilde{C} A_\odot^2}{2} \frac{S}{c} \frac{A}{m} \frac{1}{|\boldsymbol{r} - \boldsymbol{r}_\odot|} \,.$$

(3.157)

Milani et al. [74] even show explicitly that *direct radiation pressure (under somewhat generalized conditions) does not dissipate energy.*

This conclusion, on the other hand, is based on an approximation *not* taking into account the effect of light aberration, which was discovered by James Bradley (1692–1762) around 1728. In non-relativistic approximation (which is correct up to terms of order 1 in v/c, v being the velocity of the satellite relative to the Sun), the unit vector along which radiation pressure acts should actually be defined by

$$\boldsymbol{e}'_\odot \approx \boldsymbol{e}_\odot + \frac{-\dot{\boldsymbol{r}}_\odot + \dot{\boldsymbol{r}}}{c} = \boldsymbol{e}_\odot + \frac{\dot{\boldsymbol{r}}}{c} - \frac{\dot{\boldsymbol{r}}_\odot}{c} \,,$$

(3.158)

where $\boldsymbol{e}_\odot$ is the geometric direction satellite-Sun, $\boldsymbol{e}'_\odot$ the unit vector taking into account light aberration, and $-\dot{\boldsymbol{r}}_\odot + \dot{\boldsymbol{r}}$ is the velocity of the satellite w.r.t. the light source, i.e., the Sun. The difference between the correct (i.e., aberrated) and the conventional direction of radiation pressure consists of a short-period component (with the satellite's revolution period as period) and a long-period component with an annual period. The modified model for radiation pressure is now easily obtained from eqn. (3.142) by replacing the unit vector Sun $\rightarrow$ satellite according to eqn (3.158):

$$\boldsymbol{a}'_{\mathrm{rad}} = -\frac{\tilde{C}}{2} \frac{A_\odot^2}{|\boldsymbol{r} - \boldsymbol{r}_\odot|^2} \frac{S}{c} \frac{A}{m} \boldsymbol{e}'_\odot \approx \boldsymbol{a}_{\mathrm{rad}} - \frac{|\boldsymbol{a}_{\mathrm{rad}}|}{c} \dot{\boldsymbol{r}} + \frac{|\boldsymbol{a}_{\mathrm{rad}}|}{c} \dot{\boldsymbol{r}}_\odot \,.$$

(3.159)

The right-hand side represents the correct non-relativistic model of radiation pressure.

From now on, we will only consider the perturbation due to the short-period correction, because only this component leads to a significant dissipation of energy. This means that we approximate radiation pressure as:

$$a'_{\rm rad} \stackrel{\rm def}{=} a_{\rm rad} - \frac{|a_{\rm rad}|}{c} \dot{r} \ . \tag{3.160}$$

In the approximation (3.160) the radiation pressure effect is the sum of the conventional term $a_{\rm rad}$ and the term $\delta a_{\rm rad} \stackrel{\rm def}{=} - \frac{|a_{rad}|}{c} \dot{r}$ which, in view of the fact that

$$|a_{\rm rad}| \approx \frac{\tilde{C}}{2} \frac{S}{c} \frac{A}{m} = {\rm const.} \tag{3.161}$$

may be viewed as a constant drag-like acceleration which inevitably must result in a loss of energy, i.e., in a diminution of the semi-major axis a.

Assuming circular orbits one easily verifies that this effect is $\Delta a = -1$ m per 124 years for the Lageos satellites, $\Delta a = -1$ m per 142 years for Starlette, and $\Delta a = -1$ m per 1.8 years for GPS satellites using the characteristics of Table 3.4 (see [11]). The dissipative effect is very small indeed. It should, however, be taken into account, when discussing the secular behavior of the semi-major axis of satellites like Lageos. For most practical applications the model represented by eqn. (3.142) is sufficient, and one can safely say that direct radiation pressure "almost" is a conservative force.

3.7.5 Advanced Modelling for Radiation Pressure

Apart from the discussion of the effect of light aberration in the previous paragraph we assumed so far that radiation pressure acts along the line Sun $\rightarrow$ satellite and that the area-to-mass ratio A/m is constant. Both assumptions are only crude approximations in the case of active satellites like GPS satellites, TOPEX/Poseidon, etc. Such satellites have a relatively complex shape and the satellites' attitude is dictated by their mission. Usually, the Earth-observing sensors (altimeter antenna for TOPEX, transmission antennas for GPS satellites) have to point to the center of the Earth, and the solar panels' axes have to be perpendicular to the unit vector Sun $\rightarrow$ satellite (to optimize the energy absorbed by the panels).

In other words, the satellites' orientation in inertial space, i.e., their attitude, has to be known when developing refined models for direct radiation pressure. In order to demonstrate the effort necessary to meet highest demands of orbit modelling, we briefly introduce the radiation pressure models developed for GPS satellites. Only models of this kind allow it to the International GPS

Service (IGS) to come up with ephemerides for the entire system of GPS satellites which are accurate to a few centimeters.

Figure I- 3.5 illustrates the body-fixed coordinate system for a GPS satellite. The satellite's antenna axis (the z-axis) for obvious reasons always should point to the center of the Earth, whereas the solar panels axis (the y-axis) always should be perpendicular to the direction Sun $\rightarrow$ satellite. The solar panels are then rotated around the y-axis in order to be perpendicular to the direction Sun $\rightarrow$ satellite. The nominal orientation of the z- and the y-axis is achieved with four momentum wheels in the satellite. The correct attitude is controlled with so-called horizon sensors (for the alignment of the z-axis) and Sun sensors (for the alignment of the y-axis and the orientation of the panels). A feedback loop is used to maintain the nominal attitude in an iterative process.

The nominal orientation of a GPS satellite is illustrated by Figure 3.36. It shows the attitude as seen from the Sun. The projection plane is the Earth's terminator plane (the plane containing the day/night boundary on the Earth's surface). Clearly, the y-axis nominally should always be parallel to the terminator plane.

Figure 3.36 defines a Cartesian coordinate system (X, Y, Z), which has to be distinguished from the body-fixed system in Figure I- 3.5: The Y-axis of the new coordinate system is identical with the y-axis of the satellite-fixed system in Figure I- 3.5. The Z-axis of the new system represents the direction Sun $\rightarrow$ satellite, and the X-axis completes the right-handed Cartesian coordinate system. According to this definition, the X-axis also is parallel to the terminator plane.

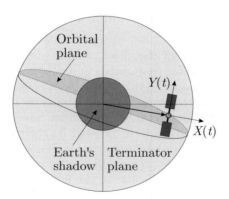

Fig. 3.36. Attitude of GPS satellite as seen from the Sun

The nominal attitude of the satellite, as illustrated by Figures 3.36, is achieved only with a limited precision in practice. One has to take into account in particular a systematic misalignment of the satellite-fixed y-axis. Two misalignment scenarios are illustrated in Figure 3.37 (Figure taken from [40]).

According to Figure 3.30 the linear momentum transferred to the satellite contains a component parallel to the misaligned y-axis. The total acceleration vector may be decomposed into the nominal Z- and Y-components. The Y-component is very small compared to the Z-component, but it may differ significantly from zero. In order to achieve highest accuracy, it has to be taken into account. (Observe that the misalignment angles will be much smaller than shown in Figure 3.37). Due to the fact that this constituent of direct radiation pressure acts along the solar panels' axis (which, according to Figure I-2.6, is the body-fixed y-axis) the effect is also called y-*bias*. When it was first introduced in the 1980s, there was a subtle alternative interpretation of this term: "Why this bias?". Meanwhile the y-bias is well established, and nobody remembers the above interpretation.

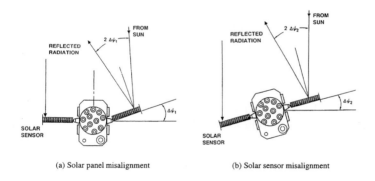

(a) Solar panel misalignment   (b) Solar sensor misalignment

Fig. 3.37. Radiation pressure along the y-axis (according to Fliegel, 1992)

As a rule of thumb the y-bias is below 1% of the primary radiation pressure acceleration. The perturbations due to the y-bias are, however, substantially different from the effects of direct radiation pressure discussed so far (see examples given below).

Figure 3.36 may be used as an argument for the existence of "biases" in the two other directions, because systematic errors in the alignments (and imperfect knowledge of reflective properties, imperfect orthogonality of axes) must exist for all axes. A thorough analysis in [12] showed that the following radiation pressure model is capable of absorbing most effects due to direct solar radiation:

$$\boldsymbol{a}_{\mathrm{rad}} = \boldsymbol{a}_{\mathrm{Rock}} + X(t)\,\boldsymbol{e}_X + Y(t)\,\boldsymbol{e}_Y + Z(t)\,\boldsymbol{e}_Z \;, \qquad (3.162)$$

where $a_{\rm Rock}$ is the a priori model as documented in [40]. e_X, e_Y, and e_Z are the unit vectors in the three orthogonal directions introduced in Figure 3.36. According to [12] the three components are defined as

$$
\begin{aligned}
X(t) &\overset{\text{def}}{=} X_0 + X_p \cos(u + \phi_X) = X_0 + X_c \cos u + X_s \sin u \\
Y(t) &\overset{\text{def}}{=} Y_0 + Y_p \cos(u + \phi_Y) = Y_0 + Y_c \cos u + Y_s \sin u \\
Z(t) &\overset{\text{def}}{=} Z_0 + Z_p \cos(u + \phi_Z) = Z_0 + Z_c \cos u + Z_s \sin u ,
\end{aligned} \tag{3.163}
$$

where u is the satellite's argument of latitude. The empirical model (3.162, 3.163) is an essential improvement over the predecessor which just allowed for a scale factor of the a priori constituent a_{Rock} and a constant y-bias.

The parameter sets $(X_0, X_p, \phi_X, \ldots)$ and $(X_0, X_c, X_s, \ldots)$ are equivalent from the mathematical point of view. From the point of view of parameter estimation the second kind of parametrization is preferable, because the parameters $(X_0, X_c, X_s, \ldots)$ occur linearly in the perturbing acceleration, thus in the equations of motion.

In [110] the empirically established parameters of the model (3.162, 3.163) were analyzed and led to a simpler, yet very accurate a priori model [40].

Program SATORB allows it to use the simplified radiation pressure model based on eqn. (3.142) or the full empirical model (3.162, 3.163). It is impossible to address all the subtleties of the more advanced tool (3.163). We confine ourselves to illustrate the effect of a constant y-bias of $Y_0 = 1 \cdot 10^{-9}$ m/s^2. For an in-depth analysis we refer to [110]. The result of two simulations covering the year 2000 is contained in Figure 3.38. The results are given for two orbits which only differ by the initial right ascension of the ascending node. In the first case the ascending node coincides with the vernal equinox, $\Omega = 0°$, in the second case the node is given by $\Omega = 180°$. The perturbations of the semi-major axis a are truly remarkable. Annual oscillations of about 55 m and 80 m are observed. When focussing on a short time interval of a few days, one gets the impression of a secular increase or decrease of the semi-major axis a. Such perturbations are similar to those due to atmospheric drag (except that the semi-major axis may also grow due to the y-bias).

The perturbation may be understood when considering the perturbation of a circular orbit for $\beta_\odot = 90°$, i.e., for the case when the Sun is in the zenith of the orbital plane. Figure 3.36 helps understanding that in this case the y-bias is equivalent to a constant along-track acceleration (the orbit lies in the terminator plane for $\beta_\odot = 90°$). The maximum elevation $|\beta_\odot|$ of the Sun over the orbital plane is limited by the inclination angle of the orbital plane w.r.t. the ecliptic, i.e., by $\beta_\odot \leq \tilde{i}$. With the special selection of the right ascension of the ascending node we have $\tilde{i} \approx 55° - 23.5° = 31.5°$ for $\Omega \approx 0°$ and $\tilde{i} \approx 55° + 23.5° = 78.5°$ for $\Omega \approx 180°$. These inclination angles explain the differences of the amplitudes in the perturbations of the semi-major axis a in Figure 3.38.

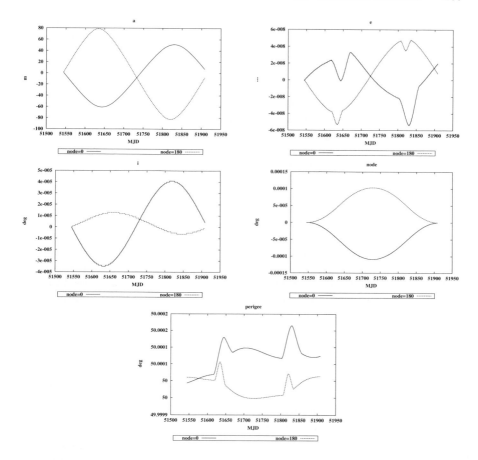

Fig. 3.38. Mean orbital elements of two GPS-like satellites ($\Omega \approx 0°$, $180°$) under the influence of a y-bias of $Y_0 = 10^{-9}$ m/s^2

If the Sun lies in the orbital plane, the perturbation term changes sign twice per revolution, and there is no net effect in a over one revolution. As the eclipse "seasons" are easily recognized in Figure 3.38, we see that actually $\dot{a} \approx 0$ during these time periods.

The periodic perturbations in the eccentricity e are small (but they grow with increasing eccentricity). They correspond to drag-like perturbations. The eclipse seasons are clearly visible in the perturbations of the eccentricity. They take place roughly at the same times within the year in both examples, but the duration is different. This difference is a consequence of the facts that the nodes are separated by $180°$ and that (as mentioned) the inclinations w.r.t. the ecliptic differ significantly.

The perturbations of the elements i and Ω due to a constant y-bias are very small and the net effect over one year is almost zero. The perturbations in the argument of perigee ω are small, as well. It should be mentioned, however, that the annual amplitudes grow when the eccentricity e of the orbit is increased.

Hugentobler [57] discussed the possibility to estimate the resonance terms J_{22}, J_{32}, J_{44}, etc. using precise GPS orbits. When such analyses are based on short time spans of observations (of a few days), it is rather difficult to separate the perturbations due to resonance and due to a constant y-bias because both effects generate drifts of very long periods in the semi-major axis.

3.7.6 Albedo of the Earth

For LEOs the radiation pressure due to the sunlight reflected or re-emitted from the Earth's surface must be taken into account. This particular kind of radiation pressure usually is referred to as *albedo radiation pressure*. Depending heavily on the actual distance of the satellite from the Earth's surface, the effect is considerably smaller than the direct radiation pressure. For LEOs it may be as much as $\sim 25\%$.

Accurate modelling of albedo radiation pressure is rather difficult. It has to be done by superposition of the accelerations due to all surface elements of the Earth visible from the satellite. One has to distinguish between radiation which is first absorbed, then re-emitted, and radiation which is reflected. The superposition of the accelerations corresponding to the absorbed/re-emitted part, is probably well taken into account by a radial component R'.

When taking into account the reflected part of the radiation, one has to distinguish between specular and diffuse reflection. For diffuse reflection it may be assumed that a surface element $d\sigma$ reflects solar radiation according to Lambert's cosine-law (named after Johann Heinrich Lambert (1728–1777)), i.e., the surface element has the same apparent brightness independently of the angle relative to the surface normal. Assuming that the portion a_s of the light actually reflected by the surface element is known relative to the incoming radiation, the acceleration due to the albedo radiation stemming from a particular surface element is easily calculated. (a_s is referred to as the albedo of the surface element). The total effect just is the sum over all surface elements of the illuminated part of the Earth. A priori models accounting for albedo radiation are of limited value, because the differences in reflectivity (albedo) of the surface elements are strongly variable with the geographical region (land, sea) and with time (e.g., due to clouds).

When estimating empirical model parameters (see Chapter I- 8), such small effects are probably absorbed by these parameters. The attempt was made to estimate albedo-scaling parameters using precise orbits of GPS satellites in [12]. No clear signals related to the albedo radiation could be found in this analysis – but it must be admitted that GPS satellites are not the ideal objects for albedo studies. Albedo radiation pressure definitely has to be considered when analyzing LEO-orbits.

3.8 Comparison of Perturbations Acting on Artificial Earth Satellites

In this chapter the discussion of perturbations was confined to the stationary part of the Earth's gravitational field and to the two prominent nongravitational perturbations, namely atmospheric drag and radiation pressure. In this summary we include in addition the perturbations due to the gravitational attraction by the Moon and the Sun and the perturbations due to the Earth tides. Third-body perturbations will be treated in Chapter 4 and the tides were discussed in Chapter 2.

Tables 3.7 and 3.8 give an overview of the order of magnitude of the perturbing accelerations and of their impact on an orbit after one day. The integration was performed with the program SATORB (see Chapter 7 of Part III). Table 3.7 deals with a LEO at a height of 500 km above the Earth's surface (semimajor axis $a \approx 6878$ km). Table 3.8 deals with a GPS-like satellite, whose inclination and eccentricity were set to $i = 55°$ and $e = 0.001$. The two tables thus refer to orbits with small eccentricities. The other three elements (Ω, ω, and T_0) were all set to zero. The initial epoch was January 1, 2001, 0^h UT.

The tables can only reflect orders of magnitude, but not subtleties. We have, e.g., seen that for short time-spans the perturbation due to the y-bias depends heavily on the initial epoch. Also, the inclination plays an important role for most perturbations included in Tables 3.7 and 3.8. Dedicated investigations are required for special orbit types, e.g., for inclinations $i \approx 0°$ and $i \approx 90°$, or for resonance studies.

The tables contain the perturbation type in the first, the absolute values of the perturbing forces in the second column. The values in the second column were actually computed as (plain) mean values over the time interval of the integration of one day. In addition the tables contain the values of the effect ("error") of the term after one day in the radial, along-track, and out-of-plane directions. The values are computed as the difference of the perturbed and the unperturbed orbit (referring to the initial epoch) after one day.

This way of comparing orbits is somewhat problematic, because most perturbations also change the mean motion of a satellite, which in turn implies a linearly growing difference in the mean anomaly. This is why the orbit error after one day is given separately for the radial, along-track, and out-of-plane direction.

There is an alternative procedure to assess the effect of individual perturbations: an ephemeris including the perturbation term is generated. Then, the generated tabular positions are considered as pseudo-observations in an orbit determination process, where the term under investigation is set to zero. This elaborate procedure gives a more realistic picture of the relevance of particular perturbations. The major disadvantage has to be seen in the dependence of the orbit effects on the arc-length. Tables of the type 3.7 and 3.8 are excellent tools for a first analysis of a perturbation problem, but they should be complemented by procedures based on parameter estimation.

Both Tables 3.7 and 3.8 show the perturbing accelerations and the corresponding impact on the orbit in decreasing order. The term C_{20} is the dominant perturbation for both, LEOs and satellites in the height of the GPS. A comparison of the two tables confirms that the influence of C_{20} decreases dramatically with the semi-major axis. Note, that the along-track error reflects (more or less) the modified Kepler law of the mean motion, and that the out-of-plane component is due to the regression of the nodes. The analytical treatment of this term using first-order perturbation theory, lets us expect a reduction of the influence of the term C_{20} by a factor of about $(a_{\mathrm{GPS}}/a_{\mathrm{LEO}})^2 \approx 15$ in the mean anomaly, and a reduction of $(a_{\mathrm{GPS}}/a_{\mathrm{LEO}})^{3.5} \approx 112$ in the node – but this effect has to be scaled by $a_{\mathrm{GPS}}/a_{\mathrm{LEO}}$ in order to get the linear effect in the out-of-plane direction. These orders of magnitude are confirmed by Tables 3.7 and 3.8. Atmospheric drag is the next important perturbation after the C_{20}-term for LEOs, whereas the gravitational attractions due to the Moon and the Sun are in this position at GPS heights. The atmospheric drag was computed with the values $A/m = 0.02$ and $C = 2$, typical values for "bulky" satellites like, e.g., navigation satellites. But even when reducing this

Table 3.7. Accelerations acting on LEOs

| Perturbation | Acceleration | Orbit Error after one Day | | |
| | | Radial | Along Track | Out of Plane |
	[m/s²]	[m]	[m]	[m]
$\frac{1}{r^2}$-Term	8.42	"∞"	"∞"	"∞"
Oblateness	$1.5 \cdot 10^{-2}$	60000	400000	900000
Atmospheric Drag	$7.9 \cdot 10^{-7}$	150	8900	1.5
Higher Terms of the Earth's Grav. Field	$2.5 \cdot 10^{-4}$	550	3400	820
Lunar Attraction	$5.4 \cdot 10^{-6}$	2	45	2
Solar Attraction	$5.0 \cdot 10^{-7}$	1	38	15
Direct Rad. Pressure	$9.7 \cdot 10^{-8}$	10	24	0
Solid Earth Tides	$1.1 \cdot 10^{-7}$	0.2	13	1
y-bias	$1.0 \cdot 10^{-9}$	0.1	4.7	0.0

value by a factor of 20 (to obtain the typical values for cannonball satellites), atmospheric drag still is the dominant perturbation after the effects of the oblateness term.

The higher-order terms of the Earth's gravitational potential are the third-important perturbation source for LEOs. LEO orbits are therefore very well suited for the determination of the gravitational field, provided the influence of drag can be eliminated (using accelerometers and thrusters) or greatly reduced (cannonball satellites). Today's knowledge of these terms is mainly based on the latter principle. The new gravity missions (GRACE and GOCE) will be based on the former principle.

Table 3.8. Accelerations acting on GPS satellites

Perturbation	Acceleration	Orbit Error after one Day		
		Radial	Along Track	Out of Plane
	$[\,\mathrm{m/s^2}\,]$	$[\,\mathrm{m}\,]$	$[\,\mathrm{m}\,]$	$[\,\mathrm{m}\,]$
$\frac{1}{r^2}$-Term	0.57	"∞"	"∞"	"∞"
Oblateness	$5.1 \cdot 10^{-5}$	2750	32000	15000
Lunar Attraction	$4.5 \cdot 10^{-6}$	400	1800	30
Solar Attraction	$2 \cdot 10^{-6}$	200	1200	400
Higher Terms of the Earth's Grav. Field	$4.2 \cdot 10^{-7}$	60	440	10
Direct Rad. Pressure	$9.7 \cdot 10^{-8}$	75	180	5
y-bias	$1.0 \cdot 10^{-9}$	0.9	8.1	0.3
Solid Earth Tides	$5.0 \cdot 10^{-9}$	0.0	0.4	0.0
Atmospheric Drag	—	—	—	—

It is interesting to note that direct radiation pressure is much more relevant for high-orbiting than for low-orbiting satellites – despite the fact that the perturbing accelerations are (almost) identical. The perturbation in the eccentricity e is the dominating effect in the case of radiation pressure. The expected ratio for the drifts in e were found to be $\dot{e}_{\mathrm{GPS}}/\dot{e}_{\mathrm{LEO}} = \sqrt{a_{\mathrm{GPS}}/a_{\mathrm{LEO}}}$. In order to obtain the ratio of the effect in meters, we have to multiply the ratio of drifts with the ratio of the two semi-major axes. The estimated growth factor from a LEO satellite to a GPS satellite (with identical satellite properties) therefore is $\sqrt{a_{\mathrm{GPS}}^3/a_{\mathrm{LEO}}^3} \approx 7.6$, which is close to the value emerging from the above tables.

Solid Earth tides and the y-bias are the least significant perturbations for both types of satellites, but both error sources need to be taken into account when aiming at high-accuracy orbits. LEOs are of course much more sensitive to Earth tides than high-orbiting satellites. The order-of-magnitude also shows, however, that satellite orbits are of limited use to develop detailed tidal models.

Table 3.8 tells that orbit determination is "in principle" a simple task for high-orbiting satellites. The influence of the high-order terms of the Earth's gravitational field is greatly attenuated. If there are no resonance problems (such as those encountered for GPS satellites), orbit determination is close to trivial for this class of satellites: only the initial state vector and, depending on the arc-length envisaged, a few parameters related to radiation pressure have to be determined.

Considering the fact that the current generation of GPS satellites allows it to the IGS to establish the polar wobble components with greatest, the UT1–UTC parameter with fair accuracy, it is not difficult to predict that a system of drag-free GPS-like satellites would allow the determination of all these parameters with unprecedented time resolution and an accuracy comparable to the accuracy achieved by VLBI.

4. Evolution of the Planetary System

Three key issues, namely

- the development of the outer planetary system, in section 4.1,
- the development of the inner planetary system, in section 4.2, and
- the orbits of minor planets, in section 4.3,

are addressed in this chapter. Extensive use is made of the concept of mean elements, which was introduced in section I-4.3. Program PLASYS, a central tool in this chapter, is documented in Chapter 10 of Part III.

The three topics are special cases of the N-body problem. The equations of motion (I-3.13) and (I-3.18) related to the first two topics and the equations (I-3.21) governing the third topic were developed in section I-3.2.

Table 4.1 shows that the total mass of all planets amounts only to about 0.15% of the planetary system's total mass. The inclinations of the planets' orbital planes w.r.t. the ecliptic of epoch $J2000.0$ are small, those of Pluto and Mercury being somewhat larger. With the exception of the same two bodies, all planets revolve around the Sun in orbits of small eccentricities. It should be pointed out, however, that the numerical values, taken from [107], are strictly valid only for the epoch $J2000.0$. It will be interesting to see how representative the values for the semi-major axes, the eccentricities, and the inclinations are over time periods of millions of years.

It makes sense to distinguish the outer planetary system, consisting of the planets Jupiter, Saturn, Uranus, Neptune, and Pluto, from the inner system consisting of Mercury, Venus, Earth, and Mars. The outer planetary system contains in essence the entire planetary mass. It also contains practically the entire energy and angular momentum of the planetary system's point mass model. As the revolution periods in the outer planetary system (ranging from about 12 years for Jupiter to about 250 years for Pluto) are one to two orders of magnitude larger than those in the inner system (ranging from about three months for Mercury to 1.9 years for Mars), it makes sense to study the outer planetary system separately from the inner system.

This separation is also indicated from the economical point of view, because the stepsize in numerical integration is essentially governed by the revolution

Table 4.1. Properties of the planetary system

Planet	Axis a [AU]	Ecc. e	Period P [Years]	Mass m^{-1} [$m_\odot^{-1}$]	Incl. i [deg]
Mercury	0.39	0.206	0.24	6023600.00	7.0
Venus	0.72	0.007	0.62	408523.50	3.4
Earth	1.00	0.017	1.00	328900.55	0.0
Mars	1.52	0.093	1.88	3098710.00	1.8
Jupiter	5.20	0.048	11.86	1047.35	1.3
Saturn	9.54	0.056	29.42	3498.00	2.5
Uranus	19.19	0.046	83.75	22960.00	0.8
Neptune	30.06	0.009	163.72	19314.00	1.8
Pluto	39.53	0.249	248.02	130000000.00	17.1

period of the innermost planet (and, to some extent, by its eccentricity). This is why the outer planetary system is studied in section 4.1 separately from the development of the inner system in section 4.2, where the development of the orbital elements of the Earth (actually the barycenter of the Earth-Moon system) is studied in particular.

Table 4.1 does not contain all members of the planetary system. Comets and minor planets have to be dealt with, as well. Their total mass is, however, negligible compared to all planetary masses (except perhaps Pluto's). It is therefore fair to assume, that these celestial bodies, despite their big number, have no significant influence on the key properties (energy, angular momentum, mass) of the total system. As the orbits of many minor planets are pretty well known today, it is, however, most exciting to study the development of these "massless" test particles in the gravitational field of the major bodies in the planetary system. Section 4.3 is devoted to this topic.

4.1 Development of the Outer Planetary System

The outer planetary system represents a six-body problem with the Sun as central point mass, and the planets Jupiter, Saturn, Uranus, Neptune, and Pluto. Subsequently, the outer system will be numerically integrated several times with program PLASYS (Chapter 10 of Part III), using the input options of Figures 10.3 and 10.4 (or slight variations thereof). It is a straight forward procedure to generate tables (files) of osculating or mean elements for all or for a selection of the celestial bodies involved. It is less trivial to interpret and visualize the almost frightening amount of data. In this section we proceed in three steps:

- The orbital elements of Jupiter, derived from an integration of the outer planetary system, are considered over relatively short time intervals in the first step. The development of the osculating elements over a time interval of a few revolutions is studied before discussing that of the mean elements over a time period 2000 years.

• The outer planetary system is then integrated over a time span of two million years (± one million years from present). The solution properties and the results of the integration (in graphical form) are discussed. Some of the key characteristics are made plausible as consequences of the conservation theorems of the N-body problem.

• In the third step the results of the integration over two million years are analyzed using the methods of spectral analysis. The associated theory and the program used for this purpose may be found in Chapter 11 of Part III.

4.1.1 The Orbit of Jupiter Over Short Time Spans

The osculating elements of Jupiter in Figure 4.1 stem from a numerical integration of the six-body problem including the planets Jupiter, Saturn, Uranus, Neptune, Pluto, and the Sun. The integration was performed using the multistep method of order 14 with a constant stepsize of $h = 30$ days. The initial conditions on January 1, 2000 were taken from the JPL's planetary ephemerides DE200 [111].

The semi-major axis a (Figure 4.1 (top, left)) shows periodic perturbation, with a peak roughly every 20 years. This period is in turn modulated by a beat period of about 60 years.

The periods of approximately 20 and 60 years may be easily explained: The perturbing effects exerted by Saturn on Jupiter are maximum when the distance between the two planets becomes minimum, which occurs when the ecliptical longitudes of the two planets coincide. Such encounters, called conjunctions, between Jupiter and Saturn occur regularly with a period of

$$P_{\mathfrak{4}\mathfrak{h}} \stackrel{\text{def}}{=} \frac{2\pi}{n_{\mathfrak{4}} - n_{\mathfrak{h}}} = \frac{P_{\mathfrak{h}} P_{\mathfrak{4}}}{P_{\mathfrak{h}} - P_{\mathfrak{4}}} \approx 20 \text{ years } (n_{\mathfrak{4}}, n_{\mathfrak{h}}, P_{\mathfrak{4}} \text{ and } P_{\mathfrak{h}} \text{ are the mean side-}$$

real motions and revolution periods of the planets Jupiter and Saturn, respectively; approximate revolution periods $P_{\mathfrak{4}}$ and $P_{\mathfrak{h}}$ for the two planets may be taken from Table 4.1). The mean interval of time between successive conjunctions of a pair of planets is called their *synodic period*. The synodic period of the planets Jupiter and Saturn is thus approximately 20 years. The period of approximately 60 years is a consequence of the near-commensurability of the revolution periods of Jupiter and Saturn ($5 P_{\mathfrak{4}} \approx 2 P_{\mathfrak{h}}$ or $5 \cdot 11.86 \approx 59$, $2 \cdot 29.46 \approx 59$). This near-commensurability implies that essentially the same relative geometry of the three bodies Sun, Jupiter, and Saturn repeats itself every sixty years.

The osculating eccentricity e (Figure 4.1 (top, right)) is dominated by the perturbation of 60 years period. Apart from signals with periods of 20 and 60 years, the osculating ecliptical longitude of the node Ω shows a linear growth

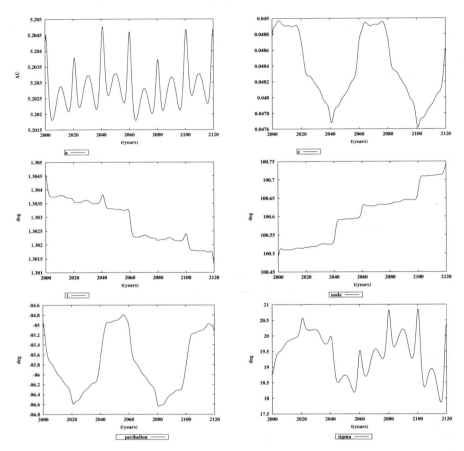

Fig. 4.1. Short-period perturbations of the orbital elements of Jupiter over a time interval of 120 years

in time, whereas the inclination i slowly decreases (middle row of Figures 4.1). Note, that on this short time scale (only about ten Jupiter and four Saturn periods) secular cannot be separated from long-period perturbations. The argument of perihelion ω is dominated by a signal of a period of about 60 years. The mean anomaly σ_0 at epoch t_0 is somewhat irregular.

In summary we may state that Jupiter's osculating elements over short periods of time are governed by short-period perturbations. Short-period perturbations have periods close to the synodical revolution periods of the planets, of fractions and of (small integer) multiples of these periods.

Graphs of the type of Figure 4.1 might be produced for each of the planets involved in the integration. It would be in particular interesting to study the correlations and particularities in the osculating elements for different planets. This discussion will be continued in the subsequent sections.

Using the definition of mean elements according to eqn. (I- 4.71) one may establish the development of Jupiter's mean orbital elements over 2000 years. The result may be inspected in Figure 4.2, where the osculating elements of Jupiter were averaged over time intervals of five revolutions

$$\Delta t(t) \stackrel{\text{def}}{=} 5\, P_{\text{2\!+}}(t) \; , \tag{4.1}$$

$P_{\text{2\!+}}(t)$ being the (osculating) revolution period of Jupiter at time t. Note, that only by averaging over five (or an entire multiple of five) sidereal revolutions (corresponding to three synodical revolutions of the pair Jupiter-Saturn), (almost) all short period effects can be eliminated.

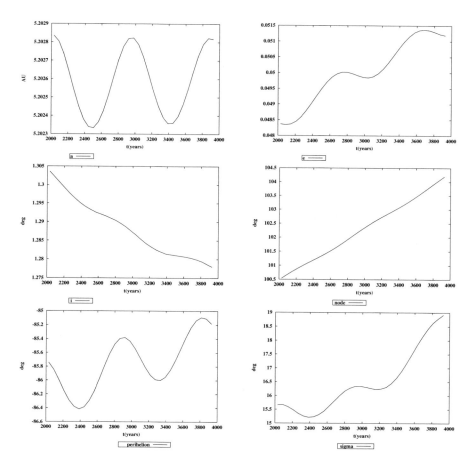

Fig. 4.2. Mean orbital elements of Jupiter over a time interval of 2000 years

Figure 4.2 clearly shows perturbations with a period of about 900 years in the semi-major axis a and in the eccentricity e. This period may be explained by the fact that the linear combination of mean motions

$$\tilde{n} = 5\, n_\saturn - 2\, n_\jupiter \approx 5 \cdot 120.45'' - 2 \cdot 299.13'' \approx 3.99''/\text{day} \qquad (4.2)$$

of Jupiter and Saturn almost vanishes. The period associated with the mean motion $\tilde{n} \approx 3.99''/\text{day}$ gives rise to a period of

$$\tilde{P} \approx 890 \text{ years} . \qquad (4.3)$$

The periodic perturbation in a in turn causes a perturbation in the longitude of the planet. This perturbation is called the *long period (or great) inequality*. It was already observed in the 18th century in the longitudes of Jupiter and Saturn and correctly explained for the first time by Laplace.

4.1.2 The Integration over Two Million Years in Overview

The general output file in Figure 4.3 summarizes the essential facts of the integration of the outer planetary system over one million years. It lists the planets involved, characterizes the averaging interval for the mean elements, and the sampling rates used to generate the files. The integration interval and the parameters of the numerical integration may be found in this general output file together with the values for the first integrals, namely the total energy and the three components of the total angular momentum.

The program output also includes the matrix of the planets' sidereal revolution periods (diagonal elements of the matrix) and the synodic revolution periods, i.e., the time interval between two subsequent conjunctions of two planets as seen from the Sun. The matrix will be required to interpret the short-period perturbations.

The power of the multistep procedure is underlined by the fact that only about 24 evaluations of the differential equations' right hand sides had to be performed per year. No mainframe computers are required nowadays to integrate the outer planetary system over millions of years. Even ten to hundredfold longer time spans may be considered as realistic integration periods on PCs. The main problem is no longer the processing time, but the organization of the results. The first attempts to numerically integrate the outer planetary system required the full computing power of the best mainframe computers available in the second half of the 20st century.

One million years sounds impressive as an absolute number, but that time interval is still a small fraction (about 0.02 %) of the estimated age of 4.5 billion years of the planetary system. Figure 4.4 indicates that the system is quite stable over the time interval considered. All planets except Pluto revolve within annuli with small inclinations w.r.t. the invariable plane (therefore

```
GENERATION OF PLANETARY SYSTEM                          22-JUN-01  07:59
*********************************************************************
NUMBER OF PLANETS:  5
     NAME      #REV(MEAN ELE)      SAMPLING FOR PLANETS-FILE
-------------------------------------------------------------
   JUPITER           5                  1000
   SATURN            2                  1000
   URANUS            1                  1000
   NEPTUNE           1                  1000
   PLUTO             1                  1000
-------------------------------------------------------------
INITIAL EPOCH:       2000  1  1.0 YEAR MONTH DAY
LENGTH (YEARS):         1000000.0 YEARS
INTEGRATION            FORWARDS
INITIAL VALUES FROM JPL DE200
NEWTON-EULER EQUATIONS USED
TABULAR INTERVAL:       50.0 DAYS
SAMPLING (PRINT):        1000
INITIAL STEPSIZE:       30.0 DAYS
INTEGRATION METHOD: MULTISTEP
INTEGRATION STEP:    FIXED
ORDER OF METHOD:          14
# ITERATIONS/STEP:         1

SIDEREAL AND SYNODIC REVOLUTION PERIODS (Means over 1 Myears)
-------------------------------------------------------------
            JUPITER    SATURN    URANUS    NEPTUNE    PLUTO
-------------------------------------------------------------
JUPITER |   11.861    19.823    13.805    12.779    12.457
SATURN  |   19.823    29.532    45.471    35.959    33.527
URANUS  |   13.805    45.471    84.247   171.907   127.632
NEPTUNE |   12.779    35.959   171.907   165.212   495.557
PLUTO   |   12.457    33.527   127.632   495.557   247.839
-------------------------------------------------------------

TOTAL ENERGY OF SYSTEM:-0.32180611D-07 M(SUN)*A.U.**2/DAY**2
TOTAL ANGULAR MOMENTUM COMP 1: 0.15953954D-05 M(SUN)*A.U.**2/DAY
                       COMP 2: 0.50749598D-06          "
                       COMP 3: 0.60717908D-04          "

CPU                   12.413 MIN
FCT CALLS          24350105.
```

Fig. 4.3. General output of integration of outer planetary system over one million years

also w.r.t. the plane of the ecliptic $J2000.0$). The diameters of the planetary annuli are about $2\,a$, their widths about $2\,a\,e$. Figure 4.4 indicates that, with the exception of Pluto, all eccentricities are and remain small over the time interval considered.

Figure 4.4 should show that the perihelion of Pluto lies within Neptune's orbit. One might suspect that strong perturbations (eventually even collisions) should occasionally occur if the time-span is long enough. Seemingly, this is not the case. Thanks to the near-commensurability $P_{♇} : P_{♆} = 3 : 2$ of Neptune's and Pluto's revolution periods, Pluto always manages to pass the perihelion at times when Neptune is far away. The close encounters between the planets always occur near Pluto's aphelion. Also, the three-dimensional illustration in Figure 4.4 shows, that during the entire time-span considered,

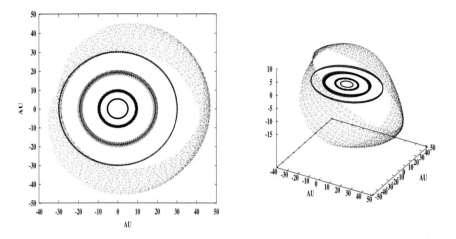

Fig. 4.4. Development of outer planetary system over one million years

the perihelion of Pluto is high above the plane of the ecliptic (ω oscillates about $90°$ with a period of about four million years and an amplitude of about $15°$).

In Chapter I-3 the invariable or Laplacian plane was defined as the plane perpendicular to the vector $\boldsymbol{h}$ of total angular momentum of the planetary system (I-3.40). It would be natural to relate planetary orbits over long time periods to this invariable plane. Despite this insight, and following astronomical tradition, we use the ecliptical system $J2000.0$ as the reference system for the integration. It is, however, easy to transform the results form the ecliptical to the "invariable system", using the (initial) value for the angular momentum vector $\boldsymbol{h}$, provided in the general output file (Figure 4.3). For the integration of the outer planetary system over one million years one finds

$$
\boldsymbol{h} \approx \begin{pmatrix} 0.160 \cdot 10^{-5} \\ 0.507 \cdot 10^{-6} \\ 0.607 \cdot 10^{-4} \end{pmatrix} = h \begin{pmatrix} \cos(\tilde{\Omega} - \frac{\pi}{2}) \sin \tilde{i} \\ \sin(\tilde{\Omega} - \frac{\pi}{2}) \sin \tilde{i} \\ \cos \tilde{i} \end{pmatrix} = h \begin{pmatrix} \sin \tilde{\Omega} \sin \tilde{i} \\ -\cos \tilde{\Omega} \sin \tilde{i} \\ \cos \tilde{i} \end{pmatrix} ,
$$

(4.4)

where

$\tilde{i}$ is the angle between the ecliptic pole $J2000.0$ and the pole of the invariable plane, and

$\tilde{\Omega}$ is the ecliptical longitude of the intersection of the invariable plane with the ecliptic.

With the above equations for $\tilde{i}$ and $\tilde{\Omega}$ the (constant) elements of the invariable plane w.r.t. the system $J2000.0$ are computed as

$$\tilde{i} \approx 1.58° \quad \text{and} \quad \tilde{\Omega} \approx 107.6° . \tag{4.5}$$

How reliable are the results stemming from the numerical integration? The integration was performed with the multistep procedure of order 14 with a stepsize of $h = 30$ days. According to the rule of thumb (I- 7.211) this stepsize makes sense, provided the integration error is governed by the error of the innermost planet Jupiter with an eccentricity of (at maximum) about $e = 0.05$. We will see below that this order of magnitude for the eccentricity of Jupiter's orbit is preserved throughout the entire time interval considered.

Figure 4.5 supports the conclusion that the accuracy of the integration is sufficient for our purpose: The relative errors in all components of the total angular momentum vector and in the total energy are very small indeed. (The units in Figure 4.5 are parts per billion (ppb); this implies, e.g., that the relative error in the energy is of the order of 10^{-13}, which seems satisfactory). Figure 4.5 shows that the classical integrals "energy" and "total angular momentum" are almost constant (as they should be). A few tests with other stepsizes, other integration orders, and other integration methods show, that the integration errors are well below the level of perturbations analyzed subsequently. The development of the polar moment of inertia (I- 3.48) which was introduced as a preserved quantity in a statistical sense shows quite a

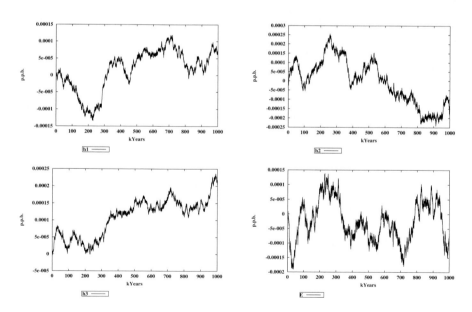

Fig. 4.5. Relative errors in components of angular momentum and in energy in parts per billion

different behavior. The polar moment of inertia oscillates between the limits

$$I(t) = 0.115 \pm 0.007 \, m_\odot \, \mathrm{AU}^2 \, , \tag{4.6}$$

where the amplitudes show periodic variations of a period of about 55000 years, which will be recognized as the precession period of Saturn's perihelion w.r.t. that of Jupiter. It is interesting to inspect the variations of the polar moment of inertia over shorter time periods, e.g., of thousand years by performing a dedicated integration over such a short time period (and storing the invariants at a high rate). One virtually sees all periods governing this six-body problem in the polar moment of inertia!

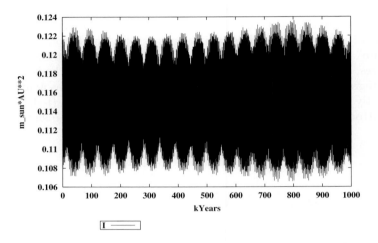

Fig. 4.6. Development of polar moment of inertia $I(t)$ over one million years

Let us now focus on the orbital elements of the five outer planets. Figure 4.7 shows the development of the semi-major axes of Jupiter, Saturn, Uranus, and Neptune, Figure 4.8 that of Pluto. The perturbations in Pluto's semi-major axis exceed those of the other four planets by one to two orders of magnitude and show a prominent period of about 19400 years. The perturbations are mainly due to Neptune. In view of the fact, that Pluto does not contribute much to the system of outer planets, we do not further investigate Pluto's orbit. For more information concerning this topic we refer to [94] and [95] and continue discussing the semi-major axes of the four planets in Figure 4.7.

We clearly see correlations between developments of the semi-major axes of Jupiter and Saturn on one hand and those between Uranus and Neptune on the other hand. The perturbations of Jupiter and Saturn basically show

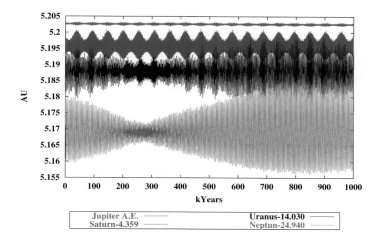

Fig. 4.7. Semi-major axes of the outer planets over one million years

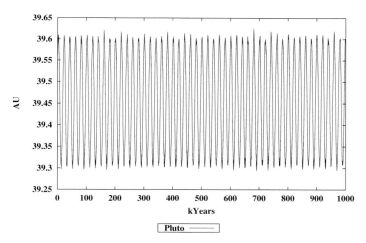

Fig. 4.8. Pluto's semi-major axis over one million years

a "high-frequency" signal (actually, the signal with the period of about 900 years mentioned in the previous section). The amplitude of the perturbations in a is modulated by a signal with a period of about 54000 years. Qualitatively a similar behavior results for the semi-major axes of Uranus and Neptune, but the modulation period exceeds one million years. Obviously, an integration period of one million years is no luxury to reveal the essential orbital characteristics of the latter two planets.

Figure 4.9 shows the detailed evolution of Jupiter's and Saturn's semi-major axis over the first 10000 years (starting January 1, 2000) of the integration. A very net correlation (with correlation coefficient of -1) is observed. The amplitudes of Saturn's perturbations in a exceed those of Jupiter by about one order of magnitude.

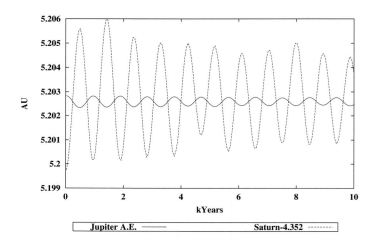

Fig. 4.9. Semi-major axes of Jupiter and Saturn over 10000 years

The characteristics revealed by Figure 4.9 may be explained qualitatively with the help of the law of energy conservation (I- 3.47). Taking into account only the Sun, Jupiter and Saturn, eqn. (I- 3.47) is reduced to:

$$\frac{1}{2} m_{\odot} \dot{\boldsymbol{x}}_{\odot}^2 + \frac{1}{2} m_{\4} \dot{\boldsymbol{x}}_{\4}^2 + \frac{1}{2} m_{\hbar} \dot{\boldsymbol{x}}_{\hbar}^2 - k^2 \frac{m_{\odot} m_{\4}}{r_{\4}} - k^2 \frac{m_{\odot} m_{\hbar}}{r_{\hbar}} - k^2 \frac{m_{\4} m_{\hbar}}{|\boldsymbol{r}_{\4} - \boldsymbol{r}_{\hbar}|} = E$$

where the index $\odot$ stands for the Sun, $\4$ for Jupiter, and $\hbar$ for Saturn.

Neglecting all terms proportional to $m_{\4} m_{\hbar}$, using $m_{\odot} = 1$, and taking the transformation equations (I- 3.36, I- 3.37) between the inertial and heliocentric system into account, we obtain the simple relation

$$m_{⚃} \left\{ \frac{1}{2} \dot{r}_{⚃}^2 - \frac{k^2}{r_{⚃}} \right\} + m_{♄} \left\{ \frac{1}{2} \dot{r}_{♄}^2 - \frac{k^2}{r_{♄}} \right\} \approx E \ .$$

Using the fact that the osculating semi-major axes of Jupiter and Saturn are defined by the astronomical version (I-4.20) of the energy conservation theorem of the two-body problem, the above equation may be written as

$$- \frac{m_{⚃}}{2\, a_{⚃}} - \frac{m_{♄}}{2\, a_{♄}} \approx E \ . \tag{4.7}$$

This equation implies that changes in the semi-major axes of Jupiter and Saturn have to obey the following rule

$$\delta a_{♄} \approx - \frac{m_{⚃}}{m_{♄}} \frac{a_{♄}^2}{a_{⚃}^2} \delta a_{⚃} \approx -11.2\, \delta a_{⚃} \ , \tag{4.8}$$

which explains the ratio of the amplitudes seen in Figure 4.9 (not the period of 900 years, however).

Let us now inspect the development of the orbital planes. Figures 4.10 and 4.11 show that there are considerable variations in the inclinations and in the longitudes of the nodes (note, that in Figure 4.10 the value $\tilde{i}_{P} \stackrel{\text{def}}{=} i_{P} - 17°$ is drawn in order to fit all inclinations into one and the same figure). Again, a rather strong correlation between corresponding elements of Jupiter and Saturn is seen in Figures 4.10 and 4.11. It is in particular remarkable, that their nodes are "only" oscillating and not precessing in the ecliptical reference system. The perturbations in the inclinations and longitudes of the nodes of Jupiter and Saturn are periodic with a period of about 49000 years.

The development of the orbital planes becomes much clearer, if the orbital poles (unit vectors normal to the orbital planes) are projected onto the plane of the ecliptic. In Figure 4.12 we see this projection for Jupiter and Saturn (left) and for all five planets (right). Obviously, all orbital poles are precessing retrograde (seen from the ecliptic pole) around the pole of the invariable plane, which, according to eqn. (4.5) has the coordinates

$$\begin{pmatrix} x_1 \\ x_2 \end{pmatrix} = \begin{pmatrix} \sin \tilde{i} \, \sin \tilde{\Omega} \\ - \sin \tilde{i} \, \cos \tilde{\Omega} \end{pmatrix} = \begin{pmatrix} 0.026 \\ 0.008 \end{pmatrix} \ , \tag{4.9}$$

which in turn corresponds closely to the center of the precession cones in Figure 4.12. Basically, all the orbital planes of all planets precess around the pole of the invariable plane of the system. The variations of the inclinations w.r.t. the invariable plane are of course smaller than those w.r.t. the plane of the ecliptic. (These facts underline the statement that, for integrations over long time-spans, the invariable plane should serve as reference plane). If the orbital poles of Jupiter and Saturn were time-tagged, we would see that the projection of the pole of the invariable plane always lies between the two orbital poles and that the three projected poles almost lie on a straight

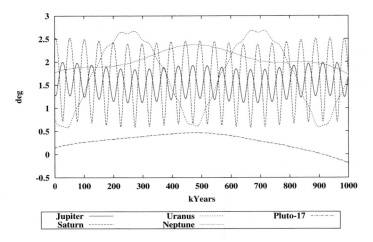

Fig. 4.10. Inclinations of the outer planets over one million years

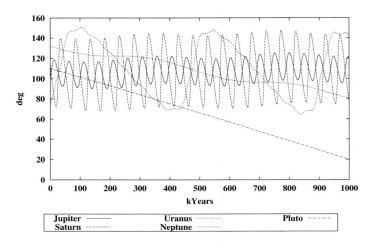

Fig. 4.11. Longitudes of the ascending nodes over one million years

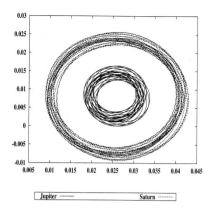 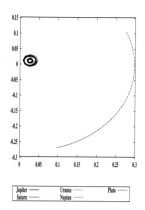

Fig. 4.12. Unit vectors of poles of the orbital planes of Jupiter and Saturn (left) and of all outer planets (right) over one million years projected onto the ecliptic

line. If the integration is performed including only the planets Jupiter and Saturn (see section I- 4.5.1, Figure I- 4.14), these empirical findings are almost perfectly true.

As in the case of the semi-major axes it is possible to explain this aspect of the motion of the orbital planes as a consequence of a conservation law, this time of the law of conservation of the total angular momentum (I- 3.40). If we include only the Sun, Jupiter and Saturn, in the definition (I- 3.40), we obtain with the same approximations as those leading to the rule (4.7) the relation

$$\boldsymbol{h} \stackrel{\text{def}}{=} m_\odot \, \boldsymbol{x}_\odot \times \dot{\boldsymbol{x}}_\odot + m_{\text{4}} \, \boldsymbol{x}_{\text{4}} \times \dot{\boldsymbol{x}}_{\text{4}} + m_\hbar \, \boldsymbol{x}_\hbar \times \dot{\boldsymbol{x}}_\hbar \approx m_{\text{4}} \, \boldsymbol{r}_{\text{4}} \times \dot{\boldsymbol{r}}_{\text{4}} + m_\hbar \, \boldsymbol{r}_\hbar \times \dot{\boldsymbol{r}}_\hbar \; . \tag{4.10}$$

In the same approximation the total angular momentum associated with the two-body motions Sun-Jupiter and Sun-Saturn may be written as

$$m_{\text{4}} \, \boldsymbol{r}_{\text{4}} \times \dot{\boldsymbol{r}}_{\text{4}} = \boldsymbol{h}_{\text{4}} \quad \text{and} \quad m_\hbar \, \boldsymbol{r}_\hbar \times \dot{\boldsymbol{r}}_\hbar = \boldsymbol{h}_\hbar \; , \tag{4.11}$$

which allows us to say that, when neglecting terms proportional to $m_{\text{4}} \, m_\hbar$, the total angular momentum $\boldsymbol{h}$ associated with the three-body motion Sun-Jupiter-Saturn simply is the sum of the angular momenta associated with the two-body motions Sun-Jupiter and Sun-Saturn, respectively:

$$\boldsymbol{h} \approx \boldsymbol{h}_{\text{4}} + \boldsymbol{h}_\hbar \; . \tag{4.12}$$

Equation (4.12) is a simplified version of eqn. (I- 4.85) leading to *Jacobi's theorem of the nodes*, which might be easily derived from the above equation as well.

Equation (4.12) says that (approximately) the vector $\boldsymbol{h}$, pointing to the pole of the invariable plane, and the orbit normals $\boldsymbol{h}_{⚄}$ and $\boldsymbol{h}_{♄}$ of Jupiter's and Saturn's orbit are coplanar. Figure 4.13 helps to reveal the relationship between the inclination angle $i_{⚄♄}$ between the orbital planes of Jupiter and Saturn, and the inclination angles $i_{⚄}$ and $i_{♄}$ of Jupiter's and Saturn's orbit w.r.t. the invariable plane. Assuming that $i_{⚄♄}$ is a small angle, we extract the following

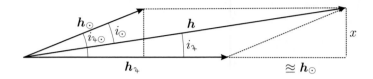

Fig. 4.13. Angular momentum geometry

relationships from Figure 4.13:

$$i_{⚄♄} = \frac{x}{h_{♄}}$$

$$i_{⚄} \approx \frac{x}{h_{⚄} + h_{♄}} = \frac{h_{♄}}{h_{⚄} + h_{♄}} \, i_{⚄♄} \tag{4.13}$$

$$i_{♄} = i_{⚄♄} - i_{⚄} \approx \frac{h_{⚄}}{h_{⚄} + h_{♄}} \, i_{⚄♄} \, ,$$

which is why the ratio of the inclination angles of Jupiter's and Saturn's orbits w.r.t. the invariable plane obeys the rule

$$\frac{i_{♄}}{i_{⚄}} \approx \frac{h_{⚄}}{h_{♄}} \approx \frac{m_{⚄}}{m_{♄}} \sqrt{\frac{a_{⚄}}{a_{♄}}} \approx 2.46 \, . \tag{4.14}$$

The latter approximation in rule (4.14) is based on the assumption of a circular orbit, where

$$|\boldsymbol{h}| = m \, |\boldsymbol{r}| \, |\dot{\boldsymbol{r}}| = m \, a \, a \, n = m \, k \, \sqrt{a} \, .$$

The rule of thumb (4.14) is pretty well confirmed by Figure 4.12. If the integration is performed as a pure three-body problem Sun-Jupiter-Saturn, the above rules are almost perfectly true (see also Figure I-4.14 in section I-4.5.1).

The orbital planes of all planets are precessing (in retrograde sense, seen from the ecliptic pole) around the pole of the invariable plane. In a fair approximation the orbital poles of Jupiter and Saturn and the pole of the invariable plane lie in one and the same plane, and the ratio of the inclination angles of the orbits of the two planets w.r.t. the invariable plane is constant. As the inclination angle between the orbits of Jupiter and Saturn does not show big perturbations, this implies that the relative geometry of the orbital planes of Jupiter and Saturn rotates in inertial space almost like a rigid body.

Figure 4.14 shows the perturbations of the eccentricities for the five outer planetary orbits (note that a value of $e_{P_0} \overset{\text{def}}{=} 0.14$ had to be subtracted from Pluto's eccentricity to fit all eccentricities into one figure). Figure 4.14 tells, e.g., that the eccentricities of all planets show periodic perturbations of considerable amplitudes. Saturn's orbit changes from almost circular (eccentricity of about $e_\hbar \approx 0.014$) to an orbit of considerable eccentricity ($e_\hbar \approx 0.085$). Even the orbit of Jupiter shows significant variations within the limits $0.03 < e_4 < 0.06$. By chance the eccentricities of Jupiter and Saturn assume more or less their mean value at present ($e_4 \approx e_\hbar \approx 0.048$, see Table 4.1). As in the case of the semi-major axes, a pronounced anti-correlation (correlation coefficient ≈ -1) is evident in the eccentricities of Jupiter and Saturn. Both planets change their eccentricities periodically with a period of about 54000 years. If Jupiter assumes its minimum eccentricity of about $e_4 \approx 0.03$, the orbit of Saturn has maximum eccentricity of $e_\hbar \approx 0.085$ and vice versa. The ratio of the amplitudes is $\frac{\Delta e_\hbar}{\Delta e_4} \approx 2.4$.

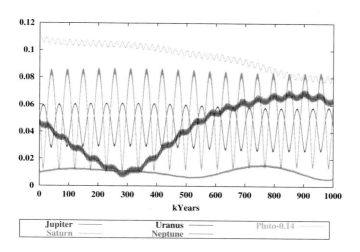

Fig. 4.14. Eccentricities of the outer planets over one million years

It is possible to explain the correlation between the perturbations of Jupiter's and Saturn's eccentricities by the conservation laws of the three-body problem. To that end we first interpret the development of the semi-latus rectum $p \overset{\text{def}}{=} a\,(1 - e^2)$ for Jupiter and Saturn in Figure 4.15. Afterwards, using the laws associated with the semi-major axis a and the semi-latus rectum p, it will be possible to derive a relationship for the eccentricity e, as well.

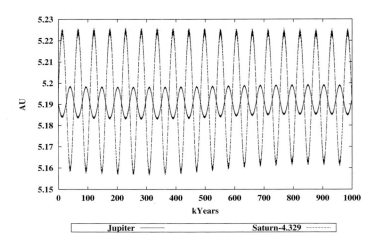

Fig. 4.15. Semi-latus rectum $p = a(1 - e^2)$ of Jupiter and Saturn over one million years

Figure 4.15 shows a net correlation (with coefficient -1) of Jupiter's and Saturn's semi-latus rectum. Despite the fact that the numerical values for the semi-major axes and the semi-latus rectum (due to the small numerical values for $e < 0.1$) are the same within 1 % for both planets, Jupiter and Saturn, the amplitudes of the variations are much larger in the semi-latus rectum p than in the semi-major axes a (compare Figures 4.9, 4.7 and 4.15). Also, the ratio $\Delta a_ħ / \Delta a_⁴$ of the amplitudes of the perturbations in the semi-major axes was established to be about -11, whereas we only see a factor 5 for the semi-latus rectum p in Figure 4.15.

This different behavior is a consequence of the fact that the variations in the axes are due to the conservation of the total energy, whereas the variations in the semi-latus rectum are due to the conservation (of the absolute value) of the total angular momentum vector: According to eqn. (I- 4.18), the semi-latus rectum of the conic section is uniquely a function of $|\boldsymbol{r} \times \dot{\boldsymbol{r}}|$.

Let us establish a theoretical relationship for the variations observed in Figure 4.15 by scalar multiplication of eqn. (4.12) with $\boldsymbol{h}$, the total angular momentum vector of the three-body problem:

$$\boldsymbol{h}_⁴ \cdot \boldsymbol{h} + \boldsymbol{h}_ħ \cdot \boldsymbol{h} = h\, h_⁴ \cos i_⁴ + h\, h_ħ \cos i_ħ = h^2 \; . \qquad (4.15)$$

By dividing both sides of the equation through the common factor $h = |\boldsymbol{h}|$ we obtain the scalar relationship

$$h_⁴ \cos i_⁴ + h_ħ \cos i_ħ \approx h_⁴ + h_ħ = h \; , \qquad (4.16)$$

where the approximation is only allowed for small inclinations i_{4} and $i_{\hbar}$ – but this is true for Jupiter and Saturn. The above relation allows us to draw the conclusion

$$\delta h_{4} = - \delta h_{\hbar} \, , \tag{4.17}$$

which essentially says that in the case of a hierarchical and almost coplanar system the conservation of the absolute value of the angular momentum only allows for an exchange of the two-body momenta between the "binaries" Sun-Jupiter and Sun-Saturn.

By virtue of eqn. (I- 4.18) we have

$$p = \frac{|\boldsymbol{r} \times \dot{\boldsymbol{r}}|}{\mu} \approx \frac{h^2}{m^2} \, , \tag{4.18}$$

from where we easily establish the relationship

$$\delta p_{\hbar} = - \frac{m_{4}}{m_{\hbar}} \sqrt{\frac{a_{\hbar}}{a_{4}}} \, \delta p_{4} \approx - 4.5 \, \delta p_{4} \tag{4.19}$$

for orbits with small eccentricities. This corresponds pretty well to the relationship in Figure 4.15. In view of the fact that the semi-latus rectum is given by $p = a \left(1 - e^2 \right)$ we have

$$\delta p = \delta a \left(1 - e^2 \right) - 2 \, a \, e \, \delta e \approx -2 \, a \, e \, \delta e \, , \tag{4.20}$$

where the latter approximation is justified by comparing Figures 4.7 and 4.15. Equation (4.20) (and using the fact that $e_{4} \approx e_{\hbar}$, which is pretty well met for the mean values of the eccentricities) allows it to provide an empirical explanation for the correlation of the perturbations in the eccentricities of Jupiter and Saturn in Figure 4.14:

$$\delta e_{\hbar} = \frac{a_{4}}{a_{\hbar}} \frac{\delta p_{\hbar}}{\delta p_{4}} \, \delta e_{4} = - \frac{m_{4}}{m_{\hbar}} \sqrt{\frac{a_{4}}{a_{\hbar}}} \, \delta e_{4} \approx - 2.46 \, \delta e_{4} \, , \tag{4.21}$$

which is numerically the same relationship as for the inclination i.

Figure 4.16 shows the development of the planets' ecliptical longitudes $\tilde{\omega} \stackrel{\text{def}}{=} \Omega + \omega$ of perihelia, reflecting the rotation of the perihelia in inertial space. With the exception of Pluto's perihelion the perihelia of the outer planets rotate prograde with different, approximately constant rates. For Saturn we observe a rotation period of its perihelion of about 45500 years, for Jupiter one of about 300000 years. This results in a synodic period of Saturn's perihelion w.r.t. that of Jupiter of about 54000 years. This is precisely the prominent period observed in the perturbations of Jupiter's and Saturn's semi-major axes and the eccentricities (see Figures 4.7 and 4.14). (Note: the jump in the perihelion longitude of Uranus is an artefact, occurring when the osculating orbit of Uranus was almost circular; it is a non-trivial affair to avoid jumps

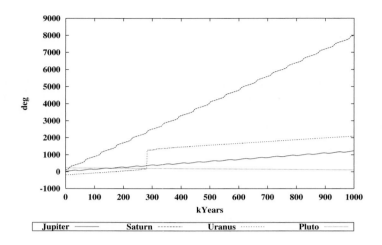

Fig. 4.16. Longitudes $\Omega + \omega$ of the perihelia of the outer planets

in the case of almost circular orbits, when the perihelion becomes almost indeterminate).

The development of the eccentricities (Figure 4.14) and of the longitudes of the perihelia (Figure 4.16) may be combined to illustrate the development of the Laplace vectors e as defined in Chapter I- 4, eqns. (I- 4.27, I- 4.32).

Figure 4.17 shows the time development of the projections of the Laplace vector onto the ecliptic for Jupiter and Saturn (left) and for all outer planets (right) in the conventional ecliptical coordinate system $J2000.0$. The figure shows that all Laplace vectors approximately revolve in circles with slowly varying radii around the origin.

Figures 4.7 to 4.17 give an impression of the time development of the orbital elements of the outer planets. The system is governed by the Sun and the two giant planets Jupiter and Saturn. Important peculiarities of Jupiter's and Saturn's orbit were explained using the invariants of the hierarchical three-body problem Sun-Jupiter-Saturn. There are of course many more fascinating aspects in the six-body problem of the outer planetary system. The discussion would differ from what we did here, however: In a very crude approximation we might now consider the dynamics of the system as defined by the Sun, Jupiter, and Saturn, and study the motion of the other planets in a predefined field. In a way, such studies are related to the studies of the orbits of minor planets (see section 4.3).

The numerical integration performed in this section might be viewed as a repetition of the work [29], [30] performed by Cohen, Hubbard, and Oesterwinter in the early 1970s. At that time numerical solutions over time periods

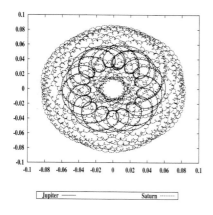 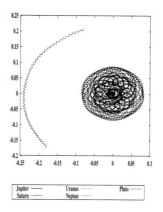

Fig. 4.17. Laplace vectors of Jupiter and Saturn (left) and of all outer planets (right) over one million years

of millions of years were very difficult to achieve. Many more numerical experiments were performed since that time. The LONGSTOP project [95], achieved by an international consortium of scientists, is a brilliant example for a large-scale numerical integration of the outer planetary system (covering over one hundred million years). These and other similar experiments are very well documented in [94].

With the advent of better and faster computers numerical experiments of the kind presented here become easier and easier. It is most important that the results are properly visualized and that the main characteristics of the results are properly understood. The figures and the orders of magnitude presented in this section hopefully serve that purpose.

4.1.3 Some Results from Spectral Analysis

The time series of osculating or mean elements, of the corresponding orbital poles, and of the Laplace vectors may be spectrally analyzed using the tools described in Chapter 11 of Part III. A full analysis (covering all elements of all planets) is out of the scope of this book. We focus on the semi-major axes a and the eccentricities e of Jupiter and Saturn on one hand, and of Uranus and Neptune on the other hand. The precession of the orbital poles and of the Laplace vectors of the same four planets will be analyzed, as well. It will be one of the goals to assign some of the spectral lines to the sidereal and/or synodic periods as given by the matrix in Figure 4.3 or to multiples of these periods.

We first address short-period effects of the order of (at maximum) few revolutions, then effects of longer periods (from tens of revolutions upwards). In principle it would be possible to use the same data set for both, the short-period and long-period, analysis. This would imply, however, that very long data files would have to be stored and subsequently analyzed. This is why a separate integration was performed for the analysis of short-period effects: The outer planetary system was integrated over ten thousand years only, and for each of the planets its osculating elements were stored at 100 day intervals. The analysis of longer periods was performed with the data set of mean elements discussed in the previous section.

Figures 4.18 and 4.19 show the amplitude spectra in the period range $0 - 70$ years of the semi-major axes and the eccentricities, respectively, for both, Jupiter and Saturn. All of the major spectral lines in the short-period range are contained in this interval. Obviously, the frequencies occurring in the spectra for Jupiter and Saturn are very strongly correlated. This indicates that all major lines are a consequence of the mutual perturbations between the two planets. The perturbations due to Uranus and Neptune seemingly do not have a major impact on the development of the system Jupiter–Saturn.

The above observation is supported by the fact that all major spectral lines in Figures 4.18 and 4.19 occur at (or near) periods derived from the basic beat frequency $P_{\uparrow\hbar} \approx 19.8$ years. In order to recognize this observation easily in the Figures 4.18 and 4.19, the periods $3\,P_{\uparrow\hbar}$, $\frac{3}{2}\,P_{\uparrow\hbar}$, $P_{\uparrow\hbar}$, $\frac{3}{4}\,P_{\uparrow\hbar}$, $\frac{2}{5}\,P_{\uparrow\hbar}$, $\frac{1}{2}\,P_{\uparrow\hbar}$, $\frac{3}{5}\,P_{\uparrow\hbar}$, $\frac{3}{8}\,P_{\uparrow\hbar}$, $\frac{1}{3}\,P_{\uparrow\hbar}$, and $\frac{3}{10}\,P_{\uparrow\hbar}$ are marked in the two spectra.

It is also interesting to note that the ratio of the amplitudes at two different frequencies are different for Jupiter and Saturn: For Jupiter, the major short-period term in the semi-major axis stems from the period $P = \frac{1}{2}\,P_{\uparrow\hbar}$, whereas in the case of Saturn it stems from the $P = P_{\uparrow\hbar}$ period. In the case of the eccentricity, Jupiter shows the biggest amplitude for the period $P = 3\,P_{\uparrow\hbar}$, whereas the $P = 0.6P_{\uparrow\hbar}$-term gives the major contribution in the case of Saturn.

Figures 4.20 and 4.21 show the amplitude spectra in the period range $0 - 250$ years of the semi-major axes and the eccentricities, respectively, for both, Uranus and Neptune. According to the matrix of synodic revolution periods the synodic revolution period of Uranus and Neptune is $P_{\delta\gamma} = 171.9$ years. Whereas the principal spectral lines for the pair Jupiter – Saturn all could be derived from the synodical period $P_{\uparrow\hbar}$ of that pair, the same is not true for Uranus and Neptune: The semi-major axis of Uranus is governed by contributions near the planet's synodic revolution period w.r.t. Jupiter (13 years) and Saturn (45.5 years). An analogue statement holds for Neptune. A mixture of periods derived from the synodic period $P_{\delta\gamma} = 171.9$ years and of the synodic periods w.r.t. the other planets may be seen in the case of the spectra for the eccentricities.

Figure 4.22 shows the spectrum of Jupiter's and Saturn's eccentricities for the range of periods between 0 years and 60000 years. The spectrum stems from the analysis of the integration over one million years. It was cut at 60000

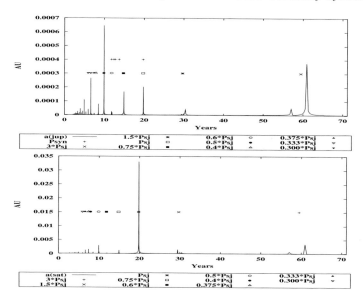

Fig. 4.18. Amplitude spectra (short periods) of the semi-major axes of Jupiter (top) and Saturn (bottom)

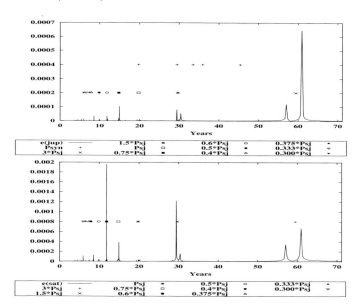

Fig. 4.19. Amplitude spectra (short periods) of the eccentricities of Jupiter (top) and Saturn (bottom)

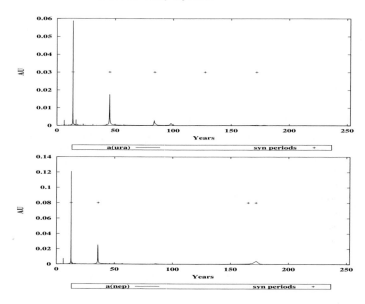

Fig. 4.20. Amplitude spectrum (short periods) of the semi-major axes of Uranus (top) and Neptune (bottom)

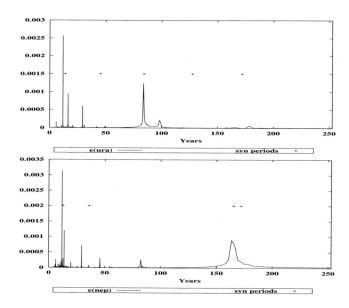

Fig. 4.21. Amplitude spectrum (short periods) of the eccentricities of Uranus (top) and Neptune (bottom)

years, because no significant contributions were observed with greater periods. The ratio of the amplitudes is (naturally) as discussed in the previous section. In essence, spectral lines are observed around 54000 years (corresponding to the strong variation of amplitudes in Figure 4.14) and at half and one third of this period. In addition, a small peak is observed near 1000 years – corresponding to the period of the great inequality.

A similar spectrum with the same frequencies results for the semi-major axes. Figure 4.23 shows a detailed view of that spectrum around the period of 900 years (the period of the long period inequality). As one can see, this spectral line has a width of about 100 years and it has a remarkably fine structure. As opposed to a line which is caused by a single period, and which has a finite width only due to discretisation (see discussion in Chapter 11 of Part III), this fine structure is real. More details would become visible, if longer time series would be analyzed.

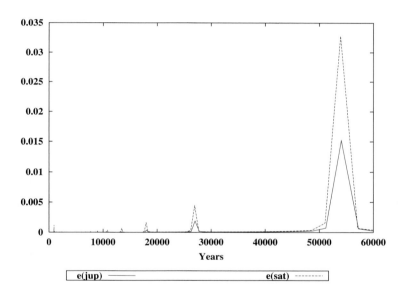

Fig. 4.22. Amplitude spectrum of Jupiter's and Saturn's eccentricities

Figure 4.24 actually contains the interesting part of the amplitude spectrum of the projections of the orbital poles of Jupiter's and Saturn's orbital planes. The method of the analysis may be found in section 11.5 of Chapter 11 of Part III. The figure documents the regression (negative sign of the frequencies) of the nodes in the invariable plane with a period of about 49000 years. The correlation between the motions of the nodes of the two planets comes out crystal-clear in this figure. The ratio of the amplitudes corresponds to the

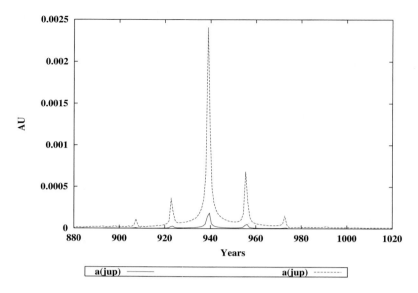

Fig. 4.23. Amplitude spectrum of Jupiter's and Saturn's semi-major axes (details around 900 years)

ratio of the inclination angles of the two orbital planes w.r.t. the invariable plane. Figure 4.25 shows the amplitude spectra of the projections of the vector *e* (the Laplace vector) onto the plane of the ecliptic *J*2000.0 for Jupiter and Saturn, using the methods specified in section 11.5 of Chapter 11 in Part III. The figure documents that the spectra are highly correlated. The motion of the two perihelia may be approximated (almost perfectly) as a superposition of two prograde circular motions with periods of about 49000 years and 330000 years. The decisive differences are the amplitudes: The amplitude of the 49000-years term is about 0.046 for Saturn, about 0.014 for Jupiter, whereas the amplitude of the 330000-years term is about 0.030 for Saturn and about 0.042 for Jupiter. These spectra illustrate the evolution of the nodes in Figure 4.16.

Let us conclude these studies of the spectral behavior of planetary orbits resulting from the integration over one million years with Figure 4.26. It shows that the long-period perturbations of the pair Uranus and Neptune are highly correlated, as well. Let us note in particular that the peak near 4300 years is the equivalent of the great inequality of Jupiter and Saturn. This *long period inequality of the planets Uranus and Neptune* is due to the near-commensurability of revolution periods of

$$\frac{P_{\text{\male}}}{P_{\text{\male}}} = \frac{165.2}{84.247} = 1.961 \approx 2 \ . \tag{4.22}$$

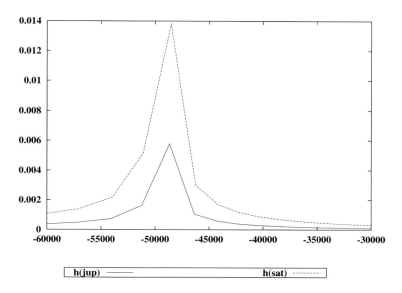

Fig. 4.24. Amplitude spectrum for Jupiter's and Saturn's lines of node

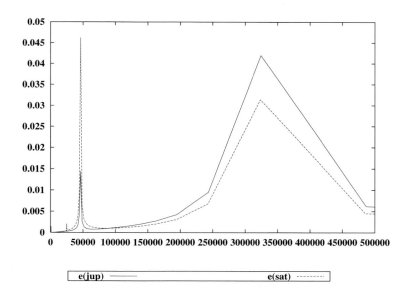

Fig. 4.25. Amplitude spectra of the Laplacian vectors e_{4}, $e_{\hbar}$ of Jupiter and Saturn

The corresponding beat period is

$$\tilde{P} = \frac{P_{\text{♃}} P_{\text{♄}}}{2\,P_{\text{♃}} - P_{\text{♄}}} \approx 4243 \text{ years} . \tag{4.23}$$

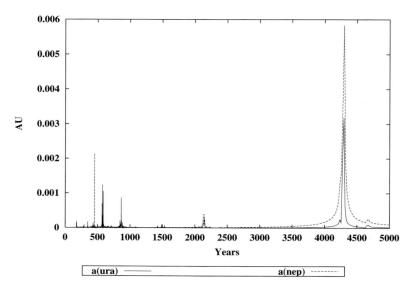

Fig. 4.26. Amplitude spectra of the semi-major axes of Uranus and Neptune

4.2 Development of the Inner Planetary System

In order to gain some insight into the development of the inner planetary system around present time, two integrations of the entire planetary system (without the planets Mercury and Pluto) were performed. The initial epoch for both integrations was January 1, 2000. The first integration covered the time interval of the next 250000 years (forward integration) from the initial epoch, the other the interval of the past 250000 years (backwards integration). Figure 4.27 contains the general output file produced by program PLASYS for the forward integration.

The numerical integration was performed with the same multistep method as that for the integration of the outer planetary system (compare Figure 4.3), but, due to the revolution period of Venus of only 225 days, a (constant) stepsize of only five days (corresponding to 45 steps per revolution for Venus) was chosen. The rule of thumb (I- 7.211) would actually ask for 77 steps per revolution. A relatively short integration interval and not too demanding accuracy requirements justify this choice.

As expected, the number of function calls is about a factor of 1.5 larger than the corresponding number for the outer planetary system (interval length reduced by a factor of four, stepsize reduced by a factor of six). The processing time required is still roughly by a factor of two higher than for the integration of the outer planetary system (where the time interval is a factor of four longer). This is caused by the circumstance that instead of integrating the orbits of five planets we are now integrating the orbits of seven bodies.

A comparison of the total energy and the total angular momentum of the systems integrated in Figures 4.3 and 4.27 shows clearly, that these quantities in essence are defined by the outer planetary system.

Figure 4.27 indicates that mean elements, with averaging periods of ten revolutions for each planet, were produced.

According to the matrix of sidereal and synodic periods in Figure 4.27 the revolution periods of Venus and Earth are almost commensurable

$$\frac{P_{\oplus}}{P_{\venus}} = \frac{1}{0.615} = 1.63 \approx \frac{5}{3} \ . \tag{4.24}$$

Therefore, the same geometry of the triangle Sun–Earth–Venus is always repeated after three sidereal years, corresponding to five sidereal revolutions of Venus. According to the experience gained one might expect "long-period" perturbations with a period of

$$\tilde{P} = \frac{P_{\oplus} \, P_{\venus}}{5 \, P_{\venus} - 3 \, P_{\oplus}} \approx 8.2 \text{ years} \ . \tag{4.25}$$

This particular commensurability might be studied by performing dedicated integrations over relatively short time intervals (let us say 10000 years) and by analyzing the resulting osculating elements stored at a relatively high rate (let us say every $50 - 100$ days).

Table 4.1 documents furthermore that Venus and Earth, with a mass ratio of $m_{\venus}/m_{\oplus} \approx 0.8$, are the dominating masses of the inner planetary system. Mars is almost by a factor of ten less massive than the Earth. Therefore, one would expect that the development of the inner planetary system is governed by the interactions between the two planets Venus and Earth.

This guess is in the first place confirmed by Figure 4.28, showing the eccentricities of Venus, Earth, and Mars in the time interval

$$[\, 250000 \text{ B.C.}, \, 250000 \text{ A.D.} \,] \ .$$

Mars, with a mean eccentricity and variations of about $e_{\mars} \approx 0.09 \pm 0.03$, at first sight seems to develop rather independently of the two inner planets.

```
GENERATION OF PLANETARY SYSTEM                                05-JUL-01  15:11
******************************************************************************
    NAME        #REV(MEAN ELE)      SAMPLING FOR PLANETS-FILE
--------------------------------------------------------------
    VENUS            10                          0
    EARTH            10                          0
    MARS             10                          0
    ...
--------------------------------------------------------------
INITIAL EPOCH:        2000  1  1.0 YEAR MONTH DAY
LENGTH (YEARS):         250000.0 YEARS
INTEGRATION           FORWARDS

INITIAL VALUES FROM JPL DE200
NEWTON-EULER EQUATIONS USED

TABULAR INTERVAL:        20.0 DAYS
SAMPLING (PRINT):      1000000
INITIAL STEPSIZE:         5.0 DAYS
INTEGRATION METHOD: MULTISTEP
INTEGRATION STEP:     FIXED
ORDER OF METHOD:         14
# ITERATIONS/STEP:        1

SIDEREAL AND SYNODIC REVOLUTION PERIODS
----------------------------------------
             VENUS     EARTH     MARS    JUPITER   SATURN    URANUS   NEPTUNE
----------------------------------------------------------------------------
VENUS   |    0.615     1.599    0.914     0.649     0.628    0.620     0.617
EARTH   |    1.599     1.000    2.135     1.092     1.035    1.012     1.006
MARS    |    0.914     2.135    1.881     2.235     2.009    1.924     1.902
JUPITER |    0.649     1.092    2.235    11.861    19.823   13.805    12.779
SATURN  |    0.628     1.035    2.009    19.823    29.532   45.469    35.959
URANUS  |    0.620     1.012    1.924    13.805    45.469   84.252   171.916
NEPTUNE |    0.617     1.006    1.902    12.779    35.959  171.916   165.223
----------------------------------------------------------------------------

TOTAL ENERGY OF SYSTEM:-0.33162740D-07 M(SUN)*A.U.**2/DAY**2
TOTAL ANGULAR MOMENTUM COMP 1: 0.15974171D-05 M(SUN)*A.U.**2/DAY
                       COMP 2: 0.50676733D-06         "
                       COMP 3: 0.60812243D-04         "

CPU                     25.131 MIN
FCT CALLS            36525105.
```

Fig. 4.27. General output of integration of planetary system over next 250000 years

The principal period of about 90000 years in its eccentricity seems to be common to the eccentricities of Venus and Mars, as well.

The orbital eccentricities of Venus and Earth are strongly correlated. Two strong signals can be distinguished: A common variation with a period of about 400000 years with an amplitude of about 0.02 and a variation of a shorter period of about 90000 years with variable amplitudes of a similar order of magnitude (the one also showing up in the eccentricity of Mars).

Figure 4.29 shows the inclinations of the three planets w.r.t. the ecliptical plane $J2000.0$. Obviously, there is a strong anti-correlation of the orbital inclinations of Venus and Earth (reminding us of the one observed in Figure 4.10 for Jupiter and Saturn).

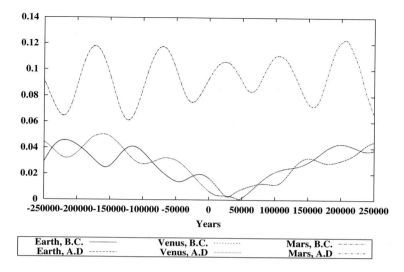

Fig. 4.28. Eccentricities of Venus, Earth, and Mars between -250000 BC and +250000 AD

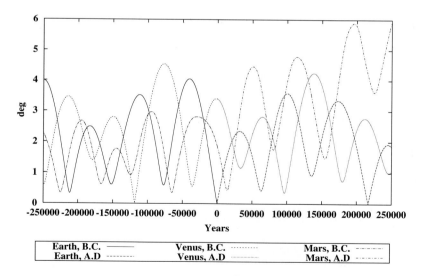

Fig. 4.29. Orbital inclinations (J2000.0) of Venus, Earth, and Mars between -250000 BC and +250000 AD

Both, the inclinations of Venus and Earth w.r.t. the ecliptic $J2000.0$, show periodic variations with amplitudes up to about four degrees in the time interval of half a million years around present time. The period of these variations is of the order of 70000 years.

In Figure 4.30 the ecliptical longitudes Ω of the nodes of the orbital planes of Venus, Earth, and Mars may be inspected. The nodes rotate prograde in the ecliptical system (as opposed to Jupiter's and Saturn's node, which "only" oscillate in this system). Obviously the rotation rates for the nodes of Venus and Earth are the same over long time periods, whereas Mars's node is rotating about 2-3 times faster.

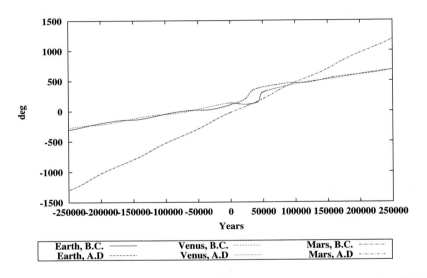

Fig. 4.30. Orbital nodes ($J2000.0$) of Venus, Earth, and Mars between -250000 BC and +250000 AD

Figures 4.31 and 4.32 give much better insight into the actual motion of the orbital planes of Earth, Venus and Mars. Obviously the orbital poles perform in good approximation a precession around the pole of the invariable plane (the coordinates of which are defined in sufficient approximation by eqn. (4.9)). The radius of the precessional motion shows strong periodic time variations.

Figures 4.31 and 4.32 should be compared to Figure 4.12. Obviously, the poles of the three orbital planes rotate (in the average over long time periods) in good approximation around the pole of the invariable plane. Clearly, the

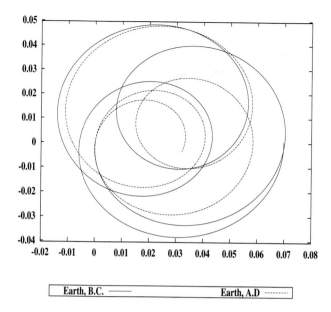

Fig. 4.31. Projection of the orbital pole of the Earth on the ecliptic $J2000.0$ between -250000 BC and +250000 AD

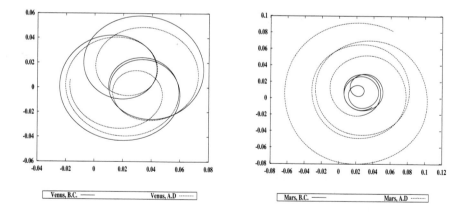

Fig. 4.32. Projection of the orbital poles of Venus and Mars on the ecliptic $J2000.0$ between -250000 BC and +250000 AD

variations of the "circles" of the orbital poles around the pole of the invariable plane is much larger than in the case of the outer planets.

In section 4.1 Figure 4.12 could be explained as a consequence of the conservation of the total angular momentum vector of the system. Is it possible to explain Figures 4.31 and 4.32 along the same lines? The answer is definitely "yes". If the outer planets are "turned off" (in the simulation), the resulting precessions of the orbital planes are much better defined: As in Figure 4.12, the orbit normals would precess in well-defined rings around the pole of the invariable plane.

In this simplified version all the statements and rules of thumb, which were established for the pair of planets Jupiter and Saturn may be transcribed to the dominating pair of planets of the inner planetary system, namely Venus and Earth. The pronounced slow variation of the radius of the precession cone is a consequence of the perturbations by the outer planetary system, by Jupiter in particular.

Let us mention one peculiarity of the orbit of Mars. Figures 4.32 and 4.29 show a significant difference in the behavior for the backward and forward integration: whereas Mars's inclination w.r.t. the ecliptic (or w.r.t. the invariable plane) was rather small $(0° - 3°)$ during the years B.C., these inclinations will increase significantly over the next 300000 years (up to about 6°). This change of the inclination is periodic with a very long period, however. The integration interval interval of half a million years is not sufficiently long to derive a reliable estimate of this period. More insight into the nature of Mars's inclination will be gained by studying the spectrum of the orbital pole of Mars (a topic addressed below).

Figures 4.33 and 4.34 illustrate the motion of the Laplace vectors in the plane of the ecliptic.   We observe a prograde motion of all three vectors of the inner planets. The motions seem to be rather complex. The spectra will reveal, however, that these seemingly complex motions may be interpreted as a superposition of only two periodic, prograde circular motions.

Let us now further refine the discussion of the development of the inner planetary system using spectral analysis as a tool. Figure 4.35 shows the spectra of the semi-major axes of the Earth and Venus generated with the entire data set between 250000 B.C. and 250000 A.D.

The first remarkable aspect of the spectra is that there are no significant contributions with periods > 250 years. This is also the reason why the time development of the semi-major axes was not illustrated by a figure. The second remarkable aspect of the spectra in Figure 4.35 is the scale. The amplitudes are of the order of a few 10^{-7} AU . Note, that mean elements (averaged over ten periods for both planets) were analyzed in Figures 4.35. If oscillating elements were studied, much larger terms would occur. Figure 4.35 is actually good news for the development of life on Earth: No climatic

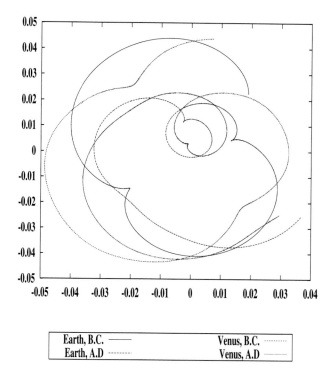

| Earth, B.C. —— | Venus, B.C. ·········· |
| Earth, A.D ------- | Venus, A.D ·········· |

Fig. 4.33. Projection of the Laplace vectors of Venus and Earth on the ecliptic
$J2000.0$ between -250000 BC and +250000 AD

changes due to long-period variations of the inner planets' semi-major axes
have to be expected.

Figure 4.36 shows the amplitude spectra associated with the projection on
the plane of the ecliptic of the orbital poles of Venus, Earth and Mars. Despite
the limited resolution – which might be improved by extending the integration interval – one clearly sees that the three spectra are of a similar and
simple structure: There is a strong contribution with a period of about 70000
years. The negative sign expresses the fact that the circular motion with the
amplitudes of 0.026 , 0.031 , and 0.033 , corresponding to inclinations of 1.5°,
1.8°, and 1.9°, takes place in the retrograde (clockwise) sense of revolution. It
corresponds to the well known linear regression of the nodes in Figure 4.30.

A term of shorter period, of about 50000 years, is barely visible in the spectrum of Venus, it clearly shows up in the spectrum of the Earth, it is prominent for the spectrum related to Mars. This spectral line is caused by Jupiter's
perturbations. It explains the strange motion of the orbital pole of Mars in

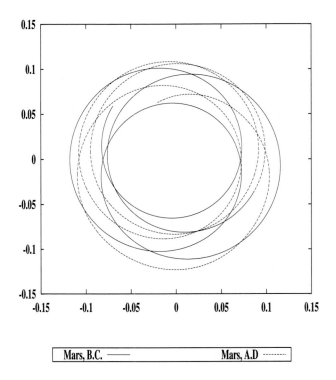

Fig. 4.34. Projection of the Laplace vector of Mars on the ecliptic $J2000.0$ between -250000 BC and +250000 AD

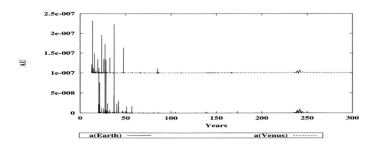

Fig. 4.35. Spectra of the semi-major axes of Venus and Earth

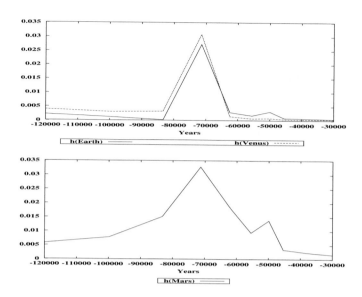

Fig. 4.36. Spectra of the projections of the orbital poles of Venus, Earth, and Mars on the ecliptic $J2000.0$

Figure 4.32 as the superposition of the two circular motions with periods of $P_7 = 70000$ and at $P_5 = 50000$ years. The beat period of the two signals is

$$\tilde{P} = \frac{P_7\,P_5}{P_7 - P_5} \approx 175000 \text{ years .} \tag{4.26}$$

This also explains Mars's seemingly different behavior in the B.C. and A.D. time spans. The ratio of the amplitudes in minimum and maximum are expected to be

$$\frac{i_{\sigma\max}}{i_{\sigma\min}} = \frac{0.033 + 0.014}{0.033 - 0.013} \approx 2.4 \;, \tag{4.27}$$

a value which underestimates the actually observed ratio in Figure 4.32. A better resolution of the spectrum would improve the situation.

The amplitude spectra for the projections of the Laplace vectors on the ecliptic are illustrated by Figure 4.37. The spectrum for Mars is clearly the simplest one. It is dominated by one spectral line with a period of about 70000 years and an amplitude of about 0.08 - in other words, this single line contains almost the entire power of the spectrum.

In view of Figure 4.33 it is not surprising that the spectra for Venus and Earth are more complex. Apart from the line around 70000 years, caused by the perturbations of the big outer planets, we observe a strong contribution around 170000 years. It must be mentioned, however, that the time series

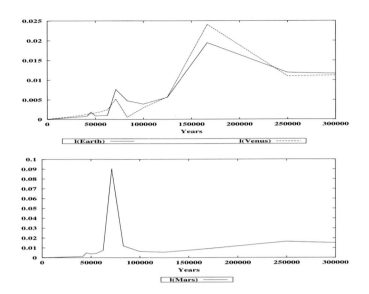

Fig. 4.37. Spectra of the projections of the Laplace vectors of Venus, Earth, and Mars on the ecliptic $J2000.0$

analyzed is too short (or it is not dense enough) to give a good resolution for such long periods (remember that the time series covers "only" an interval of half a million years). The principal characteristics in Figure 4.33 are, however, quite well explained by the superposition of the two prominent prograde circular motions in the spectra in Figure 4.37.

The above spectra show that the character of a motion is often easier to understand, if a physical phenomenon is studied in the frequency domain (period domain) and not in the time domain.

Let us conclude this section by a few remarks concerning the relations between the long-term evolution of the Earth's orbital elements and the Earth's climate.

In the first half of the 20th century Milutin Milankovitch (1879–1958) tried to explain long-term climatic variations on Earth, glaciation periods ("ice ages") in particular, by long-period variations of the (mean) orbital elements of the Earth. There is strong geological evidence, that quite a few of these glaciation periods (interrupted by warmer periods) occurred within the previous, let us say, 1–2 million years. Today, the time development of the Earth's orbital elements may be established quite well by numerical methods of the kind used here, or by advanced analytical theories, or by a combination of both.

The knowledge of the long-term behavior of these elements was very incomplete at Milankovitch's life-time (the best source of information at that time probably was the work of Simon Newcomb (1835–1909) condensed in his Tables of the Planets). Milankovitch's theory was very deductive in nature, and it was conducted in the time domain. Most of his conclusions are no longer supported by modern protagonists of this fascinating interdisciplinary field of science. Modern theories analyze the frequency domain of the climate history *and* the development of the Earth's orbital elements, and correlate the spectra of the two seemingly independent quantities. These correlations are interesting, and, at least at first sight, they seem to be quite convincing. Protagonists of this revival of Milankovitch's attempt to explain the ice ages by Celestial Mechanics are A. L. Berger and J. Imbrie. For more information we refer to their oeuvre, e.g., to [8], [6], [7], and [60].

4.3 Minor Planets

4.3.1 Observational Basis

More than 100000 so-called *minor planets* or *asteroids* are known today. Most of these objects have semi-major axes of $1.52 < a \leq 5.20$ AU, i.e., between those of Mars and Jupiter. This does not characterize the "true" population, because for an observer on Earth (or in the Earth-near space) the apparent magnitude of these objects is a function of their geocentric distance. The observation of minor planets is internationally coordinated, and the computation of orbits is performed at the *MPC (Minor Planet Center)* (60 Garden St., Cambridge MA 02138 USA) of the IAU.

The MPC offers many services. Among others, it maintains and updates the file "MPCORB.DAT", and makes it electronically available at its website (ftp://cfa-ftp.harvard.edu/pub/MPCORB/MPCORB.DAT). According to the file header, MPCORB.DAT contains "published orbital elements for all numbered and unnumbered multi-opposition minor planets for which it is possible to make reasonable predictions. It also includes published elements for recent one-opposition minor planets and is intended to be complete through the last issued Daily Orbit Update MPEC".

For a member of an institute contributing to the data base of the MPC, it is a pleasure to reproduce here the statement contained in the header of the file "MPCORB.DAT": "The work of the individual astrometric observers, without whom none of the work of the MPC would be possible, is gratefully acknowledged." The review of the observational material in this section is based on the same work.

Figure 4.38 shows the projection of all minor planets positions on the plane of the ecliptic *J*2000.0 as of July 1, 2001. The figure is based on the MPC's file MPCORB.DAT dated March 4, 2001. In order to facilitate the orientation, the (osculating) orbital curves of the planets (Earth, Mars,) Jupiter, Saturn,

Uranus, and Neptune are also drawn in Figure 4.38, and the planetary positions on July 1, 2001 are marked, as well. The positions were computed using the osculating orbital elements published in the file "MPCORB.DAT" without taking perturbations into account.

There is a clear concentration of the known minor planets between the orbits of Mars and Jupiter. An outer belt, called Kuiper-belt or Edgeworth-Kuiper-belt, mainly beyond the orbit of Saturn, is clearly visible as well (named after the discoverers Kenneth Edgeworth (1880–1972) and Gerard Peter Kuiper (1905–1973). Objects in this outer belt were only detected in the second half of the twentieth century. The first one, Chiron, as already mentioned in the introductory chapter, was detected by Charles T. Kowal in 1977. One should be aware of the fact, that due to the difficulty to observe the faint and very slowly moving objects in the outer belt, Figure 4.38 does not even approximately reflect the actual number and distribution of these objects. Most of the members of the Edgeworth-Kuiper belt have orbits outside Neptune's trajectory. This is why these objects are also referred to as TNO (Trans-Neptunian Objects).

Figure 4.39 shows the "classical" minor planets in a rectangular window of ± 6 AU centered at the Sun. The orbits of Earth, Mars and Jupiter, their positions and that of the Sun as of July 1, 2001 are also marked in Figure 4.39.

Figure 4.39 illustrates the concentration of minor planets in the asteroid belt between Mars and Jupiter. With the remarkable exception of two groups of minor planets at a heliocentric distance of about 5.2 AU (Jupiter's distance from the Sun), and separated from Jupiter by the heliocentric angle of about $60° \pm 30°$, the zones between heliocentric distances $4 < r < 5$ AU and $1.5 < r < 2$ are almost empty compared to the density in the other parts of the belt. It is safe to state that these "zones of avoidance" were created by the gravitational effects of Jupiter and Mars over long time periods.

Figure 4.40 shows the projection of the osculating aphelia on July 1, 2001 on the plane of the ecliptic of all objects in Figure 4.38. The projection of the osculating planetary aphelia and the circles with the radii of the projected planetary aphelia of the major planets should help to interpret the figure, which is oriented to have Jupiter's osculating perihelion on the positive x-axis. In this scale one can only distinguish two rings, one between Mars and Jupiter (which seems to be structured as well) and one with a radius comparable to Pluto's heliocentric aphelion distance.

The structure of the distribution of the aphelia of minor planets between Mars and Jupiter is better visible in Figure 4.41. We observe a ring-like structure and a clear asymmetry w.r.t. Jupiter's perihelion, which lies on the positive (horizontal) x-axis of Figure 4.41. This implies that Jupiter's aphelion lies on the negative x-axis. The aphelia of the minor planets seem to be concentrated near Jupiter's aphelion, as well.

The histogram of aphelia in Figure 4.42 confirms this observation. There is a clear preference of the minor planets to have their aphelia close to Jupiter's

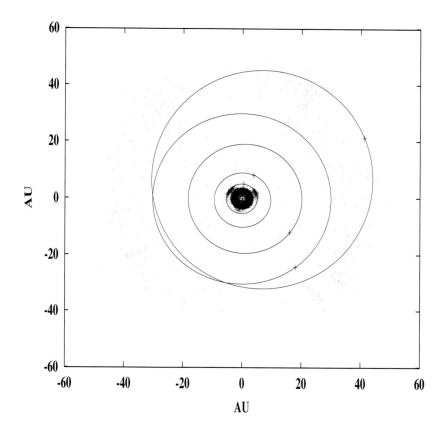

Fig. 4.38. Minor planets and planets Jupiter – Pluto (indicated by "+") on July 1, 2001 (from the MPC)

aphelion. There are about three times as many minor planets with aphelia near Jupiter's aphelion as near Jupiter's perihelion. This preference minimizes the perturbations during the times of closest approach between a minor planet and Jupiter: If the aphelion of Jupiter agrees with the minor planet's aphelion, the minimum distance $\Delta_{\min}$ between the two bodies varies within the limits

$$\Delta_{\min} = a_{\jupiter} - a \pm (a_{\jupiter}\, e_{\jupiter} - a\, e) \; , \tag{4.28}$$

under the assumption that the two orbits are elliptic and coplanar. Under the same assumption this minimum distance varies between the limits

$$\Delta_{\min} = a_{\jupiter} - a \pm (a_{\jupiter}\, e_{\jupiter} + a\, e) \; , \tag{4.29}$$

if the perihelion of one body coincides with the aphelion of the other body.

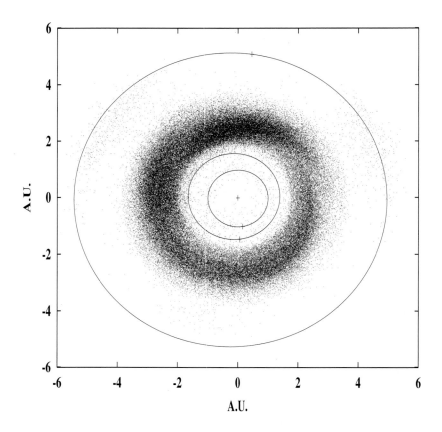

Fig. 4.39. Minor planets and Sun, Earth, Mars, Jupiter (indicated by "+") on July 1, 2001 (from the MPC)

Figure 4.42 shows that minor planets try to avoid the more violent perturbing situation. Figure 4.43 gives still more insight into the structure of the asteroid belt. The histogram of the semi-major axes (given as a function of the semi-major axis (top) and as a function of the revolution period (bottom)) shows the number of minor planets per $\Delta a = 0.002$ AU .

The first histogram shows that there are only very few minor planets with semi-major axes $a > 3.3$ AU , corresponding to a revolution period of $U = \frac{1}{2} P_{\text{4}}$, P_{4} being Jupiter's revolution period. There are only two small groups of asteroids, one at $a \approx 4$ AU corresponding to $P_{\text{4}} : P \approx 3 : 2$ and one at $a \approx 5.2$ AU corresponding to $P_{\text{4}} : P \approx 1$, which could find a "modus vivendi" with Jupiter. The former group of asteroids is called the *the Hilda*

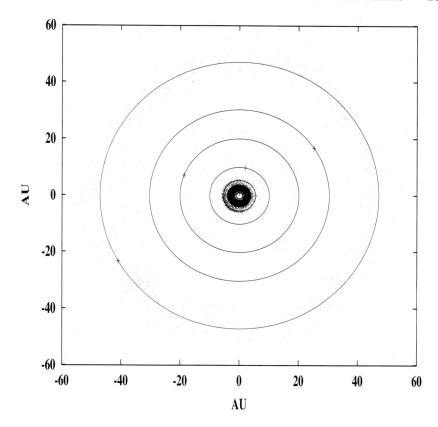

Fig. 4.40. Aphelia of the minor planets in 2001 in an ecliptical coordinate system with Jupiter's perihelion on the positive x-axis, aphelia of planets indicated by "+" (from the MPC)

group, named after one of its members. The latter group, very well visible in Figure 4.39, is called the *Trojan group*, or simply the *Trojans*, because the minor planets in this group are named after heroes of the Trojan war (the minor planets named after Greek and Trojan heroes are, by the way, with few exceptions (prisoners?) separated by Jupiter).

Figure 4.39 and the histogram 4.43 prove that the stationary triangular solutions of the planetary problème restreint Sun-Jupiter-minor planet, which was discussed in section I- 4.5.2 actually exist in reality and that the orbits associated with them are highly stable.

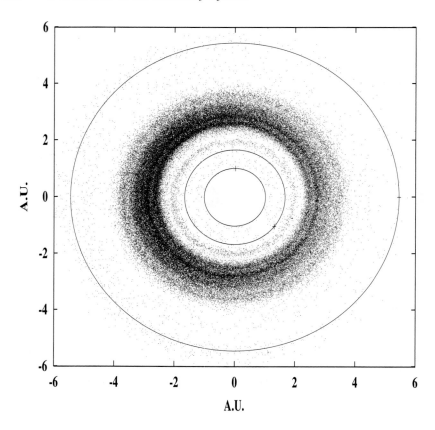

Fig. 4.41. Aphelia of the minor planets in 2001 in an ecliptical coordinate system with Jupiter's perihelion on the positive x-axis, aphelia of planets indicated by "+" (from the MPC)

The histograms 4.43 reveal other interesting characteristics: Apart from the groups mentioned there are gaps in the distribution of minor planets, the most remarkable ones being at $a \approx 2$ AU , corresponding to $P_\text{4} : P \approx 4 : 1$, at $a \approx 2.5$ AU , corresponding to $P_\text{4} : P \approx 3 : 1$, and at $a \approx 2.8$, corresponding to $P_\text{4} : P \approx 5 : 2$. The gap at $P_\text{4} : P \approx 2 : 1$ is also called the *Hecuba gap* after a minor planet very close to that gap (osculating semi-major axis $a = 3.2387771$ AU on April 1, 2001).

The data set of about 104000 minor planets available today through the MPC and used in Figure 4.43 makes it crystal-clear that the gaps in the distribution, but also the mentioned clusters (Hilda, Trojans) correspond to revolution periods which are commensurable with Jupiter's revolution period,

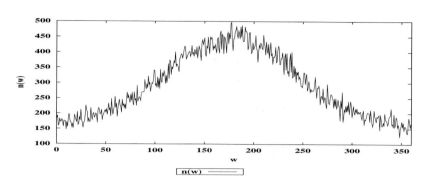

Fig. 4.42. Histogram of the angles $\tilde{\omega} \stackrel{\text{def}}{=} 180° + \Omega + \omega - (\Omega_{\jupiter} + \omega_{\jupiter})$ (number of aphelia per degree) (from the MPC)

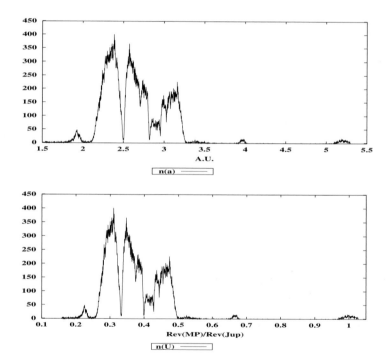

Fig. 4.43. Histogram of the minor planets as a function of the semi-major axes (top) and of the revolution periods (in units of Jupiter's revolution period)

i.e., for which

$$\frac{P}{P_{2\!\!+}} \approx \frac{k_1}{k_2},$$

where k_1 and k_2 are small integer numbers.

From samples of 88 minor planets known in 1866 and 146 in 1875, the American astronomer Daniel Kirkwood (1814–1895) concluded that there are gaps and "clusters", and he postulated their correlation with commensurabilities of the minor planets' and Jupiter's revolution periods. This certainly is a proud achievement requiring a fair portion of imagination. (Readers not agreeing with this statement are invited to randomly draw a sample of 100 minor planets in the data set "MPCORB.DAT" and to produce the histogram corresponding to Figure 4.43). In honour of Kirkwood, the gaps in the histograms in Figure 4.43 are called *Kirkwood gaps*.

The aspect of commensurability will be further addressed below. The simple and obvious question, whether the gaps may be uniquely explained through gravitational perturbations exerted by Jupiter (and possibly other planets), is not a trivial one to answer.

The osculating semi-major axes and the inclinations w.r.t. the invariable plane were plotted in Figure 4.44. The figure shows that there are clusters, called families, of asteroids in the a, i-plane. The best-known families are named in Figure 4.44. It would be preferable to use not the osculating elements but another type of elements (more representative over long time intervals), the so-called proper elements (see section 4.3.3).

When analyzing the elements of the known minor planets around 1918, the Japanese astronomer Kiyotsugu Hirayama (1874–1943) [56] identified and named some of these families, and he postulated the common origin of their members. As a matter of fact, Figure 4.44 may be very well explained by a fragmentation of larger bodies. If the fragmentation took place by collisions in the asteroid belt with moderate relative velocities, one may assume that the orbital elements of all fragments originally were very close to the elements of the proto-planetoid before fragmentation. Perturbations rather rapidly destroy the similarity of the elements Ω, ω, and σ_0 of the fragments and leave the elements a, i, and e for an identification of the families. Not all the groups visible in Figure 4.44 may be explained by fragmentation. The existence of the Hilda group and of the Trojans may be explained as a consequence of an "ordinary" orbit development. A more complete treatment of the topic of families of minor planets may be found in [42] and in [80].

Figure 4.45 shows the projection of the orbital poles (unit vectors normal to the orbital plane) on the plane of the ecliptic $J2000.0$. Rings are clearly visible. These rings seem to be (close to) concentric with the projection of the pole of the invariable plane with coordinates $(x, y) = (0.026, 0.008)$. This fact supports the existence of families: One would expect the orbital poles of

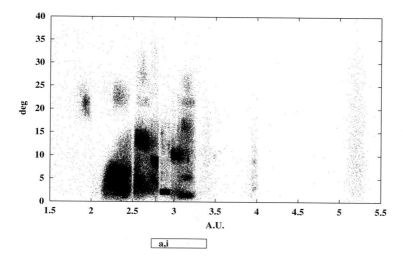

Fig. 4.44. Inclinations of the orbits of the minor planets as a function of the semi-major axes

the fragments to perform a precession in a concentric ring about the pole of the invariable plane.

Figure 4.46 gives the histogram of inclinations (w.r.t. the invariable plane) in the range $0° \leq i \leq 30°$. The ring structure, already observed in Figure 4.45, is clearly seen in this histogram as well. It is interesting that the mean inclination is not near $i_{mean} \approx 0°$, but rather $i_{mean} \approx 5°$. (Note, that the inclinations w.r.t. the invariable plane are shown in Figure 4.46).

The histogram of eccentricities in Figure 4.47 also shows that the distribution of eccentricities is rather broad. In view of the fact that the major planets all have orbits with small eccentricities, it is surprising that the mean eccentricity is somewhere close to $e_{mean} \approx 0.15$. Admittedly, it would be better to use mean elements in Figure 4.47. The main findings would, however, be the same.

4.3.2 Development of an "Ordinary" Minor Planet

In order to gain some insight into the development of the orbital elements of a presumably "ordinary" minor planet, objects with a revolution period $P = 0.385 P_4$ were integrated over a time period of one million years, starting from the initial epoch $t_0 = $ Jan 1, 2000. The integration was performed with program PLASYS (see Chapter 10 of Part III), using a multistep method of order 14 with a stepsize of 30 days.

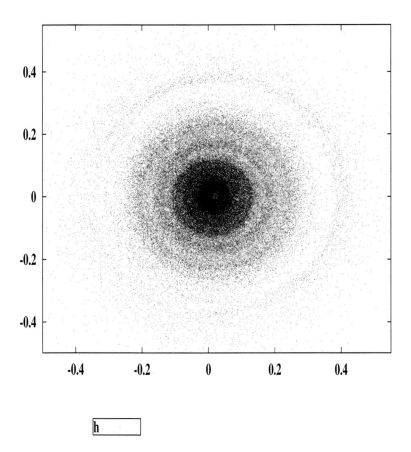

Fig. 4.45. Orbital poles of the minor planets (from the MPC)

Figure 4.48 shows the mean semi-major axes (the osculating elements were averaged over ten revolution periods) of four such bodies lying in the invariable plane (inclination $i = 1.58°$, longitude of the ascending node $\Omega = 107.6°$, see eqn. (4.5)). The osculating eccentricities were chosen to be $e = 0$ in all four cases, and the time of perihelion passage was set to zero, i.e., made to coincide with Jupiter's time of perihelion. The four simulations differed by the initial position within the orbit, which was defined by the ecliptical longitude of the perihelion (w.r.t. Jupiter's perihelion). It was set to $\tilde{\omega} = \tilde{\omega}_4 + (0°, 90°, 180°, 270°)$, where $\tilde{\omega} = \Omega + \omega$ and $\tilde{\omega}_4 = \Omega_4 + \omega_4$ are the ecliptical longitudes of the perihelia of the minor planet and Jupiter, respectively. Figure 4.48 shows that the mean values and the amplitudes of the

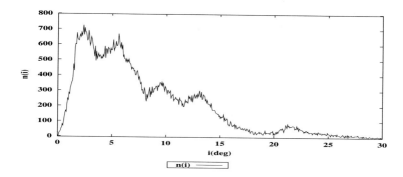

Fig. 4.46. Histogram of minor planets' inclinations w.r.t. invariable plane

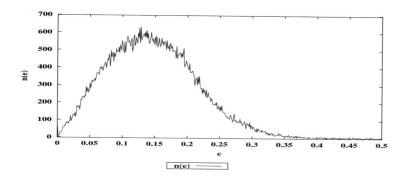

Fig. 4.47. Histogram of minor planets' osculating eccentricities

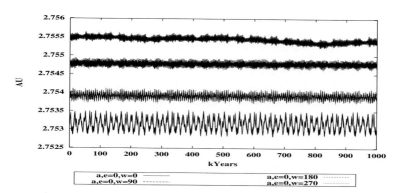

Fig. 4.48. Mean semi-major axes over one million years of four test particles with osculating semi-major axis $a_0 = 2.752$ AU ($P = 0.385 \; P_{\mars}$) and osculating eccentricity $e_0 = 0$ on Jan 1, 2000; osculating longitudes of perihelia $(\tilde{\omega} - \tilde{\omega}_{\mars})_0 = 0°, 90°, 180°, 270°$

periodic variations are functions of the initial angle $\tilde{\omega}$. The differences and the variations are small, however. With a semi-major axis of $a \approx 2.752$ AU, corresponding to a revolution period $U = 0.385\,P_4 \approx 4.57$ years, our test objects really are "ordinary" objects.

Figure 4.49, showing the development of the mean eccentricities of the same four test objects as in Figure 4.48, is more interesting. Despite the fact that all eccentricities were initially zero, the mean eccentricities show rather large periodic variations within the limits $0.0 < e < 0.085$. Similar patterns result for all four cases.

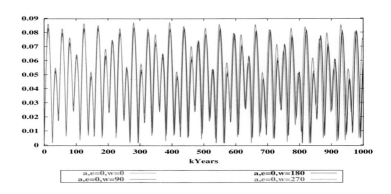

Fig. 4.49. Mean eccentricities over one million years of four minor planets with osculating semi-major axis $a_0 = 2.752$ AU $(P = 0.385\,P_4)$ and osculating eccentricity $e_0 = 0$ on Jan 1, 2000; longitudes of perihelia $(\tilde{\omega} - \tilde{\omega}_4)_0 = 0°, 90°, 180°, 270°$

Figures 4.50 and 4.51 show the development of the same orbital elements a and e as in Figures 4.48 and 4.49, the difference residing in the initial value of the osculating eccentricity, which was set to $e_0 = 0.1$ for all four simulations.

Figure 4.50 shows that the amplitudes of the periodic variations in a are much larger than in Figure 4.48 (but with amplitudes of about $\Delta a \approx 0.001$ AU they are still small).

Figure 4.51 shows periodic variations in the mean eccentricity of comparable amplitudes as in Figure 4.49, but the mean values differ substantially for the four cases (the mean values are approximately $\bar{e}_0 \approx 0.067$, $\bar{e}_{90} \approx 0.117$, $\bar{e}_{180} \approx 0.141$, $\bar{e}_{270} \approx 0.106$). For the case $\tilde{\omega} = \tilde{\omega}_4 + 180°$ the mean eccentricities reach periodically values up to $e \approx 0.18$. For the test particle with $\tilde{\omega} = \tilde{\omega}_4 + 180°$ the perturbations during the closest approach between the test object and Jupiter are initially a minimum, where the minimum distance is $\Delta_{\min} = a_4 - a - a_4\,e_4 - a\,e$, whereas this minimum distance initially is only $\Delta_{\min} = a_4 - a - a_4\,e_4 + a\,e$, if $\tilde{\omega} = \tilde{\omega}_4$.

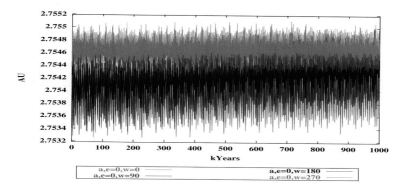

Fig. 4.50. Mean semi-major axes over one million years of four minor planets with osculating semi-major axis $a_0 = 2.752$ AU $(P = 0.385\,P_{\mathrm{4}})$ and osculating eccentricity $e_0 = 0.1$ on January 1, 2000; longitudes of perihelia $(\tilde{\omega} - \tilde{\omega}_{\mathrm{4}})_0 = 0°, 90°, 180°, 270°$

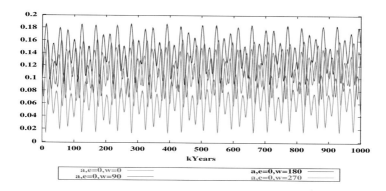

Fig. 4.51. Mean eccentricities over one million years of four minor planets with osculating semi-major axis $a_0 = 2.752$ AU $(P = 0.385,P_{\mathrm{4}})$ and osculating eccentricity $e_0 = 0.1$ on Jan 1, 2000; osculating longitudes of perihelia $(\tilde{\omega} - \tilde{\omega}_{\mathrm{4}})_0 = 0°, 90°, 180°, 270°$

Figure 4.52 shows the projection of the Laplace vectors $\boldsymbol{e}$ (see eqn. (I- 4.27)) for the eight simulations $(e = 0.0, 0.1\,,\ \tilde{\omega} = \tilde{\omega}_{\mathrm{4}} + 0°, 90°, 180°, 270°)$. The Laplace vectors rotate (in the prograde direction of rotation) around the origin $(x_0, y_0) = (0, 0)$. The Laplace vectors $\boldsymbol{e}$ are contained in (almost) circular annuli with widths of about $\Delta e \approx 0.060 - 0.085$, and they are centered at $(e \cos \tilde{\omega}, e \sin \tilde{\omega}) \approx (0, 0)$ in the average over long time intervals. Figures 4.52 only give a general impression of the actual motion of the Laplace vectors. Spectral analysis will provide more insight.

Figure 4.53 shows the development of the projections of the orbital poles on the ecliptic over one million years for two test particles with initial osculating

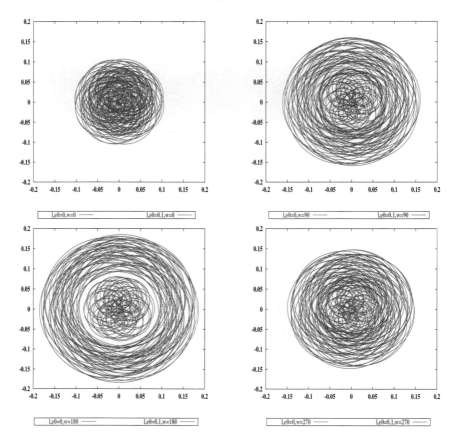

Fig. 4.52. Laplace vectors over one million years of eight minor planets with initial osculating semi-major axis $a_0 = 2.752$ AU ($P = 0.385\,P_{4}$) and osculating eccentricity $e_0 = 0.0, 0.10$ on Jan 1, 2000; longitudes of perihelia $(\tilde{\omega} - \tilde{\omega}_{4})_0 = 0°, 90°, 180°, 270°$

eccentricities $e = 0$ (and $\tilde{\omega} = \tilde{\omega}_{4}$) and inclinations of $\tilde{i} = 0°, 10°$ w.r.t. the invariable plane.

The projected vectors were multiplied by $180°/\pi$, in order to see (approximately) the inclination angle in Figure 4.53. The orbital poles move in circular annuli with radii $\bar{i} \approx i$ and with diameters $\Delta i \approx 0.5 - 1°$ around the pole of the invariable plane (see eqn. (4.5)).

Figure 4.54 superimposes the spectra for the orbital poles of the two projected vectors in Figure 4.53. When compared to the spectra of i and Ω (not reproduced here), the spectra in Figure 4.54 are amazingly simple. They are dominated by two spectral lines at periods of $-(22100 - 24300)$ and -50000 years (the sign indicating that the rotation is retrograde).

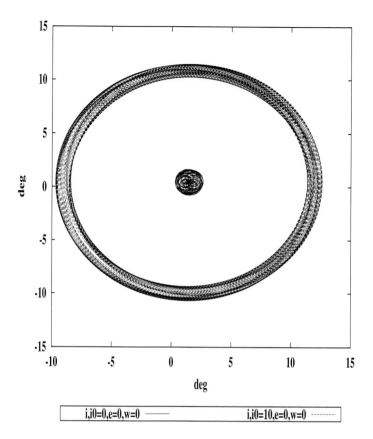

Fig. 4.53. Orbital poles over one million years of two minor planets with osculating semi-major axis $a_0 = 2.752$ AU ($P = 0.385\,P_{4}$) and osculating eccentricity $e_0 = 0.0$ on Jan 1, 2000; osculating inclinations w.r.t. invariable plane $i = 0°$, $i = 10°$; precession retrograde

The spectral line near -24000 is different in the amplitude and the precise value of the period for the two examples, whereas amplitude and period of the spectral line near -50000 years are almost identical. These empirical findings may be confirmed by additional tests starting from different initial inclinations: The resulting spectra of the projected orbital poles in essence only differ by the amplitude and somewhat by the period of the spectral line near -24000 years. The zero-order term of the harmonic series development resides (as expected) almost ideally in the projection of the invariable plane on the ecliptic.

Figure 4.55 shows the spectra (of the projections on the plane of the ecliptic) of the Laplace vectors e in the range of periods $0 < P < 600000$ years con-

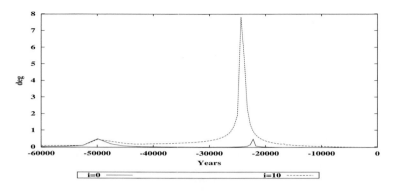

Fig. 4.54. Spectra of the projections of two orbital poles on the ecliptic. Osculating semi-major axes $a_0 = 2.752$ AU $(P = 0.385\,P_{\downarrow})$, osculating eccentricity $e_0 = 0.0\,,0.10$, osculating inclinations $i = 0°\,,10°$, $\tilde\omega - \tilde\omega_{\downarrow} = 0°$ on Jan 1, 2000

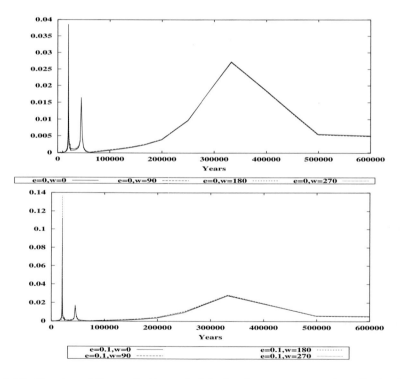

Fig. 4.55. Spectra of the projections of eight Laplace vectors on the ecliptic. Osculating semi-major axes $a_0 = 2.752$ AU $(P = 0.385\,P_{\downarrow})$, osculating eccentricities $e_0 = 0.0\,,0.1$, osculating inclinations $i = 0°$, $\tilde\omega - \tilde\omega_{\downarrow} = 0°\,,90°\,,180°\,,270°$, on Jan 1, 2000

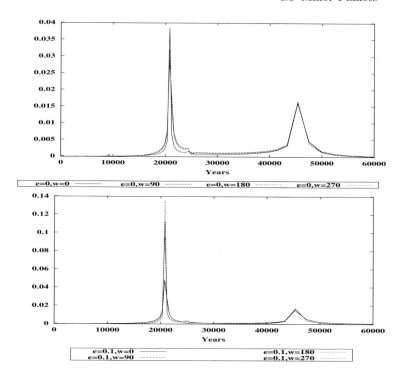

Fig. 4.56. Spectrum of the projections of eight Laplace vectors on the ecliptic. Range of periods $P = [0, 60000]$ years. Osculating semi-major axes $a_0 = 2.752$ AU ($P = 0.385\, P_{2\!\!4}$), osculating eccentricities $e_0 = 0.0\,, 0.1\,$, osculating inclinations $i = 0°$, $\tilde{\omega} - \tilde{\omega}_{2\!\!4} = 0°, 90°, 180°, 270°$, on Jan 1, 2000

taining all significant contributions. The top figure corresponds to the initial osculating eccentricity $e = 0$, the bottom figure to the initial eccentricity $e = 0.1$.

The similarity of all eight spectra is striking: There are only three major spectral lines with periods of about 20800 years, 45000 years, and 330000 years, where the latter value is not well defined, due to the fact that the length of the analyzed time series is only about three times the longest period. The retrograde part of the spectrum is not reproduced, because there is no signal in this part. Figures 4.56 and 4.57, giving details of the same spectrum as in Figure 4.55 in the period ranges $0 - 60000$ years and $15000 - 25000$ years, show that, in analogy to the findings related to the spectra of the orbital poles, the changes in the initial conditions are only reflected by one of the three spectral lines, namely by the one with a period near 20800 years.

Whereas the line near $\pm\, 20000$ years (mainly) may be attributed to the initial conditions (initial eccentricity and inclination), the other lines in the spectra

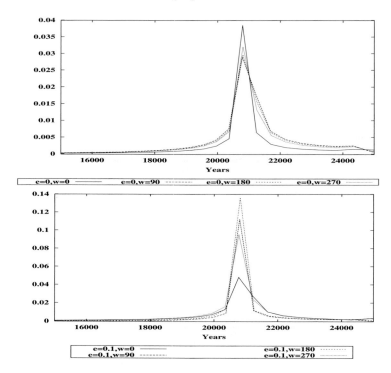

Fig. 4.57. Spectrum of the projections of eight Laplace vectors on the ecliptic. Range of periods $P = [15000, 25000]$ years. Osculating semi-major axes $a_0 = 2.752$ AU $(P = 0.385\,P_{\mars})$, osculating eccentricity $e_0 = 0.0\,, 0.1\,$, osculating inclinations $i = 0°$, $\tilde{\omega} - \tilde{\omega}_{\mars} = 0°, 90°, 180°, 270°$, on Jan 1, 2000

of the projected orbital poles and Laplace vectors are due to the perturbations by the planets Jupiter and Saturn (and to a much lesser extent by the other planets).

This statement is illustrated by Figures 4.58 and 4.59 containing the spectra of the orbital poles and the Laplace vectors of the planets Jupiter and Saturn (top spectra) and of the minor planet.

In order to improve the resolution of the spectral lines, the two figures are based on a numerical integration of the outer planetary system over ten million years (without Pluto, but including the test particle with revolution period $P = 0.385\,P_{\mars}$, $e = 0.1$, $i = 10°$, $\tilde{\omega} = \tilde{\omega}_{\mars}$). The program run took, by the way, about three hours of processing time on the PC used and produced element files of about 40 Mbytes per file for Jupiter, Saturn, and the test particle (the elements of the other planets were not saved).

The period of about -50000 years in Figure 4.58 corresponds very precisely to the eigenfrequency $f_6 = 25.73355''$/year of the secular perturba-

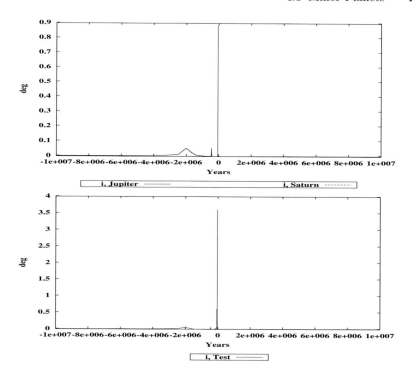

Fig. 4.58. Spectrum of the projections of the orbital poles of Jupiter and Saturn (top) and minor planet (bottom) on the ecliptic. Range of periods $[-60000, 0]$ years. Osculating semi-major axis $a_0 = 2.752$ AU ($P = 0.385\,P_{\tiny 4}$), osculating eccentricity $e_0 = 0.1$, osculating inclination $i = 10°$, $\tilde{\omega} - \tilde{\omega}_{\tiny 4} = 0°$ on Jan 1, 2000

tion theory by Brouwer and Clemence [27], the periods of about 50000 and 300000 years in Figure 4.59 correspond very precisely to the eigenfrequencies $g_6 = 27.77406''/$year and $g_5 = 4.29591''/$year of this theory.

Figure 4.60 shows an overlay of the spectral lines (detailed view of the amplitude spectra), which is defined by the initial conditions, for the projections of the orbital poles and the Laplace vectors.

In order to facilitate the comparison of the orbital poles and the Laplace vectors, the absolute value of the period was used as an independent argument. Figure 4.60 shows, that the (absolute values of the) periods roughly agree (within about 20%), but that it is somewhat keen to state that the periods are identical.

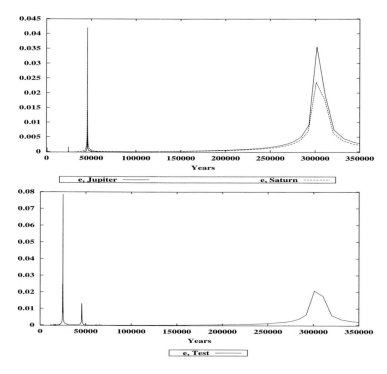

Fig. 4.59. Spectrum of the projections of the Laplace vector of Jupiter and Saturn (Top) and minor planet (bottom) on the ecliptic. Range of periods $[-60000, 0]$ Years. Osculating semi-major axes $a_0 = 2.752$ AU ($P = 0.385\,P_\text{♃}$), osculating eccentricity $e_0 = 0.1$, osculating inclination $i = 10°$, $\tilde\omega - \tilde\omega_\text{♃} = 0°$ on Jan 1, 2000

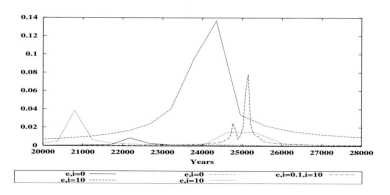

Fig. 4.60. Spectral line (as a function of the initial conditions) of the projections of the Laplace vectors and of the orbital poles for five test objects $((e, i) = (0,0), (0,10), (0,0), (0,10), (0.1,10))$. Absolute value of the periods used as independent argument

4.3.3 Proper Elements of Minor Planets

The empirical findings of the spectral analyses of the projections of the orbital poles and the Laplace vectors on the ecliptical plane may be summarized as follows:

- The spectra corresponding to the projections of the two vectors h/h (orbital pole) and e (Laplace vector) on the ecliptic are composed of only a few sizable spectral lines.

- The spectra of the orbital poles contain (practically) only retrograde contributions.

- The spectra of the Laplace vectors contain (practically) only prograde contributions.

- In both cases, only the amplitude (and, to a lesser extent, the period) of one spectral line (in the example the period P is in the range $21000 - 25000$ years) changes, if the initial (osculating) elements e and $\tilde{\omega}$ are changed.

- The absolute values of the periods P for the line depending on the initial conditions are of the same order of magnitude for the orbital poles and the Laplace vectors – but they are not identical.

These empirical findings have to be compared to the key results of the *theory of secular perturbations* as it was originally developed by Lagrange, addressed by many eminent contributors to Celestial Mechanics, and eventually brought by Brouwer and Clemence (see [27]) into the form still used today.

In secular perturbation theory all periodic terms containing the (rapidly varying) mean anomalies of the perturbing and of the perturbed body are ignored. It is customary in this theory to use the notation

$$
\begin{aligned}
h &\stackrel{\text{def}}{=} e \cos(\Omega + \omega) \ (= q_1) \\
k &\stackrel{\text{def}}{=} e \sin(\Omega + \omega) \ (= q_2) \\[6pt]
p &\stackrel{\text{def}}{=} \tan i \sin \Omega \quad (= h_1) \\
q &\stackrel{\text{def}}{=} \tan i \cos \Omega \quad (= -h_2) \ ,
\end{aligned}
\tag{4.30}
$$

where the terms in parentheses $(\ldots)$ refer to the notation used throughout this book.

Applied to minor planets, Brouwer and Clemence in [27] give the solutions:

$$h = \nu \sin(g\,t + \beta) + h_0$$
$$k = \nu \cos(g\,t + \beta) + k_0$$

$$(4.31)$$

$$p = \mu \sin(-g\,t + \gamma) + p_0$$
$$q = \mu \cos(-g\,t + \gamma) + q_0 \,,$$

where the angular velocity g is a function of the minor planet's semi-major axis a only; h_0 and k_0 are slowly varying functions of time (functions of the principal planets' secular variations).

The amplitudes ν and μ mainly depend on the initial values. They are called *proper eccentricity* and *proper inclination* of the minor planet (where the inclination was approximated by $\tan i \approx i$). The term *proper element* is justified because in the framework of secular perturbation theory, only the proper elements ν and μ (and the proper elements corresponding to the longitude of the node and of the perihelion) are affected by the initial conditions.

The results of this secular perturbation theory are approximations. The comparison with numerical results is amazingly good for the forced part of the spectrum. (The eigenfrequencies encountered here (and mentioned above) are those associated with the pair of planets Jupiter and Saturn). The periods extracted from the simulations correspond on the sub-percent level to the frequencies predicted by the secular theory. The agreement is less convincing when comparing the periods of the spectral line associated with the proper elements of the minor planet. The dependence on the inclination is not predicted by the secular theory. Also, the identity of (the absolute values of) the periods of the Laplace and orbital pole vectors is not well met in practice.

According to Brouwer and Clemence [27]:   *"... From our knowledge of the general behavior of such developments, we would expect that a rigorous treatment of the problem would give results differing from the approximate ones by considerably less than fifty percent, but there is no mathematical proof of it ..."*. In view of this statement the agreement between theory and numerical tests must be considered as excellent.

When approaching the problem not with analytical, but with numerical methods, one would simply identify the proper eccentricity ν and proper inclination μ as the amplitude of the spectral line associated with the initial conditions. The terms h_0, k_0, p_0, and q_0 at the initial epoch t_0 might be computed as the superposition of all terms (at $t = t_0$) of the harmonic series *not* related to the vector considered (either the Laplacian vector or the orbit normal). Observe that the mean values of the p_0 and q_0 must vanish over long time periods, whereas the mean values of h_0 and k_0 over very long time intervals coincide with the first two components of the projection of the unit vector normal to the invariable plane on the ecliptic.

The definition of the proper elements ν and μ proposed here differs slightly from the classical definition as it is in use today. We refer to [80] for an overview of the classical theory of proper elements.

4.3.4 Resonance and Chaotic Motion

Commensurability and Resonance. If the ratio of the revolution periods P_{4} and P of Jupiter and a minor planet (or another planet) is a *quotient of small (positive) integers k_1 and k_2*, i.e., if

$$\frac{P_{4}}{P} = \frac{n}{n_{4}} = \frac{k_1}{k_2} , \tag{4.32}$$

the revolution periods P_{4} and P of the two celestial bodies (or their mean motions n_{4} and n) are said to be *commensurable*.

Without the planetary perturbations the solution of the three-body problem Jupiter–Sun–minor planet would be periodic with period

$$P_{\mathrm{res}} = k_1 \, P = k_2 \, P_{4} . \tag{4.33}$$

In general, the perturbations due to the planets are small compared to the main term (due to the Sun), implying that the perturbing forces acting on the minor planet are (almost) periodic, or quasi-periodic, with period P_{res}.

Commensurability may give rise to *resonance*, i.e., to orbit perturbations which are orders of magnitude greater than under normal circumstances (i.e., if the revolution periods are *not* commensurable). This is why the period P_{res} also is referred to as *resonance period*.

Two bodies of the planetary system are said to be in conjunction (as seen from the Sun), if their heliocentric ecliptical longitudes are the same. In the two-body approximation the minor planet's synodic revolution period P_{syn}, the time interval between subsequent conjunctions of the minor planet and Jupiter, may be expressed as:

$$P_{\mathrm{syn}} = \frac{2\,\pi}{n - n_{4}} = \frac{P\,P_{4}}{P_{4} - P} . \tag{4.34}$$

Therefore, if the condition (4.32) holds, the resonance period must be an integer multiple of the synodic revolution period:

$$P_{\mathrm{res}} = (k_1 - k_2)\,P_{\mathrm{syn}} . \tag{4.35}$$

For commensurabilities of type $2:1$, $3:2$, $4:3$, etc., the synodic revolution period and the resonance period (4.33) are identical.

The broad scope of this book does not allow it to address resonant motion in detail. The tools developed in the previous Chapters, in particular the numerical solution methods outlined in Chapter I- 7 and the variational equations presented in Chapter I- 5, are, on the other hand, ideal for studying resonant motion in the planetary system. This is documented by the pioneer work of Wisdom in the early 1980s, e.g., in [130] and [131], who used a combination of numerical and analytical methods, and who had, last but not least, the best computer equipment for that particular purpose available at that time. Today, a comparable performance may be expected from any (reasonable) commercially available PC. Many other significant contributions to the field cannot be dealt with here. The reader is referred to [80] for a comprehensive discussion and overview.

Some important properties of the minor planets' resonant motion shall now be discussed using the numerical tools developed in this work. We will proceed in three steps:

- In the next paragraph technical problems related to the solution of the equations of motion and the associated variational equations in the special case of resonant motion will be addressed.

- The methods to distinguish between quasi-periodic (regular) and chaotic solutions of the equations of motion are introduced afterwards.

- Numerical experiments related to the (3:2)- and (3:1) resonance zones illustrate the problems.

Primary and Variational Equations of Resonant Motion. The equations of motion of a minor planet are the same whether or not a resonance condition of type (4.32) holds. The general structure of the equations is reflected by eqns. (I-5.57). If a resonance condition holds (approximately), the solution characteristics, when integrating over thousands of years, may change substantially. The development of the eccentricity is, in particular, close to unpredictable. It may very well happen that an orbit with a small eccentricity of $e < 0.1$ develops into an orbit with eccentricities $e > 0.3$. This in essence rules out all numerical solution methods which are based on constant stepsizes. Automatic stepsize control is a requirement when analyzing resonant motion. This is why the collocation method with stepsize control (see section I- 7.5.5) has to be used throughout this section for numerical experiments.

The variational equations (I- 5.58) associated with the equations of motion (I- 5.57) have already been established in Chapter I- 5. Subsequently, we will only be interested in the variational equations for the initial orbital elements, which is why only the homogeneous part of eqns. (I- 5.58) has to be considered. What was said for the associated primary equations (I- 5.57) also holds for the variational equations: the mathematical structure of these equations is the same whether or not condition (4.32) holds. If this condition holds, the characteristics of the particular solution may, however, lead to serious numerical problems, because the absolute value of such a solution may grow exponentially. Fortunately, this problem can be easily dealt with by observing that the solution of the two initial value problems

$$
\begin{aligned}
\ddot{z}_1 &= A_0\, z_1 \\
z_1(t_i) &= z_{10} \\
\dot{z}_1(t_i) &= z_{11}
\end{aligned}
\tag{4.36}
$$

and

$$
\begin{aligned}
\ddot{z}_2 &= A_0\, z_2 \\
z_2(t_i) &= \lambda\, z_{10} \\
\dot{z}_2(t_i) &= \lambda\, z_{11}
\end{aligned}
\tag{4.37}
$$

are simply related by

$$
z_2(t) = \lambda\, z_1(t) \ .
\tag{4.38}
$$

It is thus easily possible to normalize the variational equations for each integration step, e.g., by asking the absolute value of the initial value to meet the condition $|z(t_i)| = 1$. Instead of solving the original initial value problem, one may simply solve the normalized version. In order to reconstruct the original initial value problem (referred to the initial epoch t_0), one only has to keep track of the (natural or other) logarithm of the product of all normalization constants (which is the sum of logarithms of all these constants). Program PLASYS follows the procedure outlined above, if the collocation method is used.

In program PLASYS the variational equations associated with the equations of motion of the minor planet w.r.t. (one or more of) its osculating elements at the initial epoch are solved simultaneously with the primary equations (the original equations of motion). In view of the facts that the variational equations are linear and that, locally, a good analytical approximation is available (using the two-body approximation for the equations of motion), this procedure is not very efficient. It is, however, the simplest method to implement.

Let us use the notation

$$
I \in \{a_0, e_0, i_0, \Omega_0, \tilde{\omega}_0, u_0\}
$$

$$
z_I(t) \stackrel{\text{def}}{=} \left(\frac{\partial r}{\partial I}\right)(t) \ ,
\tag{4.39}
$$

introduced in Chapter I-5 to characterize the partial derivatives of the orbit w.r.t. its initial osculating elements referring to t_0. So far, we either used the time T_0 of perihelion passage or the mean anomaly σ_0 at t_0 as the sixth element. As indicated above, the argument of latitude $u_0 \stackrel{\text{def}}{=} \omega_0 + v_0$ is used here, instead. This particular element has the advantage to avoid quasi-singularities associated with orbits of small ecentricities. It is (obviously) a function of the classical elements.

The function $z_I(t)\,\Delta I$ describes the effect on the reference orbit (the considered minor planet's orbit) at time t caused by a small (infinitesimal) change ΔI in the orbital element I, provided the numerical values of all other initial osculating elements are identical with those of the reference trajectory. Obviously, the solutions $z_I(t)$ of the variational equations (I- 5.58) are very well suited to study the stability of a particular orbit w.r.t. small changes in the initial conditions.

Deterministic Chaos. A dynamical system is said to be chaotic (one also speaks of a deterministic chaos), if the absolute value of (one of) the solution(s) (4.39) of the variational equations (I- 5.58) grows exponentially with time t. Alternatively, the system is called regular or quasi-periodic. In the former case, a tiny change in the corresponding initial value causes dramatic changes of the celestial body's position vectors after a long time period.

Let us introduce the $\gamma_I(t)$-function of the partial derivative (4.39) as the following scalar function of time:

$$\gamma_I(t) \overset{\text{def}}{=} \frac{\ln |z_I(t)|}{t - t_0} \ . \tag{4.40}$$

This $\gamma_I(t)$-function related to the osculating element I is well suited to characterize the stability of the solution w.r.t. small changes in the particular initial osculating element I. It would be nice (and logical) to call this $\gamma_I(t)$-function the *Ljapunov function*, in honour of Alexander Michailovich Ljapunov (1857–1918). Unfortunately, the term has a different meaning in the theory of dynamical systems, which is why we stick to the not very inspiring term of $\gamma_I(t)$-function.

Let us assume, e.g., that the solution of the variational equation w.r.t. the osculating element I asymptotically behaves as

$$|z_I(t)| = f(t) + \lambda_0 \, e^{\gamma\,(t-t_0)} \ ,$$

where the function $f(t) \geq 0$ may be any function with a less than exponential growth in time t. In the above formula γ is a positive constant, which should not be confused with the function $\gamma_I(t)$. We will, however, associate it now with the asymptotic behavior of the function $\gamma_I(t)$.

Under the assumptions specified the asymptotic behavior of the $\gamma_I(t)$-function for $t \to \infty$ must be that of a constant, which is called the *maximum Ljapunov characteristic exponent* or simply the *Ljapunov characteristic exponent*:

$$\gamma \overset{\text{def}}{=} \lim_{t\to\infty} \left(\frac{\ln |z_I(t)|}{t - t_0} \right) \ . \tag{4.41}$$

If we integrate over sufficiently long time spans, the Ljapunov characteristic exponent will become visible as the asymptotic value of the $\gamma_I(t)$-function (4.40).

If the function $|z_I(t)|$ actually is a superposition of more than one exponential function, we are only able to see the constituent with the maximum exponent. This explains the attribute "maximum" in the expression maximum Ljapunov characteristic exponent.

The Ljapunov time is defined as the inverse of the Ljapunov characteristic exponent:

$$T_\gamma \stackrel{\text{def}}{=} \frac{1}{\gamma} \ . \tag{4.42}$$

For $|t - t_0| < T_\gamma$, the exponential growth of the solution of the variational equation will be barely visible; for $t = 10\,T_\gamma$, an initial separation of two trajectories will have grown by a factor of $e^{10} \approx 22000$, i.e., the exponential part of the error is in general the dominating feature (the precise behavior depends on the properties of the other constituent $f(t)$ of the function $|z_I|$). The Ljapunov time is a good indicator for the time-span over which deterministic predictions of a particular trajectory are reliable.

The definition (4.41) of the Ljapunov characteristic exponent and the definition (4.42) of the Ljapunov time are consistent with that adopted by Wisdom (see, e.g., [130]). According to the above developments we have to distinguish between six different $\gamma_I(t)$-functions and exponents γ for the six osculating elements referring to the initial epoch t_0. In practice, the distinction becomes immaterial, because the six solutions $z_I(t)$ are solutions of one and the same linear system of variational equations, but with different initial values. The dynamic behavior of the system is, however, governed by the structure of the system, and not by the initial values. This is why we can use simply the variational equation corresponding to one of the elements, if we only want to describe the general properties of a dynamical system. This was why we used only one symbol γ to characterize the maximum Ljapunov exponent. From now on we will also skip the index "I" for the function $\gamma_I(t)$ and simply write $\gamma(t)$. (Remark: the structure of the functions $z_I(t)$ has not yet been studied in detail over time intervals of millions of years; such studies might add considerable knowledge to the development of resonant orbits.)

The concepts outlined above, taken from the theory of dynamical systems, will be used below to characterize some of the properties of the motion of minor planets in or near the commensurability zones visible in Figure 4.43. Despite the fact that we are only studying resonant motion of the minor planet w.r.t. Jupiter, all the outer planets are included in the numerical experiments with program PLASYS. As already mentioned the collocation method with automatic stepsize control is used for all numerical experiments.

The Hilda Group. A few numerical tests with orbits of semi-major axes greater than $a \approx 3.5$ (minor planets beyond the Hecuba gap with $P_4 : P \approx 2$)

show that such orbits in general change dramatically in short time spans (thousands of years to a few hundred thousand years). According to Figure 4.43 there are only two exceptions to this rule, namely the *Hilda group* and the *Trojans*. The Trojans have already been dealt with in the context of the problème restreint in section I- 4.5.2. Here we focus on Hilda-type objects. The Hilda group (see Figure 4.43) is a small group of minor planets with orbital periods P related to the orbital period P_{4} of Jupiter by

$$P_{4} : P \approx 3 : 2 \quad \text{corresponding to} \quad a \approx 4\,\text{AU} . \tag{4.43}$$

Before performing some realistic experiments with Hilda-like objects, it makes sense to calculate two orbits in the environment of the problème restreint. Program NUMINT was used for this purpose. Two orbits with $P_{4} : P \stackrel{\text{def}}{=} 3 : 2$ were integrated over a time interval of 10000 years. Both orbits have identical initial osculating elements (eccentricity $e = 0.1$, $\Omega = i = 0°$, perihelion passing time $T_0 = t_0$) except for the longitude of perihelion ω, which was selected to be parallel to the heliocentric vector $r_{4}(t_0)$ at the initial epoch t_0 in the first case, and antiparallel in the second case. This means that the conjunctions originally took place while the test particle was at perihelion (first case) or at aphelion (second case).

The result of the simulations (in the system rotating with the system Sun-Jupiter, marked with "+" on the x-axis of the coordinate system) is contained in Figures 4.61. The first set of initial conditions (left picture, corresponding to initial conjunctions in perihelion) obviously leads to a pseudo-periodic motion, whereas the second setup leads to a rather chaotic motion. Observe, however, that the test particle is constrained (essentially) by the Hill surfaces of zero velocity in both cases. One expects that this condition (conjunctions originally taking place near the perihelion of the test object) also should be met under more general conditions.

Two types of Hilda-like orbits, both with $P_{4} : P = 3 : 2$, and an initial (osculating) eccentricity of $e = 0.1$, are numerically integrated in the force field of the Sun and the outer planets Jupiter to Pluto in order to introduce the general problem. As opposed to the problème restreint, Jupiter's orbit is elliptical and all of its orbital elements are functions of time. Both types of orbits initially were assumed to be inclined by $5°$ w.r.t. the invariable plane. The osculating inclination and longitude of node were chosen according to eqn. (4.5) as $i = 6.58°$ and $\Omega = 170.6°$. The orbit types differ only in the relative orientation of the perihelia of the test object and of Jupiter. The two types of orbits, namely *identical heliocentric longitudes of perihelia, i.e.,* $\tilde{\omega} = \tilde{\omega}_{4}$ *(case 1)* and *difference between heliocentric longitudes of perihelia of Jupiter and test object, i.e.,* $\tilde{\omega} = \tilde{\omega}_{4} + 180°$ *(case 2)*, are studied subsequently. Individual orbits do moreover differ by the initial position of the test particle within its orbit (which may, e.g., be expressed by its true anomaly v).

Let us now consider two individual orbits. In the first case we assume that $\tilde{\omega} = \tilde{\omega}_{4}$ and that the test particle is at aphelion while Jupiter passes perihelion

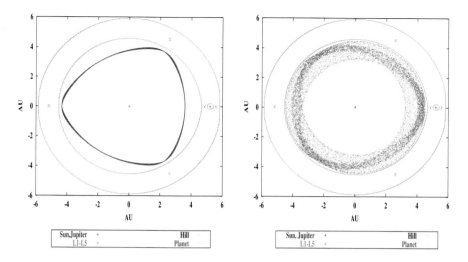

Fig. 4.61. Two Hilda-like objects over a time interval of 10000 years. Conjunctions initially near Hilda object's perihelion (left), near its aphelion (right)

(i.e., $v_{4}(t_0)0°$ and $v(t_0) = 180°$, v_{4} and v standing for the true anomaly of the Jupiter and the test particle's orbit, respectively). This means that initially the conjunctions take place while both, Jupiter and the test particle, are close to their perihelia. Based on the experiences gained with the problème restreint one would therefore expect a quasi-periodic motion of the test object. Observe that the initial true anomaly $v(t_0)$ might have been defined in such a way that the conjunctions take place at Jupiter's aphelion and the Hilda object's perihelion (this is achieved by selecting $v(t_0) = 270°$).

For the second individual orbit the perihelia of Jupiter and of the test particle are selected in opposite directions as seen from the Sun, i.e., $\tilde{\omega} = \tilde{\omega}_{4} + 180°$. Furthermore, the test object and Jupiter are assumed to cross their perihelia at the initial epoch t_0. Under these circumstances the conjunctions – at least initially – take place under the "worst" possible conditions, Jupiter being at perihelion while the test particle is at aphelion. From the experiences gained with the problème restreint one expects a rather chaotic motion in this case. As compared to the problème restreint, the situation is aggravated by Jupiter's orbital eccentricity. Observe that it is also possible to select the test object's initial anomaly $v(t_0)$ in such a way that the conjunctions (initially) take place at the minor planet's perihelion and Jupiter's aphelion (promising the "mildest" possible perturbations during the conjunction). This would be achieved by setting $v(t_0) = 90°$.

Figure 4.62 illustrates the initial positions (Jupiter $J(t_0)$ and the Hilda object $H(t_0)$) and the (approximate location of the) first conjunction between the two celestial bodies for the two individual orbits considered subsequently.

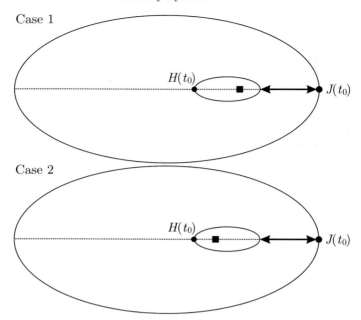

Fig. 4.62. Initial orientation of the perihelia of Jupiter and the test object, initial positions, and first conjunction ($\leftrightarrow$) of the two bodies (position of the Sun marked by a square)

Figure 4.63 gives an overview of Jupiter's and the test particle's positions during the experiment (this time in the inertial system). The first experiment (Figure 4.63 (left)) lasts for the scheduled one million years and no signs of significant orbital changes can be seen. The positions of Jupiter and of the minor planet are contained in two comparatively thin, well defined annuli, the widths being defined in essence by $2\,e_{\twoplus}\,a_{\twoplus}$ and $2\,a\,e$, respectively.

The second experiment (right) came to a close after a relatively short time span – after somewhat less than 200000 years the test particle had a very close encounter with Jupiter (perhaps even resulting in a crash on the giant planet) after quite a few previous close encounters with the planet (these encounters are documented in the general program output file). Whereas the Jupiter positions are contained – as already stated– in a rather thin annulus around the Sun, the test particle occupies a rather extended area in this particular view of the ecliptic. As opposed to the motion governed by the problème restreint (compare Figure 4.61) the test object's positions are *not* confined by a Hill surface. One can imagine that the semi-major axis and the eccentricity must have been heavily perturbed in the short "life-time" of the test particle.

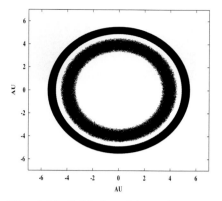 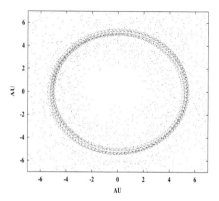

Fig. 4.63. Orbital positions of two Hilda-type objects, with initial orientation of the perihelia $\tilde{\omega} = \tilde{\omega}_{\textrm{4}}$ and $v(t_0) = 180°$ (left, case 1, over one million years)), and with the perihelion of the minor planet opposite to Jupiter's perihelion and $v(t_0) = 0°$ (right, case 2) (about 200000 years)

The pseudo-periodic solution obviously corresponds to the case of (initially) identical perihelia and conjunctions (initially) near the test object's perihelion, whereas the unstable solution corresponds to the case of (initially) opposite perihelia (as seen from the Sun) and (initial) conjunctions near the test object's aphelion.

Figure 4.64 gives an overview of the development of the semi-major axes (left) and the eccentricities (right) of the test particle for the first (top) and the second (bottom) individual orbit. The outcome of the two experiments hardly can be more different: Loosely speaking, one sees an extremely chaotic behavior in the case $\tilde{\omega} = \tilde{\omega}_{\textrm{4}} + 180°$ and (initial) conjunctions near the test object's aphelion and a pseudo-periodic behavior in the case $\tilde{\omega} = \tilde{\omega}_{\textrm{4}}$ and (initial) conjunctions near the test object's perihelion.

In the first case the semi-major axis varies over one million years only within the range of a few percent, the eccentricity between the limits of about $0.02 \leq e \leq 0.15$. In the second case the semi-major axis and the eccentricity change dramatically in a relatively short time period. Quite a few close encounters between Jupiter and test particle were also documented in the general output file. The eccentricity reaches values of about $e_{\textrm{max}} \approx 0.7$, the semi-major axis varies between about 3.2 and 20 AU (!). This explains the large scatter in the distribution of the test particle's positions.

The remarks concerning the stability of the two orbit types is confirmed by Figure 4.65, which shows the $\gamma(t)$-functions as a function of time t. The $\gamma(t)$-functions were calculated with the variational equation w.r.t. the initial

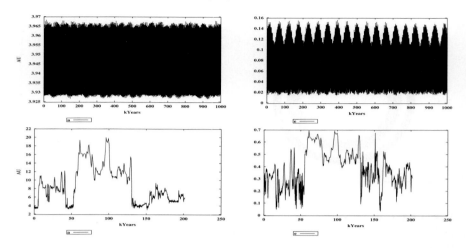

Fig. 4.64. Semi-major axes (left) and eccentricities (right) of Hilda-type objects with perihelia $\tilde{\omega} = \tilde{\omega}_{2\!\!\!+}$ and anomaly $v(t_0) = 180°$ (top, case 1) and with perihelion of minor planet opposite Jupiter's perihelion and $v(t_0) = 0°$ (bottom, case 2)

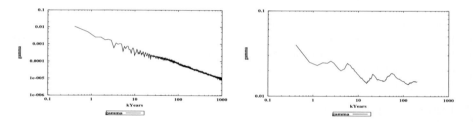

Fig. 4.65. $\gamma(t)$-functions of Hilda-type objects with perihelia $\tilde{\omega} = \tilde{\omega}_{2\!\!\!+}$, $v(t_0) = 180°$ (left, case 1) and opposite perihelia, $v(t_0) = 0°$ (right, case 2)

osculating semi-major axis a_0. Only one value of the solution of the variational equations was stored per about 410 years.

Whereas a maximum Ljapunov characteristic exponent of $\gamma \approx 0.015$, corresponding to a Ljapunov time of $T_\gamma \approx 67$ years, is observed in the second case, no chaotic behavior can be detected in the first case. A short Ljapunov time of $T_\gamma = 67$ years implies that already after 20000 years a small error of Δa in the semi-major axis will create effects of the order of $e^{20000/67}\,\Delta a \approx 4.4 \cdot 10^{129}\,\Delta a$ in the solution. As in numerical integration small errors of the order of 10^{-15} may and will occur at any integration step, and as these errors are magnified (or reduced) with the factor dictated by the variational equations, one would need more than 110 digits to perform a meaningful integration – even over such a short time interval! This also means that the solution discussed above and represented, e.g., in Figure 4.64 (bottom), should not be understood as a deterministic solution, but as the realization of an initial value problem of a stochastic differential equation system. The solution was actually generated with a collocation method of order $q = 12$ using automatic stepsize control allowing for a maximum local error of $1.0 \cdot 10^{-16}$ AU/day.

If, e.g., the integration order, or the numerical value of the error criterion, or the initial conditions, are only slightly modified, a very different realization of the same stochastic differential equation system may result. Figure 4.66 illustrates this effect by showing the development of the semi-major axis and the eccentricity of the unstable solution shown in Figure 4.64 (right) and the development of the same quantities for a solution based on identical initial conditions, the same integration method, but using a numerical value of $0.99 \cdot 10^{-16}$ instead of $1.0 \cdot 10^{-16}$ AU/day to control the local errors in the velocity components. In this second solution the crash on Jupiter does occur already after about 50000 years, and even before, after an initial phase of a few hundred years, the two solutions differ substantially!

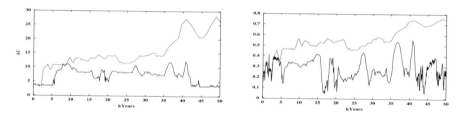

Fig. 4.66. Semi-major axes (left) and eccentricities (right) of two Hilda-type objects with perihelia opposite to Jupiter's perihelion using different integration error criteria

It was mentioned initially that for both initial orientations of the perihelia (see Figure 4.63) the initial true anomaly $v(t_0)$ of the test object may be selected in such a way that the initial conjunctions take place either at the test object's perihelion or aphelion. It is of course even possible to find an initial orbital position $v(t_0)$ of the test object to let the initial conjunctions take place near any chosen value v_c of the true anomaly v of the test object. When performing experiments of this kind (with either case 1 or case 2 of the initial orientation of the orbital planes) one realizes that a regular motion results if the initial conjunctions take place near the test object's perihelion and that a chaotic motion results when the initial conjunctions take place near the test object's aphelion. These findings are the same as in the case of the problème restreint.

Figure 4.67 shows the $\gamma(t)$-functions for six individual orbits, three per case of the initial orientation of the orbital plane, where the initial conditions were selected to let the initial conjunctions take place at $v_c = (0°, 90°, 180°)$.

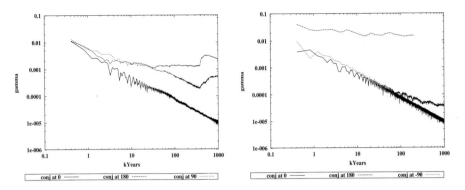

Fig. 4.67. $\gamma(t)$-function of three Hilda-type with conjunctions at $v(t_0) = (0°, \pm 90°, 180°)$. Longitudes of perihelia initially identical (left), opposite of Sun (right)

When the conjunctions initially take place near the test object's perihelion there is no sign of of a chaotic behavior in the time interval considered. Also, when the conjunctions initially take place near the test object's aphelion, a finite asymptotic value γ for $\gamma(t)$ results, indicating a chaotic motion. The asymptotic value of $\gamma(t)$ is larger for case 2 (perihelia initially opposite of the Sun), which is plausible because the distances between Jupiter and the test object are smaller during the conjunctions in this case. The orbits, where the initial conjunctions take place at anomaly values of $v \approx \pm 90°$, indicate that the regularity zone is rather small.

Figure 4.68 shows the development of the longitudes of perihelia in the six cases documented in Figure 4.67. The perihelia rotate (in the retrograde sense) relatively rapidly in those cases leading to a pseudo-periodic motion, whereas the rotation of the perihelia is rather slow when a chaotic behavior is observed. The rotation of the test object's aphelion is synchronized with the rotation of Jupiter's perihelion under these circumstances.

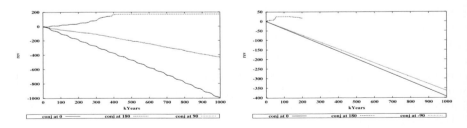

Fig. 4.68. Longitudes of perihelia of three Hilda-type objects with conjunctions at $v(t_0) = (0°, \pm 90°, 180°)$. Longitudes of perihelia initially identical (left), opposite of Sun (right)

Figure 4.69 (left), the histogram of conjunctions as a function of the test objects' true anomaly, gives more information. In both cases there is a pronounced preference to avoid the test objects' aphelion for conjunctions. For the initial orientation $\tilde{\omega} = \tilde{\omega}_4 + 180°$ and $v(t_0) \approx 90°$ (which leads to "mildest possible" conjunctions), the test particle even manages to let *all* conjunctions take place near its perihelion. Figure 4.69 (right) also shows that for the regular case there is a clear preference to avoid Jupiter's perihelion for conjunctions.

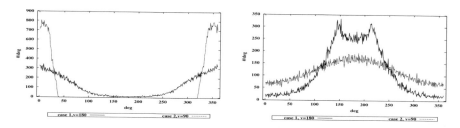

Fig. 4.69. Histogram of conjunctions, as a function of the minor planet's true anomaly (left), and as a function of the minor planet's aphelion w.r.t. Jupiter's perihelion (right); initial conjunctions near minor planet's perihelion

The above simulations let us expect that the perihelia of the real Hilda objects follow more or less the pattern of Figures 4.69. This hypothesis is confirmed by Figures 4.70, which shows the histograms of the conjunctions of all Hilda objects closest to the osculation epoch contained in the file MPCORB.DAT dated March 4, 2001, which was retrieved from the MPC homepage. Figure 4.70 (left) illustrates the distribution of the actual conjunctions as a function of the minor planets' true anomalies of the 419 Hilda objects found in the file MPCORB.DAT. Obviously the surviving Hilda objects follow the expectations rather closely. The preference to let the conjunctions take place near Jupiter's aphelion is also confirmed by Figure 4.70 (based on the real data). The histograms have a resolution of $5°$.

More information about the dynamical properties of the Hilda-group of asteroids may be found in [104] and [105]. Comparative studies of the Hilda group with the Hecuba gap (the (2:1)-commensurability of the revolution of test objects with that of Jupiter) may be found in [39].

The (3:1)-Commensurability. Wisdom studied the (3:1)-commensurability with analytical and numerical experiments. He showed (see, e.g., [130] and [131]) that a large chaotic zone is associated with this commensurability. His work was a breakthrough in the way of interpreting simulation-type results: Whereas previous attempts to explain the gaps directly interpreted the perturbations in the semi-major axes, Wisdom considered the perturbations in all orbital elements for that purpose. Figures of type 4.43 seemingly asked

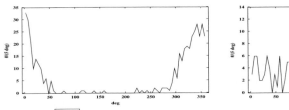 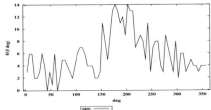

Fig. 4.70. Histogram of conjunctions of real Hilda objects, as a function of the minor planet's true anomaly (left), and as a function of the minor planet's aphelion w.r.t. Jupiter's perihelion (right)

for the classical way of interpreting the simulations, but Wisdom's numerical experiments showed that the perturbations in the eccentricities e are far more spectacular than those in the semi-major axes a. Even when starting with osculating (better: proper) eccentricities of the order $e \approx 0$, the eccentricities $e(t)$ eventually assume rather high values, in particular values $e \geq 0.3$, which make asteroids in or near the (3:1)-commensurability Mars orbit-crossing. This in turn allows for a "brute force" explanation, a collision with (or heavy perturbations by) Mars, to eliminate the resonant minor planets – obviously a very efficient way to generate gaps.

This pioneer work shall now be illustrated with a few numerical experiments using numerical tools, by studying the evolution of a few trajectories in or near the (3:1)-commensurability zone.

In the subsequent experiments the revolution periods of the test particles were varied according to the following scheme:

$$P_i \stackrel{\text{def}}{=} \left(\tfrac{1}{3} + i \cdot 0.001\right) , \quad i = -7, -6, \ldots, +7 . \tag{4.44}$$

Using the value of $P_4 = 11.86$ years (Table 4.1) this implies that the osculating semi-major axes at the initial epoch t_0 were

$$P = 0.326\bar{3}\,P_{\textrm{4}} \; ; \quad a = 2.465$$
$$P = 0.327\bar{3}\,P_{\textrm{4}} \; ; \quad a = 2.470$$
$$P = 0.328\bar{3}\,P_{\textrm{4}} \; ; \quad a = 2.475$$
$$P = 0.329\bar{3}\,P_{\textrm{4}} \; ; \quad a = 2.480$$
$$P = 0.330\bar{3}\,P_{\textrm{4}} \; ; \quad a = 2.485$$
$$P = 0.331\bar{3}\,P_{\textrm{4}} \; ; \quad a = 2.490$$
$$P = 0.332\bar{3}\,P_{\textrm{4}} \; ; \quad a = 2.495$$
$$P = 0.333\bar{3}\,P_{\textrm{4}} \; ; \quad a = 2.500 \tag{4.45}$$
$$P = 0.334\bar{3}\,P_{\textrm{4}} \; ; \quad a = 2.505$$
$$P = 0.335\bar{3}\,P_{\textrm{4}} \; ; \quad a = 2.510$$
$$P = 0.336\bar{3}\,P_{\textrm{4}} \; ; \quad a = 2.515$$
$$P = 0.337\bar{3}\,P_{\textrm{4}} \; ; \quad a = 2.520$$
$$P = 0.338\bar{3}\,P_{\textrm{4}} \; ; \quad a = 2.525$$
$$P = 0.339\bar{3}\,P_{\textrm{4}} \; ; \quad a = 2.530$$
$$P = 0.340\bar{3}\,P_{\textrm{4}} \; ; \quad a = 2.535$$

With this selection of test orbits we expect to scan through the commensurability in steps of about 0.005 AU .

The osculating eccentricity was initially set to $e = 0.1$, the initial inclinations i and the longitudes Ω of the ascending node (w.r.t. the ecliptic $J2000.0$) were defined as

$$\Omega = 107.6°$$
$$i \ = 6.58° \,, \tag{4.46}$$

i.e., all test particles have an initial inclination of $i = 5°$ w.r.t. the invariable plane, and the node initially lies in the intersection of the invariable plane with the ecliptic $J2000.0$ (see eqn. (4.5)).

The solutions were generated with a collocation method of order $q = 12$ with automatic stepsize control. The tolerance in the components of the velocity vector was set to $10^{-16}\,\textrm{AU/d}$.

Two series of solutions were produced: In the first series, hereafter called *Series A*, the perihelia of the test particles were initially opposite to Jupiter's perihelion, in the second series, hereafter called *Series B*, the two perihelia longitudes were initially identical:

$$\tilde{\omega}_{(\textrm{Series A})} = \tilde{\omega}_{\textrm{4}} + 180°$$
$$\tilde{\omega}_{(\textrm{Series B})} = \tilde{\omega}_{\textrm{4}} \,. \tag{4.47}$$

The time of perihelion passage of the test particles was defined to agree initially with that of Jupiter in both series:

$$T_0 \stackrel{\textrm{def}}{=} T_{\textrm{4}_0} \,. \tag{4.48}$$

With this definition of the longitudes of perihelia and of the associated times of perihelia passage, the conjunctions initially take place *either* while both celestial bodies are in their perihelia *or* aphelia in Series B. For series A the conjunctions initially take place at distances of approximately ±90° from the perihelia (or aphelia) of the two planets. Figure 4.71 illustrates the initial positions $T(t_0)$ and $J(t_0)$ of the test particle and of Jupiter, respectively, and the initial conjunction geometries ($\leftrightarrow$).

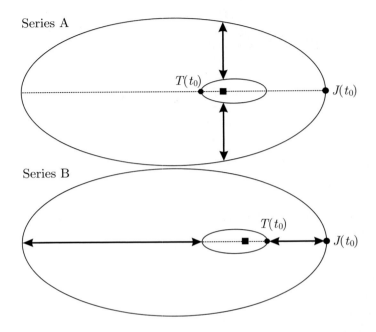

Fig. 4.71. Initial orientation of perihelia of Jupiter and test object, initial positions, and first conjunction ($\leftrightarrow$) of the two bodies (position of the Sun marked by a square) for the experiments in (3:1)-commensurability

Violent instabilities of the kind encountered in the experiments with Hilda-type objects (with Ljapunov-times of the order of one century) do not occur in the case of the (3:1)-commensurability. The key difference between the two series A and B becomes apparent in Figure 4.72, showing the projection of the Laplace vectors onto the plane of the ecliptic for the two first solutions of series A and B (with $P = \left(\frac{1}{3} - 0.007\right) P_{24} \approx 0.326333\,P_{24}$).

The examples used in Figure 4.72 are obviously not yet in deep, but in shallow (3:1)-resonance. We see essentially the same behavior as in the case of an "ordinary" minor planet. This statement is confirmed by the spectra of the motion of the Laplace vector and in Figure 4.73: Only three strong spectral lines are observed. In analogy to the interpretation of Figure 4.59 two lines (centered at periods of about 45000 and 350000 years) may be associated with Jupiter perturbations, the one near 31000 years with the proper eccentricity.

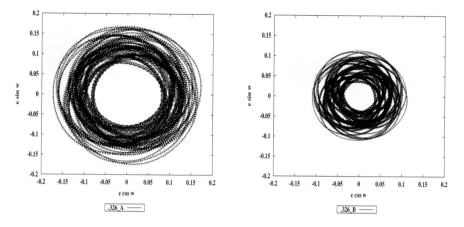

Fig. 4.72. Projection of the Laplace vectors of two test particles in (3:1)-resonance with Jupiter over one million years ($P = 0.326\overline{3}\, P_{\mathrm{4}}$). Series A (left), series B (right)

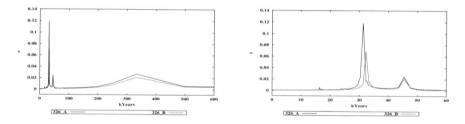

Fig. 4.73. Spectra of the projection of the Laplace vectors of two test particles in shallow (3:1)-resonance with Jupiter over one million years ($P = 0.3263\overline{3}\, P_{\mathrm{4}}$)

The main difference between the two data sets thus consists of the power of the spectral line associated with the proper eccentricity. From Figure 4.72 and Figure 4.73 the mean eccentricities over long times are seen to be

$$\bar{e}_{\text{(Series A)}} \approx 0.13$$
$$\bar{e}_{\text{(Series B)}} \approx 0.07 \ . \tag{4.49}$$

These mean values over the longest possible time span should be defined as the proper eccentricities of the test particles.

The development of the mean semi-major axes over one million years of the two test series is illustrated by Figure 4.74. The differences between the test particles in deep and shallow resonance are truly remarkable: Whereas the variations are constrained to a few thousands of an AU for objects in shallow resonance, variations up to $0.05 - 0.07$ AU are observed in deep resonance.

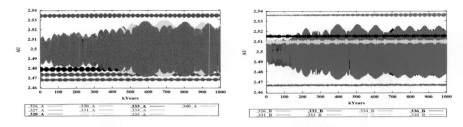

Fig. 4.74. Mean semi-major axes of test particles in (3:1)-resonance with Jupiter over one million years. Series A (left), Series B (right)

According to eqns. (4.45) the test particles were originally equally distributed in the interval $[a_{-7}, a_{+7}] \approx [2.465, 2.535]$. This distribution is preserved (more or less) only near the upper and lower limits of the resonance zone. Figures 4.74 illustrate that the minor planets try to avoid the zone of deep resonance, which would support the statistical argument to explain the particular Kirkwood gap. The gap created is, however, much too narrow to explain the actual distribution of the minor planets' semi-major axes (see Figures 4.43).

Figures 4.75 illustrate the development of the mean eccentricities for test series A. The first five members of the series are in the top row, the last four in the bottom row. The figures on the left-hand side cover the entire time span of one million years, the figures on the right-hand side only a time slot of 100000 years.

A clear distinction can be made between the test particles in shallow resonance and those in deep resonance: Whereas the test objects in shallow resonance show more or less regular periodic variations within the limits $0.08 < e < 0.16$, irregular variations with maximum values of up to $e_{max} \approx 0.7$ are seen for the test particles in deep resonance. From Figure 4.75 one may conclude that particles in deep (3:1)-resonance are characterized, at least temporarily, by orbits of eccentricities of the order $e \geq 0.3$. This fact implies that the perihelia of such orbits are at heliocentric distances of $r = a(1 - e) = 1.75$ AU , which may come close to the orbit of Mars. Orbits with an eccentricity of $e = 0.6$ have their perihelia in the vicinity of the Earth's orbital curve.

Similar statements may be made for the test Series B (Figure 4.76), where the zone of deep resonance is somewhat narrower, which might indicate that the size of resonance zone is a function of the (proper) eccentricity.

Figures 4.75 and 4.76 prove that asteroids in deep (3:1)-resonance may penetrate deeply into the inner planetary system. Wisdom [130] proposes that all asteroids in the (3:1)-commensurability which develop eccentricities of the order $e \geq 0.3$ are removed by the planets of the inner planetary system. When adopting this hypothesis one obtains a good agreement of the observed

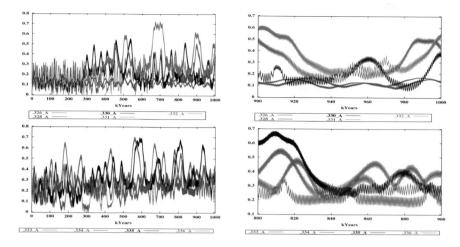

Fig. 4.75. Eccentricities of test particles in (3:1)-resonance with Jupiter over one million years (left), over 100000 years (right). Objects with $P = (0.326, 0.328, 0.330, 0.331, 0.332) P_{2\!\!\!\!\!\;}$ in top, objects with $P = (0.333, 0.334, 0.335, 0.336) P_{2\!\!\!\!\!\;}$ in bottom row; Series A

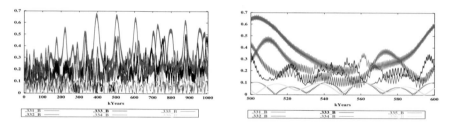

Fig. 4.76. Eccentricities of test particles in (3:1)-resonance with Jupiter over one million years (left), over 100000 years (right), Series B

histogram of minor planets (see Figure 4.43) and the results of Monte-Carlo type simulations of test particles in the (3:1)-commensurability of the kind performed here.

Figures 4.77 indicate that the chaotic aspects are not confined to the semi-major axis and the eccentricity, but that they are visible in the development of the orbital poles, as well. Only four examples are given in Figure 4.77. With $P = 0.32633 P_{2\!\!\!\!\!\;}$ (top, left), $P = 0.33133 P_{2\!\!\!\!\!\;}$ (top, right), $P = 0.33333 P_{2\!\!\!\!\!\;}$ (bottom, left), and $P = 0.33633 P_{2\!\!\!\!\!\;}$ (bottom, right) in essence the entire chaotic zone of test series A is spanned. The curve for $P = 0.34033 P_{2\!\!\!\!\!\;}$ would be (almost) undistinguishable from the first one ($P = 0.32633 P_{2\!\!\!\!\!\;}$ (top, left)). In deep resonance we see excursions of the orbital pole of the order of 15°, i.e., about three times the value expected for the regular case (top, left).

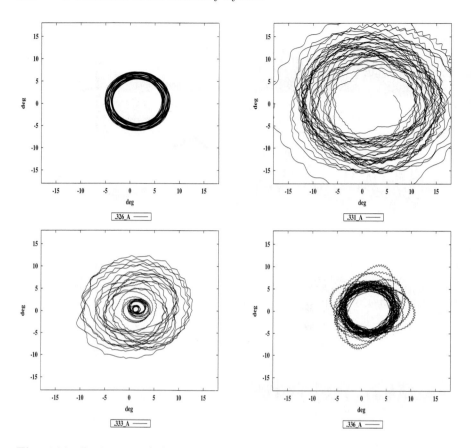

Fig. 4.77. Projection of the orbital poles of the test particles in (3:1)-resonance with Jupiter on the ecliptic over one million years; Series A

Figure 4.78 shows the development of the Laplace vectors for the same four test particles. The results in Figure 4.78 are consistent with those of Figure 4.77: The first curve (top, left) corresponds to what we expect from an ordinary minor planet, whereas the other three curves, corresponding to the zone of deep resonance, seem to be chaotic in nature.

For both, Figures 4.77 and 4.78, we might produce the corresponding spectra. As both spectra show in essence the same features, only the spectra associated with the Laplace vectors are reproduced in Figure 4.79.

Eight amplitude spectra are shown in Figure 4.79. In order to improve the visibility, two spectra with the same initial osculating revolution period P are given per sub-figure, one corresponding to test series A, the other to test series B. The spectra associated with $P = 0.326\,P_4$ are given in the top row (these

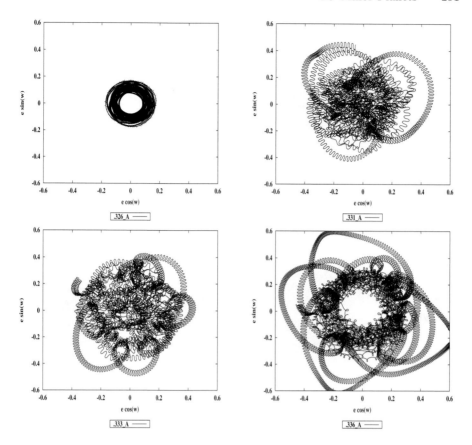

Fig. 4.78. Projection of the Laplace vectors of test particles in (3:1)-resonance with Jupiter on the ecliptic over one million years. Initially, perihelia opposed to Jupiter's perihelion (series A)

spectra are identical with those in Figure 4.73), those with $P = 0.331\,P_{2\!\!4}$ in the second, with $P = 0.333\,P_{2\!\!4}$ in the third, and with $P = 0.336\,P_{2\!\!4}$ in the last row. The first column of Figures 4.79 shows the spectra in the range $[0, 600000]$ years, the right column in the range $[0, 60000]$ years.

It is striking that in Figure 4.79 the simple pattern of three spectral lines, one associated with the proper elements, two with the Jupiter perturbations, completely disappears for the test particles in deep resonance. The spectral line corresponding to the proper element becomes much weaker, or even disappears completely. The lines corresponding to the Jupiter perturbations are always visible in deep resonance, in most cases they are even much enhanced compared to the case of shallow resonance.

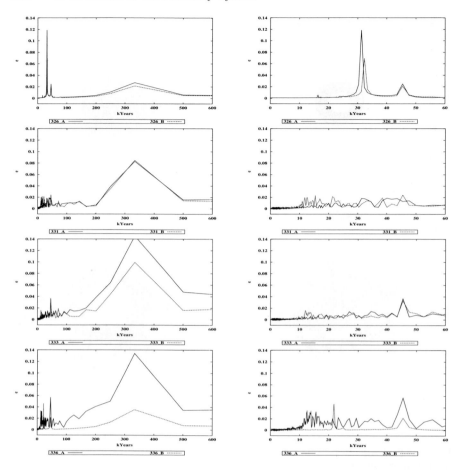

Fig. 4.79. Spectra of the projection of the Laplace vectors on the ecliptic of eight test particles in (3:1)-resonance with Jupiter ($P = 0.32633\,P_{2\!\!\!+}$, $P = 0.33133\,P_{2\!\!\!+}$, $P = 0.33333\,P_{2\!\!\!+}$, and $P = 0.33633\,P_{2\!\!\!+}$), one from test series A and B per picture; left column: over 600000 years, right column: over 60000 years

From such experiments one may conclude that the "three spectral line model" is in essence correct for all minor planets between Mars and Jupiter, except for objects in deep resonance. The top and bottom row (only in the case of series B) of Figure 4.79 illustrate this fact. It is interesting to keep track of the location of the line associated with the proper elements.

Only the prograde part of the spectrum was provided in Figure 4.79. This is fully justified for the spectra corresponding to orbits in shallow resonance, because there is no significant power in this part of the spectrum. It cannot be justified for the spectra corresponding to deep resonance. Figure 4.80

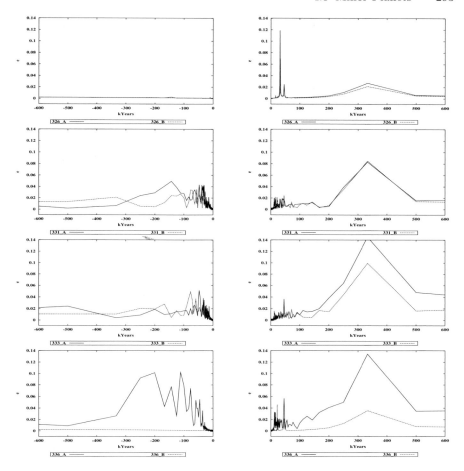

Fig. 4.80. Retrograde (left) and prograde (right) part of the spectra of the projection of the Laplace vectors on the ecliptic of eight test particles in (3:1)-resonance with Jupiter ($P = 0.32633\,P_{2\!\!\!+}$, $P = 0.33133\,P_{2\!\!\!+}$, $P = 0.33333\,P_{2\!\!\!+}$, and $P = 0.33633\,P_{2\!\!\!+}$), one from test series A and B per picture

underlines this statement. In order to facilitate the comparison, all spectra in Figure 4.80 are reproduced in the same scale. The figures nicely document that the only common features are the lines with periods near 45000 and 350000 years. Apart from that the spectra are typical for a non-periodic, random motion of the Laplace vectors.

Figures 4.81 and 4.82 show the $\gamma(t)$-functions associated with test series A and B in the logarithmic representation. (In order to improve the visibility, not all of the members of the two series were included.)

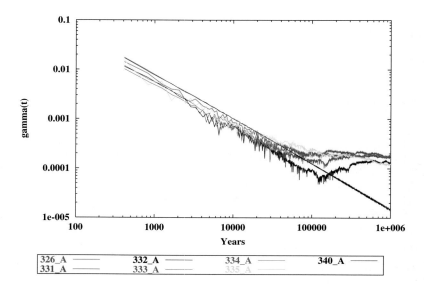

Fig. 4.81. $\gamma(t)$-function over one million years for test particles in or near the (3:1)-resonance. Initially, the perihelia were opposed to Jupiter's perihelion (series A)

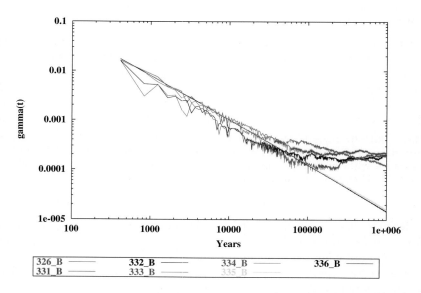

Fig. 4.82. $\gamma(t)$-function over one million of test particles in (3:1)-resonance. Initially, the perihelia were aligned with Jupiter's perihelion (series B)

In essence we may distinguish two regimes in Figures 4.81 and 4.82, one corresponding to the regular motion, with

$$\lim_{t \to \infty} \gamma(t) \to 0 \ ,$$

and one corresponding to chaotic motion, where

$$\lim_{t \to \infty} \gamma(t) = \gamma$$

with a finite value $\gamma > 0$ for the Ljapunov characteristic exponent γ. Typical values for the maximum Ljapunov characteristic exponent γ in the (3:1)-resonance vary between the limits

$$0.0001 < \gamma < 0.0002 \ , \tag{4.50}$$

corresponding to Ljapunov times of about 5000 to 10000 years.

Figures 4.81 and 4.82 should be interpreted cautiously: we are only able to safely detect chaotic components with Ljapunov times $1/\gamma < \Delta T_{\text{int}}$, where ΔT_{int} is the length of the integration interval. The cases classified as non-chaotic might very well show a chaotic behavior with Ljapunov characteristic exponents around $\gamma \approx 10^{-7}$, corresponding to Ljapunov times of about $T_\gamma \approx$ 10 million years, which still is a short time-span compared to the age of the planetary system.

Let us complete the experiments related to the (3:1)-commensurability with the histograms of the location of the aphelia w.r.t. Jupiter's perihelion and of the true anomaly of the test particles at the moment of a conjunction, in Figure 4.83, which shows that for all examples in both test series A and B there is a clear preference for aligning the perihelia of the test particles with that of Jupiter. This alignment is much better developed for test objects in deep resonance. The bottom figure (left), for the initial revolution period of $P = 0.333 \, P_4$ with the corresponding objects from test series A, test series B, and one object with an initial osculating longitude of perihelion $\tilde{\omega} = \tilde{\omega}_4 + 90°$ (with label "333_C"), nicely demonstrates this behavior.

The right column of figures shows that there is a clear preference for the conjunctions to take place midway between perihelia and aphelia, i.e., to avoid the line of apsides for conjunctions. This preference is much more pronounced for orbits in deep resonance than for the other orbits.

4.3.5 Summary and Concluding Remarks

The observational basis concerning the orbits of minor planets in the planetary system was reviewed in section 4.3.1 with the help of the archive of the MPC. Today one has to distinguish two belts of minor planets, the classical

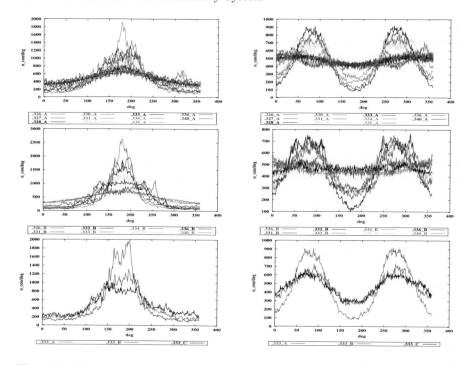

Fig. 4.83. Histogram of the aphelia of test particles (at conjunctions) w.r.t. perihelion of Jupiter (left column), and of the true anomalies at conjunctions (right column); series A in top row, series B in middle row, example for $P = 0.33\overline{3} P_{4}$ in bottom row

belt between Mars and Jupiter and the Edgeworth-Kuiper belt with aphelia in the region of Pluto's aphelion. Despite the fact that the observational basis for the objects in the outer belt is steadily growing and is already now considerable, we were essentially concerned with the classical belt of minor planets between Mars and Jupiter in this section.

Quite a few structural elements can be extracted from the data base of the MPC:

- The Kirkwood gaps and groups of planetoids are clearly seen in the histogram of the semi-major axes (or of the associated revolution periods) in Figure 4.43.

- A ring structure becomes visible in Figure 4.41 of the osculating aphelia of the minor planets. As a celestial body with an eccentric orbit (according to Kepler's law) spends more time near the aphelion (than near perihelion), this ring structure would become visible in a figure of the time averages (over one revolution) of the minor planets' positions.

- A clear asymmetry of aphelia, already apparent in Figure 4.41, is confirmed by the histogram 4.42.

- Figure 4.44 reveals more of the structure of the belt of minor planets. In the (a, i)-plane so-called families of minor planets, presumably of common origin, can be identified. Such families were first described and named by Hirayama after members of the families in 1918 [56].

- Figure 4.45 showing a ring structure of the projections of the osculating orbital poles on the ecliptic also supports the existence of groups of minor planets of common origin – provided the orbital poles of the fragments of a collision remain in the same precession cone of the "mother planetoid". This ring structure is also clearly reflected by the histogram 4.46 of inclinations w.r.t. the invariable plane.

- The histogram of eccentricities 4.47 does not show a clear structure. It is interesting, however, that, as opposed to the eccentricities of the major planets (see Table 4.1), the mean values for the eccentricities are not close to $e = 0$. (Admittedly, it would be preferable to produce a histogram of mean or proper eccentricities, but the main characteristics would remain the same).

In sections 4.3.2, 4.3.3 and 4.3.4 some of the observed peculiarities of the minor planets' orbits and of their distribution between Mars and Jupiter were addressed form the theoretical point of view. The essential elements of the analysis were:

- Consequent use of the tool of numerical integration in program PLASYS (see documentation in Chapter 10 of Part III).

- Analysis of mean orbital elements and of functions thereof, to reveal essential characteristics of the orbits and their evolution in time.

- Generation of spectra of the series of mean elements (and functions thereof) to extract the essential periods and amplitudes of the mean orbital elements.

- Solution of the variational equations associated with (one or some of) the osculating elements at the initial epoch, to further characterize the orbit, in particular to distinguish between regular and chaotic motion.

The analysis was performed in two steps: We first looked at "one million years in the life of an ordinary minor planet" (in sections 4.3.2 and 4.3.3) then we focused on minor planets in resonant motion with Jupiter (in section 4.3.4), in particular on the (3:2)- and (3:1)-commensurabilities, as prominent representatives of a group of asteroids and a the Kirkwood gaps.

The key findings of the analysis of the orbits of "ordinary" minor planets were:

- The mean semi-major axes are remarkably stable (see Figures 4.48 and 4.50).

- The mean eccentricities are rather stable, as well (no secular terms or terms with very long periods). The amplitudes of the periodic variations are, however, considerable (of the order of several hundredths, see Figures 4.49 and 4.51). The amplitudes of these periodic variations are (mainly) a function of the semi-major axis. This explains the shape of the histogram 4.47 of osculating eccentricities of real minor planets.

- A change of the initial osculating eccentricity does not affect the amplitude of the long-period variations of the eccentricities, but the mean value of the eccentricity (see Figures 4.49 and 4.51).

- The orbital poles precess in the retrograde sense about the pole of the invariable plane (see Figure 4.53). The angle of the orbital pole w.r.t. the pole of the invariable plane is in essence given by the initial inclination w.r.t. the invariable plane.

- The inclination also shows long-period variations of the order of up to about one degree. The stability of the mean inclinations (Figure 4.53) very strongly supports Hirayama's arguments in [56] concerning the common origin of families of minor planets.

- The spectra of the projections of the Laplace vectors and of the orbital poles onto the ecliptic reveal that the motion of these vectors may be represented as the linear combination of three circular motions, corresponding to the three dominant spectral lines. The amplitude of only one spectral line may be altered by the initial conditions. It makes thus sense to consider the amplitude of this spectral line as the proper eccentricity or the proper inclination, respectively.

- The zero-order term of the harmonic series describing the motion of the orbital poles in all cases agrees very well with the projection of the pole of the invariable plane, whereas the zero-order term, when spectrally analyzing the Laplace vectors, in general has the value $(0, 0)$.

- The proper elements emerging from numerical methods correspond quite well to the proper elements of the theory of secular perturbations as formulated by Brouwer and Clemence in [27]. There are, however, also important differences (due to the non-modelled inclination dependence, differences of the frequencies for proper inclination and proper eccentricity, etc. in the theory by Brouwer and Clemence).

- In view of the availability of fast computers and of the differences seen between the classical secular theory and the numerical results, it would make sense to complement the analytical theory of secular perturbations with a numerical analogue.

In the analysis of the motion of minor planets in resonant motion we introduced the variational equations as an essential tool to distinguish between regular and chaotic motion. It is much simpler to use the solution of the variational equations as indicators of stability than to actually integrate two different orbits. All problems associated with the exponential growth of the difference between trajectories may be handled very easily when using this technique (see Chapter I-5). All problems associated with the "re-normalization" may be avoided when using the variational equation approach. The computation of the maximum Ljapunow characteristic exponent γ becomes a very inexpensive tool which should accompany each numerical test.

Only few experiments related to the resonant motion of minor planets could be performed in section 4.3.4. We were nevertheless able to identify stable and chaotic solutions associated with the (3:2)-commensurability (the Hilda group) and with the (3:1)-commensurability.

The analysis of the spectra of the orbital poles and of the Laplace vectors with the tools of spectral analysis gives additional insight into resonant motion. A clear distinction can be made between the spectra of regular motion (in shallow resonance) and of chaotic motion (in deep resonance). Whereas the regular case is very closely related to the case of an ordinary minor planet (with three dominating spectral lines, one associated with the initial values, two with Jupiter), the spectra are completely different in the case of chaotic motion: There is little or no power left in the line associated with the initial values, whereas the other two lines become far more important. Also, as expected for spectra corresponding to a random motion, a multitude of additional lines show up in the "resonant" spectra. No consistent pattern could be seen for these additional lines. One might make an attempt to classify the spectra corresponding to resonant motion. Some aspects (like the strength of the lines associated with the two lines attributed to Jupiter and Saturn, and dis- and reappearance of the line corresponding to the proper elements) might be good indicators for a classification. It might be more difficult to find a meaningful way to characterize the "chaotic lines" of the spectrum.

It is remarkable that the key results of pioneer work, such as, e.g., performed by Wisdom in the early 1980s, may nowadays be reproduced with few numerical experiments without too much analytical or numerical work. One should keep in mind, however, that a complete statistical treatment of a commensurability or even of the entire belt of minor planets requires much more systematic work. Advanced analyses of this kind do not fit within the scope of this book. The methods developed in this section would, however, be ideally suited for further investigations in this fascinating field of science.

Part III

Program System

Part III

Feynian system

5. The Program System Celestial Mechanics

5.1 Computer Programs

The two volumes of this work offer a thorough overview of the theory of modern computational Celestial Mechanics. The author is, however, convinced that, apart from the theoretical background, a "hands-on" introduction to a broad variety of applications is of equal importance. Solving practical tasks in the field of Celestial Mechanics means using (and sometimes even developing) efficient numerical algorithms and computer programs. The close relation between theory and applications, realized by means of specially developed computer routines, is a (if not the) central aspect of this book; the computer programs represent an integral part of this compendium of Celestial Mechanics. The program should help to better understand the theory, stimulate experiments and (perhaps) give additional pleasure when reading and digesting the main body of the book.

Several aspects matter in connection with computer program systems. In the field of science mathematical correctness, numerical stability and the efficiency of the algorithms are the key issues. The development of a scientific program system is a rather lengthy process, where usually "old" and reliable pieces of software are reused and incorporated into a new systems. This approach may "ruin" the technical level of the programs (from the point of view of computer science and of professional program developers) and their user-friendliness.

Eight computer programs accompanying this book were written by the author, who made extensive use of routines written by a team of the Astronomical Institute, University of Bern. Some of the routines (quite a lot of them, actually) stem from the so-called Bernese GPS Software [58], a program system designed for the processing of data of the Global Positioning System. The Bernese GPS Software is sold as a source code package. It would have been desirable to follow the same approach here. Time and resource limitations excluded this approach. The program system described in Part III is "only" available in the form of executable programs.

Computer programs accompanying a book should have a certain level of user-friendliness – comprising the actual usage of the programs as well as the installation of the system on a given computer platform. Prof. Leoš Mervart from the Technical University of Prague, who wrote his Ph.-D. thesis at the Astronomical Institute of the University of Bern, is an experienced designer of scientific software systems. He already developed the user surface of the latest version of the Bernese GPS Software. The entire design of the program system's user interface including the incorporation of the accompanying programs into a user-friendly environment were performed by Leoš Mervart. The graphical user interface has been developed using the Qt library (see `http://www.trolltech.com`). A brief description of this menu system is provided in the following sections.

5.2 Menu System

The first impression of the graphical user interface of CelestialMechanics is obtained by studying Figure 5.1 showing the primary menu of CelestialMechanics. The menu system allows the user to

- prepare the input options,
- start the programs,
- browse the output files,
- graphically display some of the more important results, and
- consult the help panels (in the HTML-format) "in real time", while preparing a program run. Figure 5.2 shows the start of the help file available with the primary menu shown in Figure 5.1.

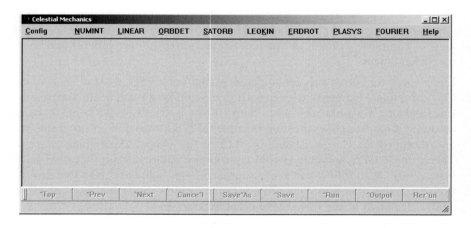

Fig. 5.1. Graphical user interface of system CelestialMechanics

The control elements (menus, buttons, "editable" fields, checkboxes etc.) are standard. It is assumed that Microsoft-Windows users feel familiar with these elements.

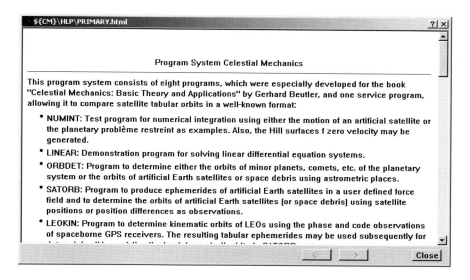

Fig. 5.2. Primary help panel of the system CelestialMechanics

5.2.1 Installation

The program system CelestialMechanics is distributed on a CD and can be installed on any computer with an operating system from the Microsoft Win32 family (Windows 95, 98, 2000, NT, and XP). There are no special requirements considering the hardware. The programs themselves take approximately 11.8 Mb of the disk space. One has to keep in mind, however, that the program output files may become very large. Some of the programs perform very CPU-intensive tasks. The integration of the planetary motion over millions of years may, e.g., result in a rather long program run. Machines equipped with Pentium IV processors should, however, not experience major difficulties related to the performance of the system.

The installation is initiated by clicking the `setup` program icon on a CD. The user must say where the system shall be installed. The default location is `C:\CelestialMechanics`. According to the location specified the setup program sets the environment variable `%CM%` that contains the name of the CelestialMechanics's root directory (i.e., `C:\CelestialMechanics` per default). Then the setup program copies all the necessary files and creates the following directory structure:

```
%CM%\BIN
     \GEN
     \HLP
     \INP
     \SKL
     \NUMINT
     \LINEAR
     \ORBDET
     \SATORB
     \LEOKIN
     \ERDROT
     \PLASYS
     \FOURIER
```

There are eight program-specific directories where the programs look for data and store results. Some of these directories contain further subdirectories. Apart from that, there are five general directories:

BIN contains the executable files of the eight accompanying programs and the CelestialMechanics menu system itself.

GEN contains various files for general use and of general interest (Earth orientation parameters, constants, etc.).

HLP contains the on-line help files (in HTML format) that can be displayed directly from within the menu, and

INP contains nine input files of the programs and the file CM_MENU.INP, where the configuration of the CelestialMechanics menu system is stored. The input files are ASCII files that may be viewed in any text editor. We do not recommend you, however, to edit these files unless you exactly know what you do. Corrupted input files may cause a malfunctioning of the entire program system.

SKL contains a copy of the subdirectory INP. Used to restore initial parameter settings.

5.2.2 Running a Program

After selecting an appropriate menu item (e.g., *Fourier*→*Run program*) using the computer-mouse or (alternatively) pressing the combination Alt-f-r on the keyboard, the menu system allows it to edit the input options of the program. Figure 5.3 shows Panel FOURIER 1, the first panel displayed, when starting to prepare a program run.

For each program, the options are contained in one or more input panels. Each program panel contains the name of the program and a panel number in the title line. In addition there may be some comments characterizing the principal purpose of the panel. The panels are presented in the order of

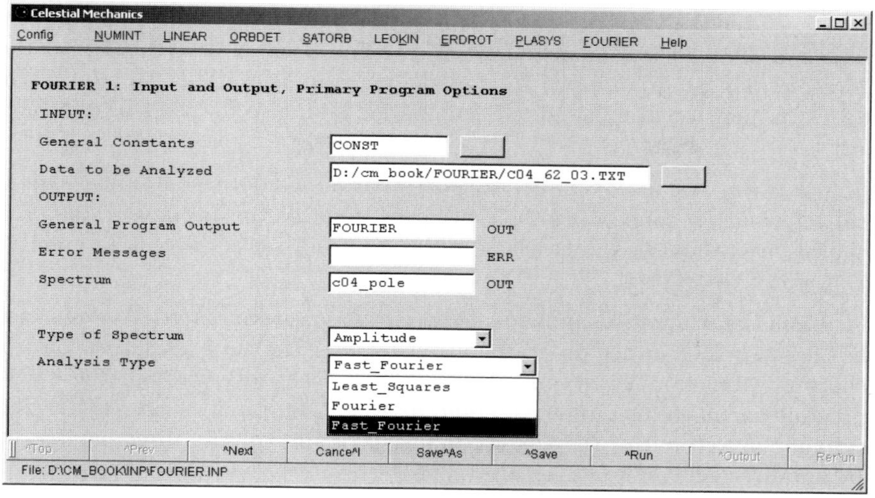

```
· Celestial Mechanics                                                    _ |□| x|
 Config   NUMINT   LINEAR   ORBDET   SATORB   LEOKIN   ERDROT   PLASYS   FOURIER   Help


    FOURIER 1: Input and Output, Primary Program Options
    INPUT:

    General Constants           |CONST                 |      |

    Data to be Analyzed         |D:/cm_book/FOURIER/C04_62_03.TXT       |      |
    OUTPUT:

    General Program Output      |FOURIER               |   OUT

    Error Messages              |                      |   ERR

    Spectrum                    |c04_pole              |   OUT


    Type of Spectrum            |Amplitude          ▼|

    Analysis Type               |Fast_Fourier         ▼|
                                |Least_Squares          |
                                |Fourier                |
                                |Fast_Fourier           |
||  ^Top    ^Prev      ^Next   |  Cance^l  |  Save^As |  ^Save  |  ^Run  |  ^Output  |  Re^run  |
    File: D:\CM_BOOK\INP\FOURIER.INP
```

Fig. 5.3. Program FOURIER - panel FOURIER 1

increasing panel numbers. Depending on the options selected, some numbers may not show up. Usually one number, e.g., "3" characterizes one panel. Panel numbers like "3a" and "3b" do also occur, in which case they characterize similar tasks in different program options. The user may browse through the panels using the *Next Panel, Prev Panel* buttons (or the corresponding shortcuts Ctrl-n and Ctrl-p) as often as required or wished. The options are specified by means of the following control elements:

- Editable fields are used, e.g., for specifying output file names.
- Editable fields with a selection button (e.g., the field "General Constants" in Figure 5.3) are used primarily for specifying input file names. After pressing the selection button a standard open-file dialog appears that allows the user to select an existing file.
- Checkboxes are used if the corresponding options require the simple yes/no answer.
- So-called "comboboxes" and "spinboxes" are used if the option has to be selected from a given set of alternatives or from a given range of values. An example is provided in the last line of panel FOURIER 1 in Figure 5.3, where one of the three analysis strategies has to be selected.

The menu system guides the user by reporting some invalid options (e.g., non-existing input files, missing output file names, etc.). Some options are only meaningful in connection with a particular setting of other options. These may be set "inactive" in certain panels if their value is irrelevant under the given circumstances.

After specifying all options the user may start the program by pressing the *Save and Run* button (shortcut `Ctrl-r`). The options are stored into the corresponding program input file and the program is started. An alternative to that is just to save the edited options for later use by pressing the *Save* button.

Programs run as console applications in separate windows containing in general information concerning their progress. One of the programs (`ORBDET`) requires even the user's interaction during runtime – it is the only interactive program of the system.

If the console window disappears, the program was terminated. At that time the user may look at the program output by pressing the *Last Output* button (shortcut `Ctrl-o`). All output (log) files are accessible through the corresponding menu items *Browse Output*, as well. The log files are ASCII files and they are stored in the corresponding program-specific directory (or its `OUT` subdirectory).

5.2.3 Visualizing the Results

Visualization of results is a very important step of data analysis. In the CelestialMechanics system there are two ways of visualizing results. The first way is provided in the menu system in the *Display Results* menu item related to the program used. An example of such a plot is given in Figure 5.4. It was produced by program FOURIER, after having spectrally analyzed the data set "c04_62_03.txt", which is provided as a startup example in the FOURIER subdirectory of the menu system. The figure is available (in color) in the primary on-line help panel. When activating the figure, one obtains initially the full spectrum (except, if a range of periods was specified). Figure 5.4, containing only a small (but interesting) portion of the entire spectrum, may then be generated by making use of the "zoom-option". Alternatively, one may specify a range (either in period or in frequency) in the corresponding panel. This method is preferable, if one would like to have precise interval boundaries, like e.g., $I = [300, 500]$, as in Figure 5.4.

The described way of visualizing results is very user-friendly. The authors of the CelestialMechanics program system tried to give the user the opportunity to display the most interesting and the most important results without the need to know anything about the structure of the output files and their formats. However, this part of the menu system has two limitations: (1) The plot facilities are programmed only for few typical examples. (2) The graphical tool may become rather slow when dealing with large data sets. The advanced users are therefore encouraged to use their own plotting programs to meet their own, perhaps more special, requirements. This is possible because the output files of all the programs are formatted (ASCII) files (as opposed to binary) and because the file content is described in detail in the subsequent

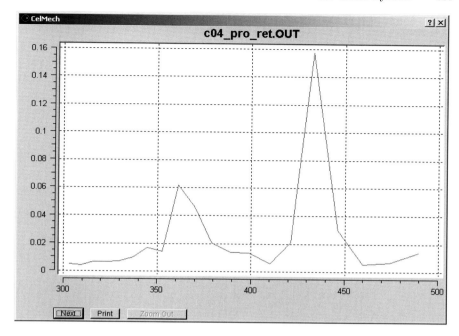

Fig. 5.4. Plot example: Prominent spectral lines in the polar motion data series (C04 pole of the IERS)

chapters of Part III. The on-line help panels contain, moreover, examples for most of the input and out data sets.

Most figures contained in the two volumes of this work on Celestial Mechanics were prepared with the Gnuplot utility (see http://www.gnuplot.info for more information). Gnuplot is a freeware software package, which may be installed from the internet.

The latest news concerning this software system are available under the URL http://www.aiub.unibe.ch/CelestialMechanics

6. The Computer-Programs NUMINT and LINEAR

NUMINT and LINEAR are test programs for numerical integration. NUMINT allows it in addition to generate the Hill surfaces of zero velocity of the problème restreint. Both programs are extensively used in Chapter I-7 to illustrate the performance of numerical integration algorithms. NUMINT numerically solves the non-linear equations of motion associated with satellite and minor planet orbits, LINEAR solves a limited set of typical linear differential equations and of systems of linear differential equations using special collocation methods exploiting the linearity of the systems.

6.1 Program NUMINT

The Panel NUMINT 1 in Figure 6.1 contains primary selections of the program NUMINT. The program may be used for

- integrating the orbit of an artificial satellite in the gravitation field of the Earth composed of the main term and the term C_{20} (see section I-3.4.1),

- integrating the orbit of a minor planet in the gravitation field of the problème restreint (see section I-4.5.2), and for

- the generation of files allowing it to "draw" the *Hill surfaces of zero velocity* in the framework of the problème restreint (see section I-4.5.2).

The program may be used to extract rather detailed CPU timing information. If the CPU time spent in the subroutines calculating the right-hand sides of the differential equations is measured each time this subroutine is called, this may (on certain platforms) significantly reduce the efficiency of the integration. It is therefore possible to by-pass this detailed timing procedure (last input line in Panel NUMINT 1, shown in Figure 6.1) and to measure only the total CPU time, instead.

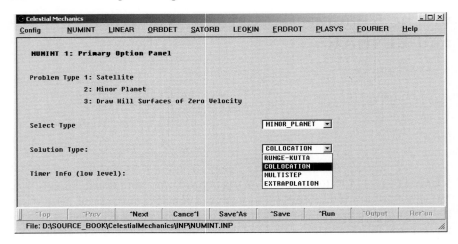

Fig. 6.1. Primary menu of program NUMINT

6.1.1 The Use of Program NUMINT for Numerical Integration

Panel 1 in Figure 6.1 shows that four types of algorithms suitable for numerically solving initial value problems associated with non-linear ordinary differential equations systems are implemented in program NUMINT, namely

- Runge-Kutta methods (see section I- 7.4.4),
- extrapolation methods (see section I- 7.4.5),
- collocation methods (see sections I- 7.4.1 and I- 7.5),
- multistep methods (see sections I- 7.4.2 and I- 7.5.6).

The solution method and the problem type may be selected in Panel NUMINT 1 in Figure 6.1

For all methods, except for the Runge-Kutta methods, the order of the method (equivalent to the order of the approximating truncated Taylor series development) may be defined within reasonable limits by the program user. The Runge-Kutta methods of the orders 4, 7 and 8 are available in program NUMINT.

For all methods, except for the multistep method, the stepsize h may be either selected as fixed or defined automatically by the program using certain error criteria. The error control is comparatively refined in the case of the collocation method, whereas rudimentary methods are underlying the Runge-Kutta and the extrapolation methods. For these methods better criteria may be found, e.g., in [88].

The menu system associated with program NUMINT is relatively straight forward. It does not make sense to reproduce here all the panels for all the methods implemented. We confine ourselves to provide and discuss the panels associated with the extrapolation method when applied to the integration of the problème restreint and assume that the program user will find his/her way through the panels for the other options.

Panel NUMINT 2 in Figure 6.2 is common to all integration methods. The program makes use of the constant-file (residing in the "GEN-subdirectory of the directory "CelestialMechanics). The content of the ASCII-file may be inspected, but should not be altered by the program user. Not counting the error file (which, when left blank, is identical with the general output file), three output files are generated in this program mode: the general output file with general information and statistics (timing information and number of function calls, etc.), the file with the osculating elements (relative to the initial osculating elements), and the file containing the position vectors and the differences of the numerically integrated orbit w.r.t. the initial two-body orbit. The contents of the file with the osculating elements may be viewed directly in the menu-system. The intersections of the Hill surfaces with the coordinate planes may also viewed in the menu-system. The contents of the *.osc- and the *.res-files may, of course, also be used to generate figures with the gnuplot package (see http://www.gnuplot.info for more information).

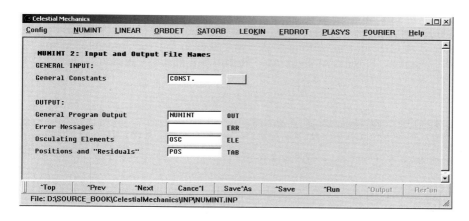

Fig. 6.2. Input- and output-files in program NUMINT

The *.osc-file contains the time argument in the first column, the differences of the osculating elements $a(t)$, $e(t)$, $i(t)$, $\Omega(t)$, $\omega(t)$, and $\sigma_0(t)$ (the mean anomaly at the initial epoch t_0) w.r.t. the corresponding initial osculating elements (referring to the initial epoch t_0) in the columns $2-7$ and the angle $\Delta\sigma(t) \stackrel{\text{def}}{=} n\ (t-T_0(t))-n_0\ (t-T_0(t_0))$ in column 8 (T_0 is the pericenter passing time). Columns $9-11$ contain the differences in radial (R), along-track (S), and out-of-plane directions (W) of the numerically integrated solutions at time t w.r.t. the two-body orbit (elements referring to t_0) referring to the same time t. When the perturbations are set to zero, these columns contain the integration errors in the three directions. Column 12 contains the step-size h (only of interest when this parameter is adjusted automatically in the program).

The *.res file contains the time argument in the first column and the corresponding three Cartesian coordinates in the columns $2 - 4$. The underlying coordinate system is either the inertial system (where the fundamental plane is *either* the orbital plane of the two finite bodies when a minor planet is considered *or* the equatorial plane when a satellite is considered) or the rotating coordinate system of the problème restreint (in the case of a minor planet) or the equatorial, Earth-fixed system (in the case of a satellite orbit). The coordinate system is defined in Panel NUMINT 6 in Figure 6.6. The columns $5 - 7$ of the *.res file contain the position differences (in the coordinate systems described above) of the numerically integrated orbit w.r.t. the two-body orbit with elements referring to the initial epoch t_0.

Panel NUMINT 3 in Figure 6.3 is only displayed if the orbit of a minor planet is integrated (or if the primary option HILL_Zero_V was chosen). In the case of integrating the orbit of a minor planet the program computes the Jacobi-constant (I- 4.109) associated with the orbit to be integrated using eqn. (I- 4.110) and it produces the files containing the information for the Hill surfaces. Three options related to the Hill surfaces may be selected:

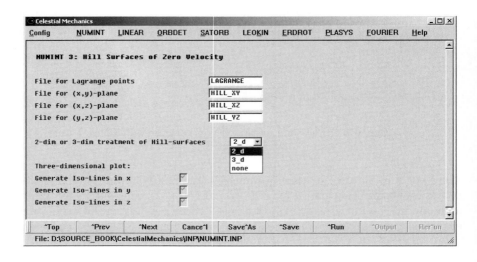

Fig. 6.3. Definition of Hill surfaces

- *none*: No Hill-related files are generated,
- *2-d*: the intersections of the Hill surfaces with the coordinate planes are generated and stored in three files, and
- *3-d*: the intersections of the Hill surfaces with the selected coordinate planes and the corresponding parallel planes are calculated and stored into one file.

The latter option is used to produce three-dimensional representations of the Hill-surfaces (examples were included in section I- 4.5.2).

Either no (option *none*), one (option *3-d*), or three (option *2-d*) Hill-file(s) may thus be generated when a particular problème restreint is solved. In addition the program generates one file containing the coordinates of the two finite bodies in the rotating system *and* the coordinates of the five stationary solutions associated with the particular problème restreint considered.

The first two lines in the file called "Lagrange" Panlel NUMINT 3 in Figure 6.3 correspond to the two finite point masses in the rotating system (only the first column, specifying the corresponding x-coordinates is relevant). Lines $3 - 7$ correspond to the five stationary solutions, where column 3 corresponds to the x-, column 4 to the y-coordinate. The somewhat peculiar format is optimized for using the *gnuplot*-package.

When using option *2-d* the files *HILL_XY, HILL_XZ*, and *HILL_YZ* contain only two columns (the (x, y)-, (x, z)- and the (y, z)-coordinates) of the intersections of the Hill-surfaces with the corresponding coordinate planes. When using option *3-d*, the first Hill-file contains three columns, where the columns correspond to x-, y- and z-coordinates of the intersections of the Hill-surfaces with the coordinate planes and parallel planes to these coordinate planes. The file may, e.g., be visualized with the gnuplot command *splot "hill_xyz" u 1:2:3 w d* (assuming that the first of the Hill-files was named accordingly).

Panel 4 in Figure 6.4 allows it to refine the definition of the problem type. The panel is only active if either the option MINOR_planet or satellite was selected in Panel NUMINT 1 of Figure 6.1. Depending on this selection either the first three or the latter three defining constants may be altered. The values supplied on the CD correspond to the Sun-Jupiter problème restreint and to the Earth. The panel also allows it to neglect the perturbations (internally either the mass of planet or the term C_{20} are set to zero).

Panel NUMINT 5, shown in Figure 6.5, allows it to define the initial orbital elements. Depending on the selection of the problem type in Panel NUMINT 1 in Figure 6.1 the semi-major axis is defined either directly as such (satellite) or via the revolution period in units of the revolution period of the planet with finite mass (minor planet).

Panel NUMINT 6 in Figure 6.6 serves to select the algorithm-independent specifications of the numerical integration, namely the length of the integra-

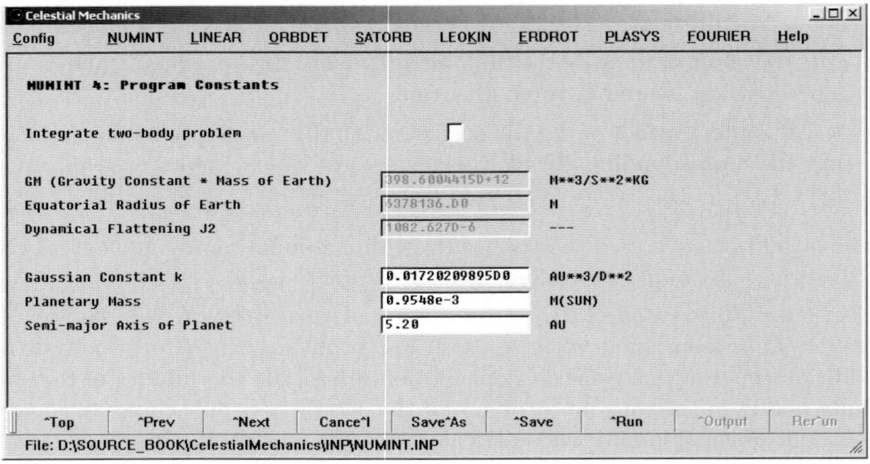

Fig. 6.4. Fine-tuning of the problem type

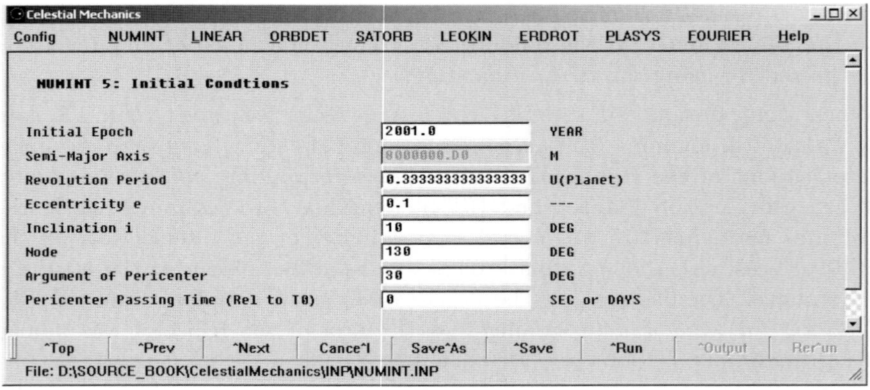

Fig. 6.5. Selection of the initial orbit

tion interval, the (initial) stepsize h and the sampling rate for the output files. Also, one may select the type of the coordinate system for the *.res file: either the rotating system (in the sense described above) or the inertial system may be chosen.

The algorithm-dependent integration specifications for the extrapolation method are defined in panels of the kind shown in Figure 6.7. They are numbered NUMINT 7a, 7b, 7c, or 7d, depending on the integration method. Only the order of the method (even orders between 2 and 16) may be defined in the example shown in Panel NUMINT 7d, characterizing extrapolation methods (see section I- 7.4.5). A constant stepsize was chosen in the example. If the

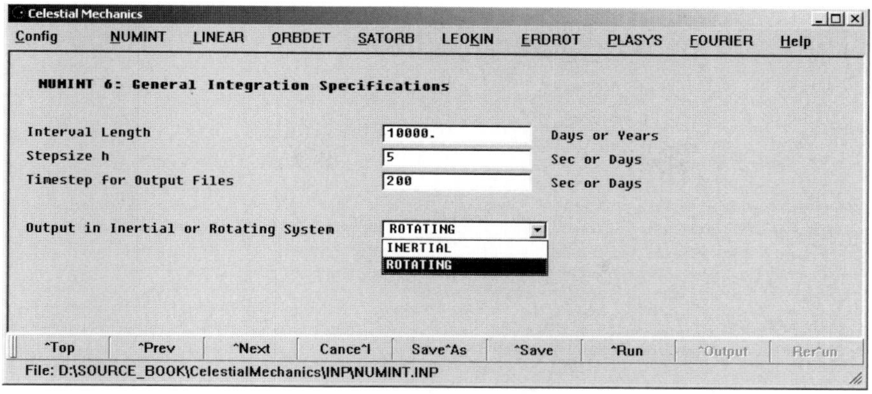

Fig. 6.6. Numerical integration: Algorithm-independent quantities

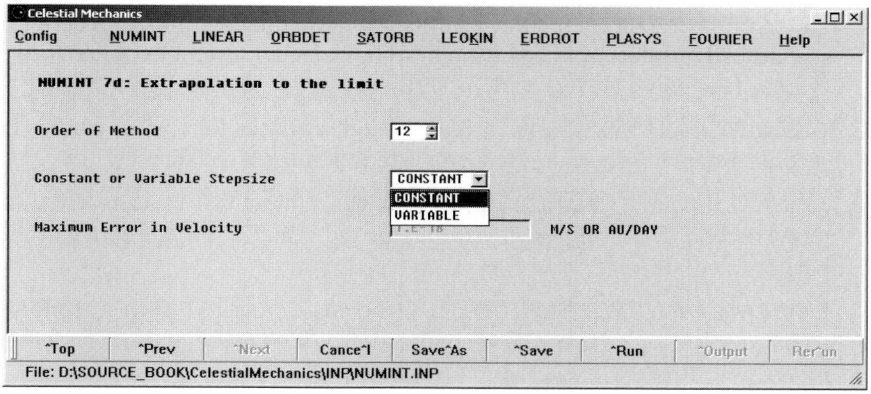

Fig. 6.7. Numerical integration: Algorithm-dependent quantities for extrapolation method

stepsize is to be automatically defined by the program, the maximum allowed error in the velocity must be specified in addition.

Panel NUMINT 7a in Figure 6.8 shows the menu corresponding to panel NUMINT 7d in Figure 6.7 for the collocation method. There are many more options available in this case – indicating that collocation is viewed by the author as a central method for orbital motion. The following sub-methods may be selected under the first option "Method":

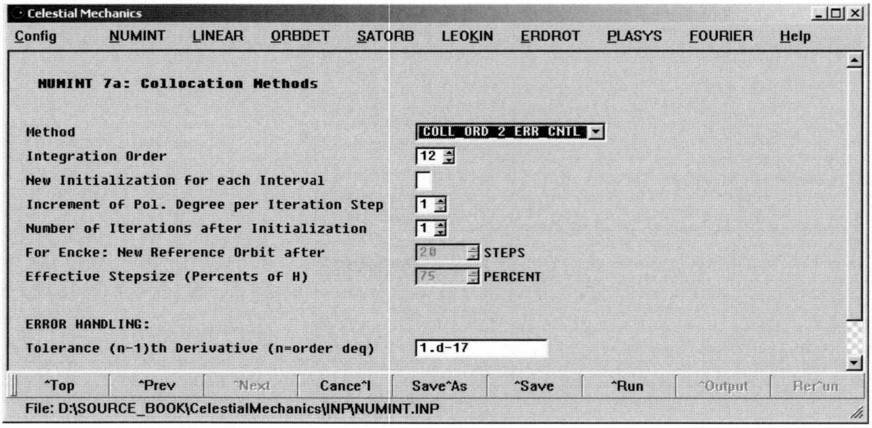

Fig. 6.8. Numerical integration: Algorithm-dependent quantities for collocation method

1. *COLLOC_ORDER_1*: Collocation method with fixed stepsize, where the second-order differential equation system is decomposed into a first-order system (see eqns. (I-7.7) and (I-7.9)).

2. *COLLOC_ORDER_2*: Collocation method with fixed stepsize, where the second-order differential equation system is integrated directly.

3. *ENCKE*: Encke's equations (I-6.4) for the differential motion w.r.t. an unperturbed two-body orbit are integrated using a collocation method with a fixed stepsize.

4. *COLLOC_ORDER_2_ERR_CNTL*: Collocation method with automatic stepsize selection, where the second-order differential equation system is integrated directly.

5. *ENCKE_ERR_CNTL*: Encke's equations (I-6.4) for the differential motion w.r.t. an unperturbed two-body orbit are integrated using a collocation method with automatic stepsize selection.

As usual, the order q of the collocation method may be chosen (between the limits $2 \leq q \leq 14$). With the option "new initialization after ..." one may enforce *not* to make use of the approximative solution of the previous integration step. The integration method becomes rather inefficient in this case, the result, however, may become slightly better. For the initial subinterval one may select the order of the system to increase the integration order (from initially $q = 1$ or $q = 2$) by 2 or only 1 order per iteration step. The number of iteration steps (after the initial subinterval) is chosen in the next input line. When integrating Encke's equations one must define the number n_{encke} of integration steps after which a new reference orbit has to be selected (the instantaneous osculating two-body orbit is chosen). A value between 1 and 40 may be chosen for n_{encke}.

For the methods with constant stepsize one may also decide *not* to use the approximating function over the entire subinterval, but to define the new initial values within this subinterval. Only reductions of 25%, 50% and 75% may be selected. The accuracy of the solution improves when the stepsize is reduced, but the efficiency is reduced, as well.

If a method with automatic error control is selected one has to define the maximum allowed errors in velocity. The on-line help panel offers hints and tps in this context. For more information concerning automatic stepsize control we refer to section I- 7.5.5.

6.1.2 The Use of Program NUMINT to Generate Hill Surfaces

With each orbit integrated (in the case of the problème restreint) the corresponding value of the *reduced Jacobi constant* is provided in the general output file. Also, files containing the intersections of the Hill surface with the coordinate planes (or with parallel planes) may be generated (see Panel NUMINT 3 in Figure 6.3 and the discussion associated with it). Program NUMINT allows it also, however, to generate the same files without actually performing an integration. The program option is invoked by selecting the primary option HILL_ZERO_V in Panel NUMINT 1, reproduced in 6.1. In this case, after having defined the options in the panels NUMINT 2 and NUMINT 3 (in Figures 6.2 and 6.3), additional options have to be defined in Panel NUMINT 3(cont) shown in Figure 6.9. If the two-dimensional Hill surface option was selected in Panel NUMINT 3 in Figure 6.3, up to five different reduced Jacobi constants may be selected in Panel NUMINT 3(cont) of Figure 6.9. The intersections with the coordinate planes may then be inspected using the "Display Results" option of program NUMINT or they may be drawn with the gnuplot-package (see http://www.gnuplot.info for more information). The on-line help file of program NUMINT contains the figure corresponding to the selection in Panel NUMINT 3(cont).

The numerical values in Panel NUMINT 3(cont) of Figure 6.9 correspond to Figure I- 4.16 (left column). The right column of the same figure was produced using the masses $m_1 = 1$ and $m_2 = 1/1047.35$ (see Table 4.1) to characterize the three-body problem Sun-Jupiter-minor planet.

If the "3-d" version is selected in Panel NUMINT 3 in Figure 6.3. Only one Jacobi-constant may be specified in this case, because otherwise the resulting figure would be "too busy". Consequently, only one (the first) Jacobi-constant may then be specified. Figure 6.10 shows two examples for the three-dimensional representation of the Hill surfaces, for $J = 4.0$ and $J = 2.5$. The mass ratio is the same as that underlying Figures I- 4.16 (left). The example 6.10 (left) correspond to the innermost case in the example of Figure

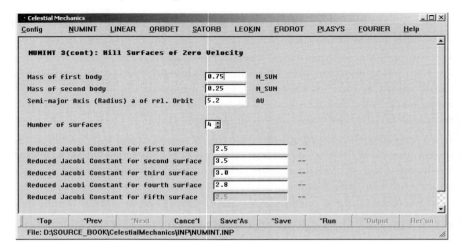

Fig. 6.9. Generation of several Hill surfaces of zero velocity

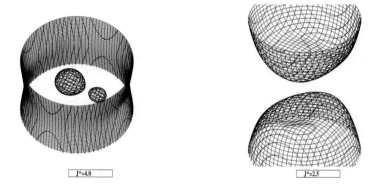

Fig. 6.10. Three-dimensional view of Hill's surfaces of zero velocity for $m_0 = 0.75 \, m_\odot$, $m_1 = 0.25 \, m_\odot$, $a_{01} = 5.2$ AU and $J = 4.0$ (left), $J = 2.5$ (right)

I- 4.16 (top, left). The test body is either constrained to the spheroidal regions around the bodies m_0 and m_1 or outside the cylindrical regions in Figure 6.10 (left). The test body may be move everywhere in Figure 6.10 (right) except within the bowl-shaped areas.

6.2 Program LINEAR

Panel LINEAR 1 in Figure 6.11 shows the primary menu of program LIN-EAR. The program allows it to solve the nine linear problems discussed in section I-7.6.6.

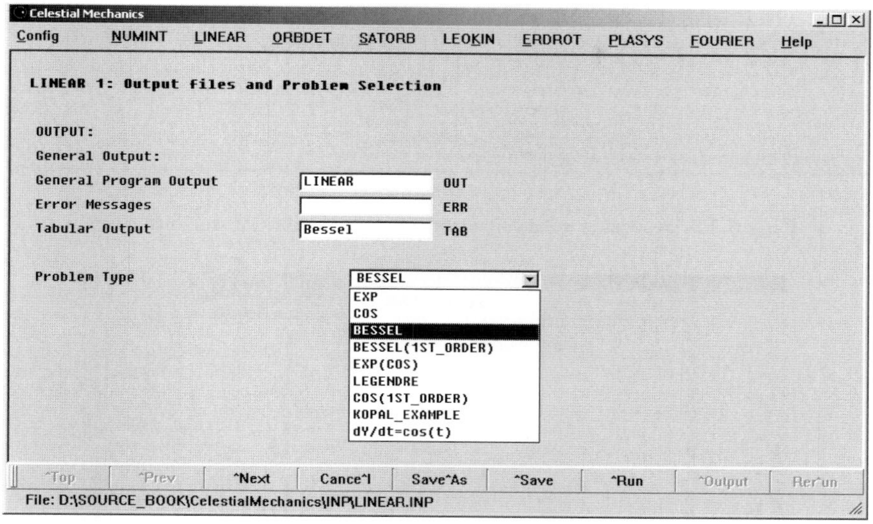

Fig. 6.11. Primary menu of program LINEAR

In Panel LINEAR 2 in Figure 6.12 the integration and output specifications are defined. This panel is problem-independent. The integration order may be selected between the limits $2 \leq q \leq 30$. The left interval boundary for the integration is assumed to be known, the right interval boundary may be selected (minor differences concerning these assumptions are automatically taken care of). The number of output points (file "Bessel" specified in the Panel LINEAR 1 6.11) should not exceed 2000 for the graphical display associated with the menu system to work rapidly. The last option to be set in this panel concerns the definition of the collocation epochs. The three options (EQUIDISTANT, CHEBYSHEV and LEGENDRE) are discussed in section I-7.6.

Depending on the problem type, a third panel may follow. In the case of the Bessel functions the pointer (index) of these functions may be defined in the third panel.

Figure 6.13 shows the result of numerically solving Bessel's differential equa-tion in the interval [0,10] using a collocation procedure of order $q = 30$, when selecting the collocation epochs as the roots of the Legendre polynomial of

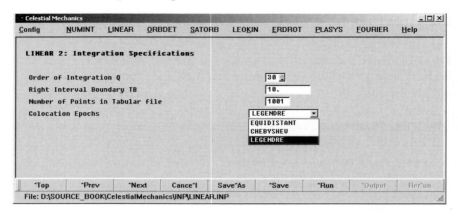

Fig. 6.12. Selection of integration specifications (Bessel functions)

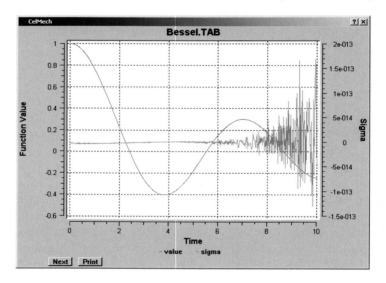

Fig. 6.13. Bessel function of pointer "0" in interval [0,10] with integration error

degree $q = 30$. The figure is obtained by using the "Display Results" option of program LINEAR. The scale for the function is given at the left ordinate, the scale for the integration error at the right one. The performance is impressive!

7. The Computer-Programs SATORB and LEOKIN

7.1 Program SATORB

Program SATORB may be used to generate satellite orbits or to determine the orbits of satellites using (a) tabular satellite positions as pseudo-observations or (b) astrometric positions as real observations. More precisely the following problems may be solved:

1. Generation of satellite ephemerides using a wide variety of models.

2. Orbit determination using tabular positions or position differences as pseudo-observations. Two concrete problems may be addressed:

   a) Orbit determination for GPS or GLONASS satellites using satellite positions in the SP3-format (or a special tabular format used in the Bernese GPS Software (see [58])) as pseudo-observations. No a priori information concerning the orbits other than the tabular positions is required to solve this task.

   b) Orbit improvement for LEOs using satellite positions (and possibly position differences) as pseudo-observations. The positions and position differences stem from program LEOKIN (see section 7.2).

3. Orbit determination using astrometric positions following the use of program ORBDET (option "satellites"). Approximate osculating elements referring to the initial epoch are required in addition to the astrometric positions. The information stems from the program ORBDET.

Panel SATORB 1 in Figure 7.1 shows the primary menu of the program. The concrete program option is selected in the first input field. The three problem types are addressed in the next three paragraphs.

The following input files have to be defined in program SATORB:

- The file with constants shared by all programs of the package.

- The file with different geodetic datum definitions, which was already described in Chapter 8 (content see Figure 8.3).

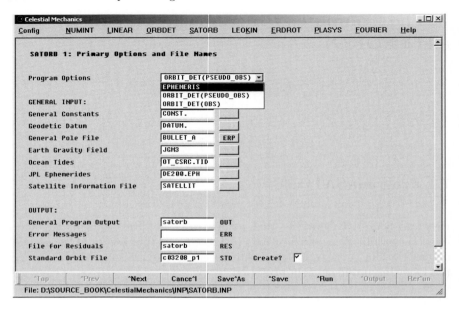

Fig. 7.1. Primary menu panel of program SATORB

- A file containing the erp-parameters. The file may be retrieved from the address of the CODE analysis center http://www.aiub.unibe.ch/download/ BSWUSER/GEN/ under the file name BULLET_A.ERP. An example is provided in the subdirectory /GEN of the directory "CelestialMechanics" on the CD accompanying this book.

- A file containing the coefficients of the gravity field. Two examples (JGM3 and GEMT3) are contained in the subdirectory /GEN of the CD.

- A file with the coefficients of the UTCSR Ocean tide model (based on the Schwiderski model). The model is described and referenced in the IERS conventions [70].

- The JPL development ephemerides DE200 [111] are read from the binary file "de200.eph" in the same subdirectory. The file covers the time interval between 1981 and 2025. Other DE-files might be attached to the program to cover other time periods.

- The satellite information file is only needed when processing GPS positions. The file is described in [58].

7.1.1 Generation of Satellite Ephemerides

The initial epoch, the length of the integration interval, and the initial osculating elements a (semi-major axis), e (eccentricity), i (inclination w.r.t. the

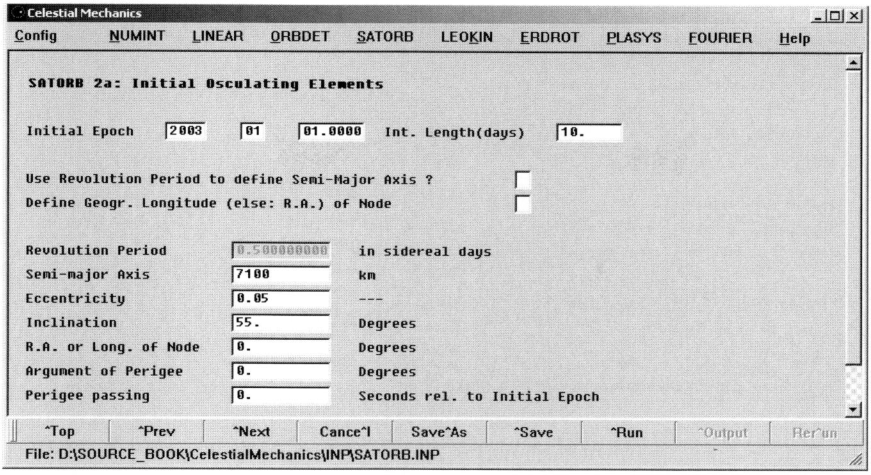

Fig. 7.2. Integration interval and initial osculating elements for ephemeris generation using SATORB

equatorial plane $J2000.0$), Ω (node), ω (argument of perigee), T_0 (time of perigee passage in seconds relative to the initial epoch) are selected in Panel SATORB 2a of Figure 7.2 presented to the program user when invoking the option "EPHEMERIS". The semi-major axis a may be defined either directly or via the revolution period (expressed in sidereal days). The latter option is in particular useful when studying resonance problems. One may either define the initial right ascension of the ascending node Ω or, alternatively, the initial geographical longitude of the ascending node. Again, the latter option is useful when studying resonance problems.

In Panel SATORB 3a (Figure 7.3) the output file names for the osculating or mean elements and the tabular positions must be defined. Moreover the averaging process for forming the mean elements is defined here (observe that the unperturbed initial revolution period is used for this purpose). For long integration intervals it may be advisable or even necessary to replace the correct computation (based on the file *.erp selected in the previous panel) of the transformation between the inertial and the Earth-fixed system by an approximation, where the polar motion and the Earth's variable rotation rate is neglected.

The "*.ELE"-file defined in Panel SATORB 3a (Figure 7.3) may be considered as the principal result file when generating ephemerides with program SATORB. It contains the modified Julian date in the first column and the corresponding osculating or mean elements $a(t)$, $e(t)$, $i(t)$, $\Omega(t)$, $\omega(t)$ and $\sigma_0(t)$ in columns 2 – 7.

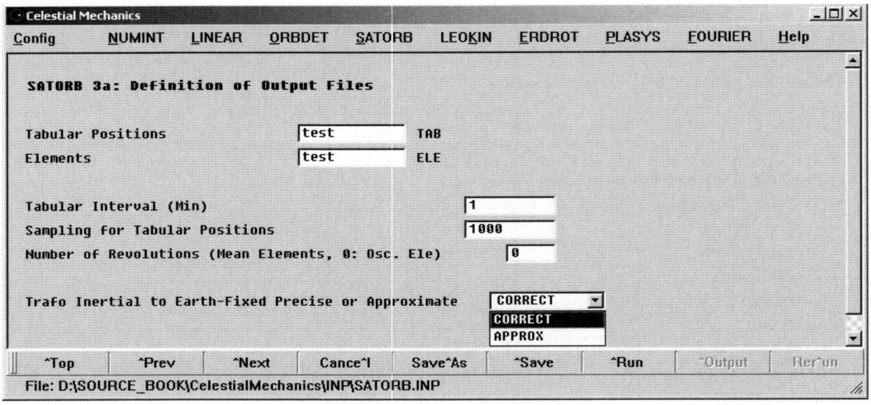

Fig. 7.3. Names of files with osculating or mean elements and tabular positions in SATORB; definition of mean elements and of transformation between inertial and Earth-fixed systems

The "*.TAB"-file, defined in Panel SATORB 3a (Figure 7.3), as well, may prove to be rather useful for particular purposes. It contains, as a function of the modified Julian date (MJD) in the first column, the Earth-fixed geocentric, equatorial coordinates $X(t)$, $Y(t)$, and $Z(t)$ of the satellite in columns 2, 3 and 4, the differences in radial, along-track, and out-of-plane directions of the true orbit w.r.t. the osculating Keplerian orbit (defined by the osculating elements at the initial epoch t_0) in columns 5, 6 and 7, and the perturbing accelerations (in m/s^2) in radial, along-track, and out-of-plane directions in columns 8, 9, and 10. The argument of latitude u is provided in the eleventh and last column. The file may, e.g., be used to produce figures of the subsatellite track (radial projection of the satellite's radius vector on the Earth's surface).

The force model (for all of the three principal options in Panel SATORB 1 (Figure 7.1)) is selected in Panel SATORB 4 (Figure 7.4). The upper limits for degree n and order m are defined first. Observe that $m \leq n$. In the case $m < n$ all terms $C_{ik} = 0$ and $S_{ik} = 0$ for $k > m$. This option may, e.g., be used to study the impact of the zonal terms of the Earth's gravity field on the orbit of the satellite.

One may then decide to take the direct gravitational attractions exerted by the Sun and Moon into account or not. The tidal attraction and the "perturbations" due to the theory of relativity may be activated as well.

The remaining force model constituents are of non-gravitational origin. One may use a simple direct radiation pressure model or alternatively the ROCK-IV model for GPS satellites (the option may, however, only be activated when determining the orbits of GPS satellites) and a drag force. An albedo

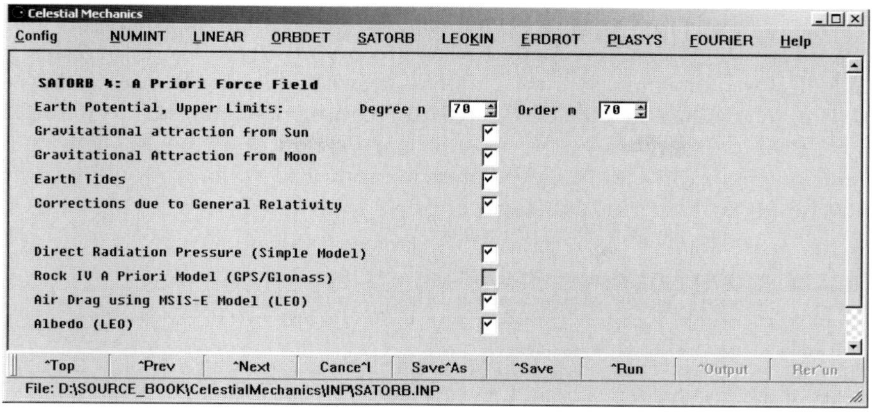

Fig. 7.4. Force model definition in SATORB

acceleration may be activated as well (a mean reflectivity of the Earth of 0.3 is adopted in this case). If either the simple radiation pressure model or the drag was selected, Panel SATORB 4(cont) (Figure 7.5), where the area-to-mass ratios and related information for the two surface forces have to be defined, is produced as well. The models implemented are simple: Constant (but possibly different) area-to-mass ratios and values for the coefficients C may be specified for radiation pressure and drag. In the latter case one may also define the values for the solar flux and for the planetary magnetic index. We refer to Chapter 3.5 for an explanation of these terms.

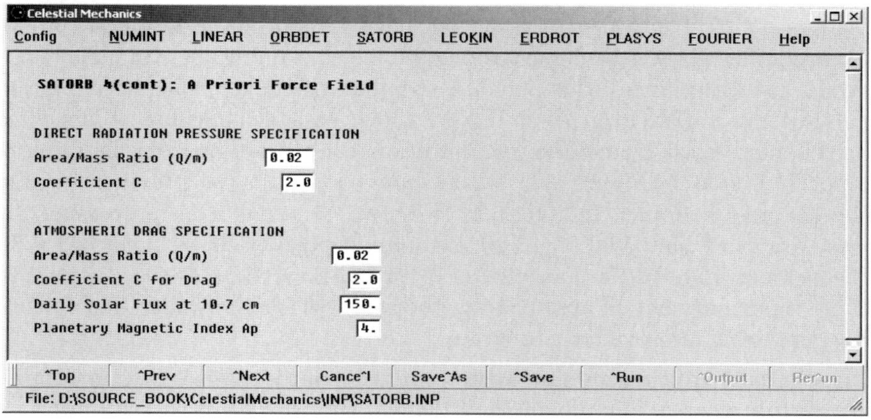

Fig. 7.5. Parameters of surface forces in SATORB

The selection of input options is completed in Panel SATORB 5 (Figure 7.6), where the integration specifications have to be provided. The integration in SATORB is performed with the collocation method with automatic stepsize control. The values in Panel SATORB 5 (Figure 7.6) make sense for LEOs. For GPS-type satellites or for geostationary satellites an initial stepsize of about one hour (3600 s) would be more appropriate. No harm would be done by choosing a shorter initial stepsize.

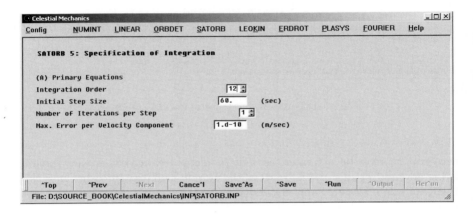

Fig. 7.6. Integration parameters

7.1.2 Determination of Orbits Using Astrometric Positions

This option is selected in the first input line of the primary Panel SATORB 1 (Figure 7.1). The program user then has to define in addition to the options and files already discussed above the file containing the residuals in the second-last input line of this panel. A so-called standard orbit file (definition see [58]) may be written as well – in which case a name has to specified for this file. Apart from these two additional input options the handling of Panel SATORB 1 (Figure 7.1) is the same as that in the previous section. The example in Panel SATORB 2c (Figure 7.7) stems from processing the observations of MeteoSat 7 (COSPAR number 97049B00, see Chapter I- 8). The element input file is that written by program ORBDET (see Chapter 8). These input parameters are necessary because SATORB (under this option) is a pure orbit improvement program.

The standard orbit represents the determined orbit piecewise by polynomials of a certain degree. The polynomial degree, the lengths of the sub-intervals (governed by one and the same set of polynomials) and the length to be covered by the entire standard orbit have to be defined in Panel 7.8. The

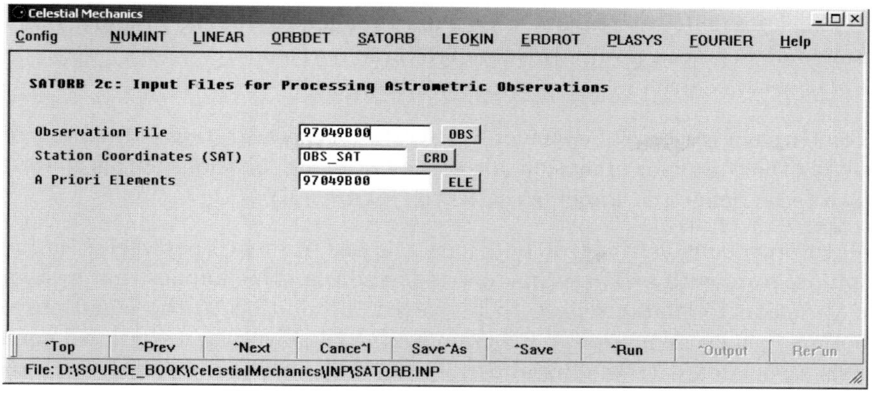

Fig. 7.7. Input files (observations, station coordinates, a priori elements)

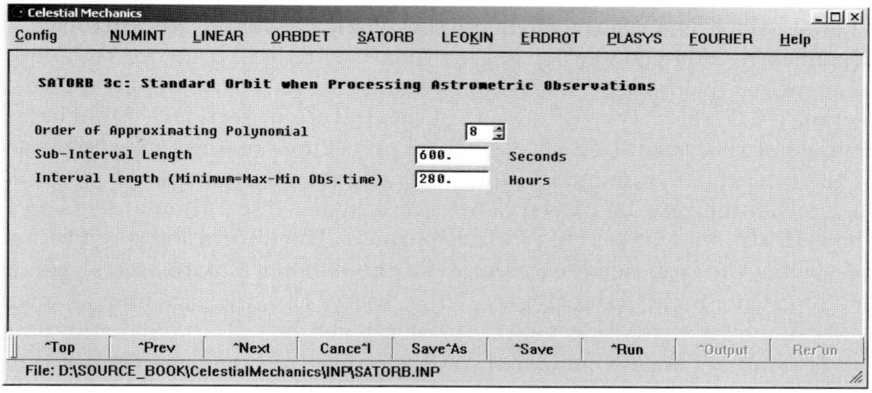

Fig. 7.8. Definition of standard orbits

panel is only activated, if the decision was made to generate a standard orbit (in Panel SATORB 3c, Figure 7.1). Observe that the standard orbit will cover at least the time interval containing the observations. The standard orbit file is reserved to the expert user of the program.

The orbit force model has to be defined in Panel SATORB 4 (Figure 7.4). As opposed to the example shown in that panel, the ROCK-IV model may not be selected (it is not assumed that GPS satellites are observed optically). The albedo radiation pressure model is also deactivated in this option – in view of the limited accuracy of astrometric positions this seems to be justified. When either a (simple) radiation pressure model or a drag model were selected in Panel SATORB 4 (Figure 7.4), Panel SATORB 4(cont) (Figure 7.5) is presented to define the parameters for these surface forces (an example for this panel was given in the previous section).

Exactly as in the case of ephemeris generation the integration parameters for solving the equations of motion, also called primary equations in this context, have to be defined in Panel SATORB 5 (Figure 7.6).

When processing astrometric positions, one has to solve a parameter estimation program with orbit parameters as unknowns. This implies that systems of variational equations have to be solved in addition to the equations of motion (primary equations). Program SATORB solves the variational equations associated with the initial osculating elements simultaneously with the equations of motion. The solutions of the variational equations accompanying the so-called pseudo-stochastic parameters may be formed as linear combinations of the solutions associated with these six initial osculating elements (see section I-8.5.4).

The variational equations referring to the dynamical parameters are solved independently, after the solution of the primary equations, using the techniques of numerical quadrature. The theory behind this technique was presented in section I-5.2, the theory of numerical quadrature in section I-7.6. The advantage of this separation of integration procedures resides in the efficiency of the procedures of numerical quadrature. Typically intervals of about 0.5 to 1 revolution may be chosen. Also, much higher integration orders (up to about thirty) may be selected for this purpose. The integration specifications for solving the variational equations (for the dynamical parameters) are defined in Panel SATORB 6 (Figure 7.9). Observe that the sub-interval length is about half a revolution period. In this panel one also defines the number of iteration steps. One may also decide to screen the observations using the standard 3σ criterion (σ being the rms a posteriori of the observation), where observations with bigger residuals (in absolute value) are not used for the adjustment. It is prudent to apply this criterion only after a few iteration steps (otherwise observations might be left out because of the poor quality of initial orbit parameters – stemming from program ORBDET in this case).

The orbit parameters other than the initial osculating elements (which are estimated under all circumstances) are defined in Panel SATORB 7 (Figure 7.10). The following parameters may be introduced and estimated in program SATORB:

- Initial osculating elements (selected automatically).

- Any combination of the nine radiation pressure model parameters (3.163) (where the $X-$, $Y-$, $Z-$ decomposition of perturbing forces has to be selected in the first input field of Panel SATORB 7, Figure 7.10).

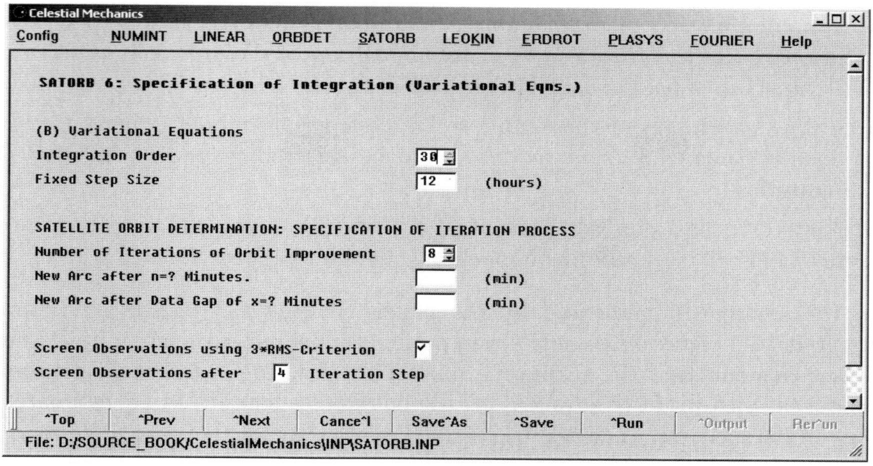

SATORB 6: Specification of Integration (Variational Eqns.)

(B) Variational Equations
Integration Order   30
Fixed Step Size   12   (hours)

SATELLITE ORBIT DETERMINATION: SPECIFICATION OF ITERATION PROCESS
Number of Iterations of Orbit Improvement   8
New Arc after n=? Minutes.   (min)
New Arc after Data Gap of x=? Minutes   (min)

Screen Observations using 3*RMS-Criterion   ✓
Screen Observations after   4   Iteration Step

File: D:/SOURCE_BOOK/CelestialMechanics\INP\SATORB.INP

Fig. 7.9. Solving the variational equations for dynamical parameters, define iterative orbit improvement process

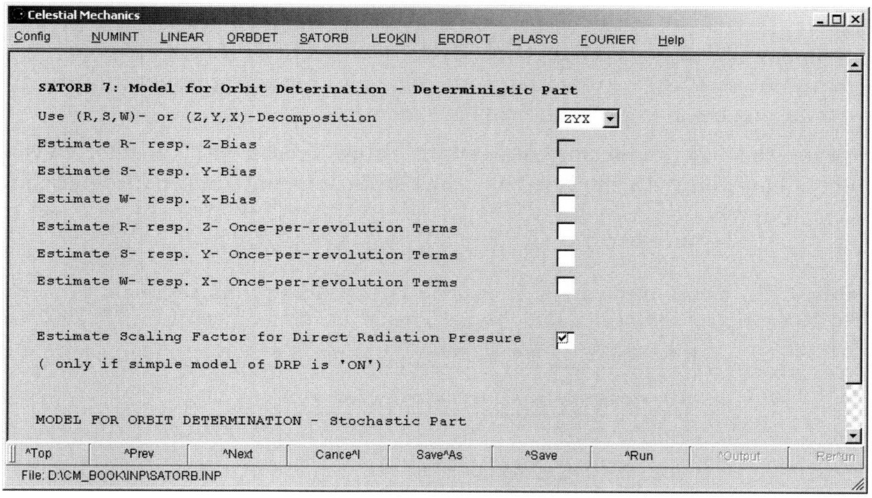

SATORB 7: Model for Orbit Deterination - Deterministic Part

Use (R,S,W)- or (Z,Y,X)-Decomposition   ZYX
Estimate R- resp. Z-Bias   □
Estimate S- resp. Y-Bias   □
Estimate W- resp. X-Bias   □
Estimate R- resp. Z- Once-per-revolution Terms   □
Estimate S- resp. Y- Once-per-revolution Terms   □
Estimate W- resp. X- Once-per-revolution Terms   □

Estimate Scaling Factor for Direct Radiation Pressure   ✓
(only if simple model of DRP is 'ON')

MODEL FOR ORBIT DETERMINATION - Stochastic Part

File: D:\CM_BOOK\INP\SATORB.INP

Fig. 7.10. Parametrization of the Orbit

- Any combination of nine empirical parameters (constant, once per revolution) using the $R-$, $S-$, $W-$ decomposition of the perturbing forces in the first input field of panel SATORB 7, Figure 7.10).

- Scaling parameter of the simple radiation pressure model (where, for obvious reasons, this orbit model has to be selected in panel SATORB 4, Figure 7.4).

- Pseudo-stochastic pulses (instantaneous velocity changes in predetermined directions at predetermined epochs) as explained in section I- 8.5.4.

In the example documented by the panels SATORB 2c (Figure 7.7) and SATORB 7 (Figure 7.10) only seven parameters, namely the six initial osculating elements and the scaling parameter for the simple radiation pressure model (a constant force acting along the line Sun $\rightarrow$ satellite) were estimated.

The resulting parameters for this program run may be inspected in Table 7.1, which should be compared to Table I- 8.14, documenting a similar run based on the same 155 astrometric observations. The difference resides in the selection of a constant acceleration in the direction Sun $\rightarrow$ satellite in the case of Table I- 8.14, whereas a scaling parameter for the the simple radiation pressure model was estimated in the alternative case. Obviously, the product $C\,(Q/m) = 2 \cdot 0.02$ (see Panel SATORB 4(cont), Figure 7.5) had to be scaled a factor of ≈ 0.707. From the point of view of the representation of the observations the two program runs are identical.

Observe that the osculating elements in Tables I- 8.14 and 7.1 differ considerably. This is due to the fact that the osculation epoch are *not* identical in

Table 7.1. Orbital elements of *MeteoSat 7* (97049B00), and residuals w.r.t. the best-fitting perturbed orbit (155 observations in 11 days, January 2 - 13, 2002)

```
ORBIT DETERMINATION FOR OBJECT 97049B00          DATE: 20-FEB-03 TIME: 05:11
---------------------------------------------------------------------------

ORBIT DETERMINATION USING *.OBS-FILES FOR   1 SATELLITE(S)
**********************************************************
SATELLITE   1    ARC        =        1
                 FROM (MJD) = 52276.778
                 TO (MJD)   = 52287.792
                 # OBS-EPOCHS =      155
                 # ITERATIONS =        8
-----------------------------------------

ORBITAL ELEMENTS AND THEIR RMS ERRORS
*****************************************************
OSCULATION EPOCH = 52276.7777778 MJD
SEMIMAJOR AXIS  = 42167155.460 M   +-        0.187 M
REV. PERIOD U   =     1436.221 MIN
ECCENTRICITY    = 0.0002640861 --- +-0.0000000673
INCLINATION     =       0.1441911 DEG +- 0.000007362
R.A. OF NODE    =       5.9132629 DEG +- 0.002311448
ARG OF PERIGEE  =     -72.3902282 DEG +- 0.045008633
ARG OF LAT AT TO =     16.7096479 DEG +- 0.002310933
*****************************************************

NUMBER OF DYNAMICAL PARAMETERS   : 1
*****************************************************
PARAMETER = DRP  VALUE =0.707106D+00 +-0.738183D-02
*****************************************************
SAT   1 : RMS=   0.21"  # OBS =  310 # PARMS =  7 BETA= -22.74 GRAD
```

both cases. The osculation epoch t_0 is adjusted in the case a standard orbit is produced: The osculation epoch is set to $t_0 \stackrel{\text{def}}{=} int(t_1) + k\,h_{\text{std}}$, where t_1 is the first observation epoch, k the biggest integer for which $T_0 \leq t_1$, and h_{std} the tabular interval of the standard orbit. If no standard orbit is produced, the osculation epoch is defined as $t_0 \stackrel{\text{def}}{=} t_1$.

The residuals (in right ascension and declination) are contained in the general output file, but they are also available in the SATORB.RES. The five columns contain the following information:

1. Observation time in the modified Julian date (MJD).

2. Number of astrometric positions.

3. Residuals in right ascension $\alpha \cos \delta$ in arcseconds.

4. Residuals in declination δ in arcseconds.

5. Mark '*', if an observation was not used in the adjustment.

No header is written. The file may be easily used to plot the residuals.

7.1.3 Determination of GPS and GLONASS Orbits

The program SATORB may be used as an orbit determination program using tabular satellite positions as pseudo-observations. It is in particular possible to analyze one or several precise orbit files of GPS (and/or GLONASS) satellites produced by one of the IGS Analysis Centers or by the IGS Analysis Coordinator in the SP3-format. One IGS precise orbit file contains (in general) 96 tabular positions per day at 15-minutes intervals for each active GPS satellite. The tabular positions of one or several contiguous precise orbit files may be analyzed in the same SATORB program run; one particular or all available satellites may be selected for analysis.

After having defined the appropriate primary option ("ORBIT_DET (PSEU-DO_OBS)"), the Panel SATORB 2b (Figure 7.11) serves to select in the first input field GPS and/or GLONASS satellites for analysis. The alternative "LEO", used for processing tabular positions and/or position differences as established by program LEOKIN, will be dealt with in the concluding section. Three file formats (SP3, TAB, PPD) are supported for the input tabular positions. The SP3-format is, as already mentioned, the official IGS format for providing positions (and clock information) of the GPS satellites in the Earth-fixed coordinate system. TAB-files contain the satellite positions in the inertial coordinate system in a format supported by the Bernese GPS Software [59]. The PPD-format (Positions and Position Differences) may only be selected when processing LEO-positions (see section 7.3). The value for the a priori rms error of the observed satellite position is "only" used to define the correct constraints when solving for pseudo-stochastic pulses. The

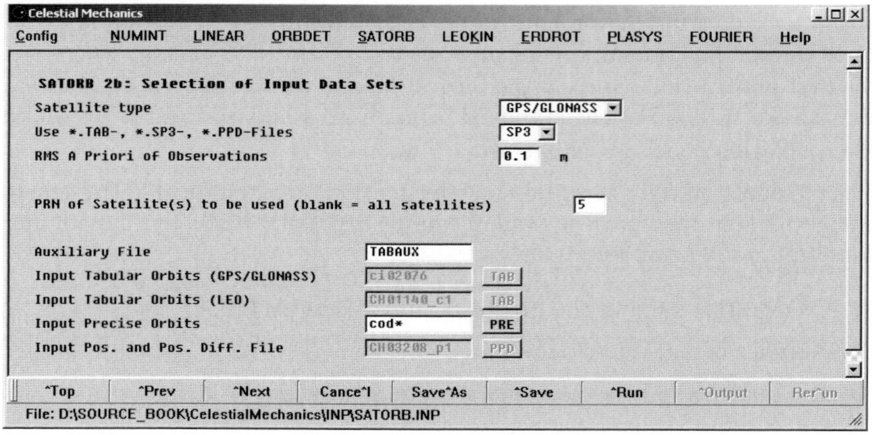

Fig. 7.11. Selection of pseudo-observations

rather pessimistic value for SP3-orbits of 0.1 m was selected above. Either one satellite (to be defined by its PRN number) or all satellites (as in the example of panel SATORB 2b, Figure 7.11) may be selected in the fourth input field of Panel SATORB 2b (Figure 7.11).

The pre-selection "cod*" for the SP3-files ("*" serves as a wild card) indicates, that only precise files produced by the code processing center will be presented for inclusion in the list of Figure 7.12, which is activated by pressing the button "PRE" in Panel SATORB 2b (Figure 7.11). The files are assumed to be available in the sub-directory "/LEOKIN/ORB". After having selected one, several, or all COD*-files available, COD* will be replaced by "SELECTED" in the corresponding input field of panel SATORB 2b (Figure 7.11). For subsequent runs this input selection must only be altered, if a different set of files shall be analyzed. In the example documented by the list in Figure 7.12 the seven official CODE files of the GPS week 1151 (Jan 27, 2002 - Feb 02, 2002) were selected. Observe that the corresponding file "COD11517.ERP" was selected to describe the transformation between the ITRF and the inertial reference frame $J2000.0$ (as opposed to the setting in Panel SATORB 1, Figure 7.1).

The force field model is defined in Panel SATORB 4 (Figure 7.4). The option ROCKIV may bow be activated, where the input file "SATELLIT", defined in Panel SATORB 1 (Figure 7.1), is required to obtain the necessary information for the ROCK-IV radiation pressure acceleration. One may either use the ROCK-IV or the simple model (or none) to take radiation pressure into account – but not both. Air drag and albedo radiation pressure are not available when analyzing GPS orbits.

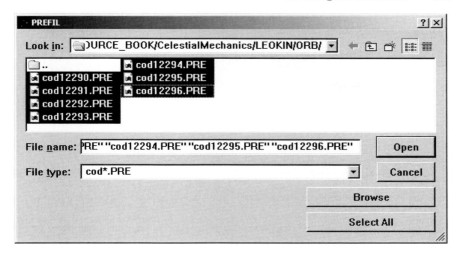

Fig. 7.12. Selection of SP3-files

The integration characteristics are defined in panels SATORB 2c (Figure 7.8) and SATORB 7 (Figure 7.9). It makes sense here to ask the program to set up a new arc after a long gap: if no observations are available for a particular satellite for a time period longer than (let us say) four hours (240 minutes), the observations before and after the gap are assumed to lie in two different arcs. This assumption makes sense, because a long gap often indicates that a satellite manoeuvre took place within the gap. It would also be possible to set up automatically new arcs after certain time periods. The two options to define new arcs are active in this program option.

When analyzing one week of GPS positions it makes sense to solve for all parameters of the Bernese radiation pressure model. An analysis of this kind was actually performed in [12] and led to the creation of the Bernese radiation pressure model. Panel SATORB 7 (Figure 7.13) indicates how such long arcs may (should) be parametrized: On top of the six initial osculating elements the nine parameters of the Bernese model are selected, implying that each orbital arc is modelled by fifteen free parameters.

Program SATORB produces a summary table for the entire program run in the general output file. Table 7.2 contains the result for the example prepared in the above panels. The table shows the mean rms errors per coordinate of the orbital positions for all active GPS satellites in January 2002 using the initial osculating elements *and* the nine parameters in eqn. (3.163) as orbit parameters. Data of one week (GPS week 1151, corresponding to January 27 – February 2, 2002) were used in the orbit determination process. In the case of satellites 15 and 17 long data gaps occurred, which led to a splitting of the one week arc into two shorter arcs (one of one day and one of five days).

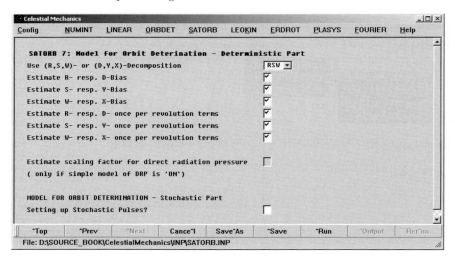

Fig. 7.13. Parametrization of long GPS arcs

Let us mention that Table 7.2 also shows the elevation β of the Sun above the orbital plane, the right ascension Ω of the ascending node of the orbital plane. A flag "*" is set to indicate that the sunlight is eclipsed (as seen from the satellite) by the Earth for some time during the saellite's revolution around the Earth.

The residuals are stored in the file defined in Panel SATORB 1 (Figure 7.1). The eight columns in the *.RES file contain the following information (in the program options processing tabular satellite information):

- *Column 1:* Time in days relative to the first observation epoch.
- *Column 2:* Argument of latitude.
- *Columns 3 – 5:* Residuals (in meters) in Cartesian coordinates in the inertial system.
- *Columns 6 – 8:* Residuals (in meters) in radial (R), along-track (S) and out of plane (W) directions.

Generally, the rms errors a posteriori per satellite coordinate (column 5 of Table 7.2) are of the order of few centimeters. This orbital consistency is typical for "well-behaving" satellites. Because the satellite positions are not real measurements, but derived from a particular solution of the equations of motion, the residuals of these rectangular orbit positions w.r.t. the best-fitting orbit are not randomly distributed. Figure 7.14 (top) shows a typical example.

In contrast to this normal case, the orbital fit for certain satellites may be much worse (showing rms values of more than, let us say, 10 cm). Figure 7.14

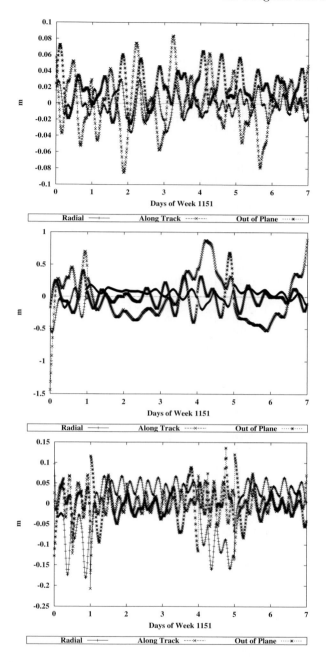

Fig. 7.14. Residuals of long arc orbit fit for $SVN09$ (top), $SVN02$ (middle), $SVN02$ including stochastic pulses (bottom)

Table 7.2. Long-Arc Analysis of GPS-Week 1151 (Jan 27, 2002 - Feb 02, 2002) of CODE/IGS Orbits

PRN	Arc	n_o	n_p	rms [cm]	β	Ω	Shadow
1	1	2016	15	2.3	19.8	87.5	
2	1	2016	15	24.4	−66.2	201.6	
3	1	2016	15	5.0	−47.4	263.4	
4	1	2016	15	2.7	4.0	327.7	*
5	1	2016	15	3.5	−67.0	202.8	
6	1	2016	15	3.6	−45.4	266.1	
7	1	2016	15	4.8	−46.7	264.4	
8	1	2016	15	5.1	−25.7	148.1	
9	1	2016	15	2.6	−23.2	144.9	
10	1	2016	15	2.1	36.2	26.5	
11	1	2016	15	2.8	−0.8	322.9	*
13	1	2016	15	3.5	20.7	86.4	
14	1	2016	15	4.0	20.7	86.1	
15	1	288	15	1.0	6.1	330.4	*
15	2	1440	15	8.6	4.8	330.3	*
17	1	288	15	0.7	7.8	332.7	*
17	2	1440	15	160.5	6.6	332.6	*
18	1	2016	15	2.6	35.7	29.1	
20	1	2016	15	2.5	35.2	26.1	
21	1	2016	15	95.1	36.2	26.8	
22	1	2016	15	4.6	−66.6	202.4	
23	1	2016	15	3.7	36.9	29.3	
24	1	2016	15	8.2	5.0	328.8	*
25	1	2016	15	2.9	−21.2	142.4	
26	1	2016	15	2.8	20.5	86.6	
27	1	2016	15	2.1	−22.3	143.8	
28	1	2016	15	3.3	−69.9	206.4	
29	1	2016	15	99.0	21.4	84.9	
30	1	2016	15	2.9	−68.4	204.9	
31	1	2016	15	3.8	−46.7	264.4	
2	1	2016	93	4.7	−66.2	201.6	

(middle) shows the residuals of the orbital fit for satellite $SVN02$. Instead of showing typical orbit errors well below a 10 cm limit as in the above example, maximum residuals of up to almost one meter occur for $SVN02$. Observe, that there are no big "jumps" in the residuals.

Obviously, it does not make sense to represent the entire arc by one set of only 15 orbit parameters under such circumstances. Considering the facts that (1) the orbits of all but a few satellites may be very well represented by deterministic orbits characterized by only few parameters and that (2) all satellites are more or less identical in construction, one may safely conclude that the reason for bad behavior is related to satellite manoeuvres and/or to errors related to attitude control.

One radical method of curing such problems is to break up the original arc into shorter arcs. This well known method usually is referred to as *short arc method*. One should be aware of the fact, however, that through such a procedure the number of parameters is multiplied by the number of arcs generated – a fact which considerably weakens the solutions. Also, an old Latin proverb says *natura not facit saltus*: It is thus preferable to allow from time to time for instantaneous velocity changes δv_i in pre-defined directions. The program user has the option in program SATORB to introduce such instantaneous velocity changes, called *pseudo-stochastic pulses* at regular intervals. He may set up pulses either in the radial, the along-track, or the out-of-plane directions (or any combination of the three options).

In addition, these velocity changes may be constrained to "reasonable" values by introducing artificial observations of the velocity changes with a weight proportional to $\sigma^{-2}(\delta v_i)$, where $\sigma(\delta v_i)$ is the user-specified rms-scatter (root of variance) of the pulse.

The last row in Table 7.2 shows the success of setting up pseudo-stochastic pulses every six hours in radial, along-track and out of plane directions (with $\sigma(\delta v_i) = 2$ cm/s) for satellite $SVN02$. By comparing the rms-values for satellite $SVN02$ with and without pseudo-stochastic pulses in Table 7.2 we can see that the orbit representation was improved by a factor of about 5 – at the price of increasing the number of orbit parameters from 15 to 93 (most of them constrained, however). The corresponding input Panel SATORB 8 (Figure 7.15) is activated by selecting the corresponding (last) option in panel SATORB 7 (Figure 7.13). The estimated pulses may be written into a file (in the example of Panel SATORB 8, Figure 7.15 called "prn_2.pls"), which is stored in the subdirectory "OUT" of the directory "SATORB". The file contains the MJD in the first column, the corresponding argument of latitude

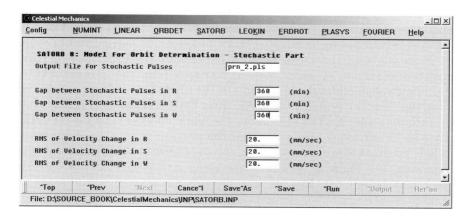

Fig. 7.15. Definition of pseudo-stochastic pulses

in the second column, and the pulses (in units of m/s) in the third, fourth, and fifth columns for the radial (R), along-track (S), and out-of-plane (W) pulses, respectively.

An analysis very similar to that described above is performed every week by the IGS Analysis Coordinator. The analysis corresponding to Table 7.2 performed by the IGS Coordinator is contained in the IGS Report No. 9034 (available under http://igscb.jpl.nasa.gov/). For a description of the IGS procedures we refer to [13]. Additional references are provided in the IGS Report No. 9034.

7.2 Kinematic LEO Orbits: Program LEOKIN

Program LEOKIN may be used to determine a so-called *kinematic orbit* of a LEO using the observations of a spaceborne GPS receiver. A kinematic orbit is nothing but a series of satellite positions, an ephemeris of the satellite's center of mass, which was established (almost) *without* making use of the equations of motion. In essence only the GPS code and phase measurements were used to determine the satellite ephemeris. The satellite positions may then be used to determine a so-called *reduced-dynamics* orbit by using the tabular positions as pseudo-observations in an orbit determination process in program SATORB. The procedure is very similar to the ones described in section 7.1. The peculiarities of analyzing LEO positions will be dealt with in the concluding section 7.3.

Program LEOKIN was developed by Heike Bock in context with her Ph.D. thesis written at the University of Bern [24]. The version described here is a strap-down version of this original LEOKIN program. The primary menu of LEOKIN is contained in panel LEOKIN 1 (Figure 7.16). The "usual" input files required for GPS processing have to be defined first. The only files that *really* has to be adapted when processing a particular LEO data set is the file with the Earth rotation parameters (BULLET_A.ERP), the most recent version of which may be retrieved from the internet address http://www.aiub.unibe.ch/download/ BSWUSER/GEN/. One may also retrieve the "satellite problem file" SAT_ 200x.CRX and the most recent satellite definition file "SATELLIT" from the same address. The file "*.CRX" may also be left blank – in which case the program does *not* automatically exclude satellites which were set "bad" by the CODE Analysis Center during processing. If a name is set blank, the menu system generates a warning, which may be ignored. Experts may use this file to exclude bad satellites themselves.

Program LEOKIN does – at least initially – *not* require a priori information concerning the LEO orbit. LEOKIN may, however, make use of an approximate a priori orbit in the (very important) step of pre-processing the

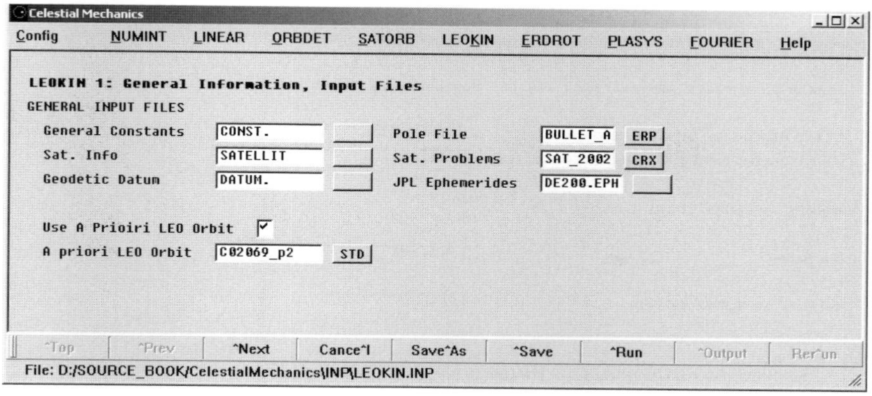

Fig. 7.16. Primary menu of program LEOKIN

GPS observations. The a priori orbit information must be made available to LEOKIN in a special binary format, the so-called *standard orbit format* of the Bernese GPS software [58]. The file is generated by the SATORB program. In a first analysis of the LEO GPS data of a particular RINEX-file it is thus *not* possible to select an a priori LEO orbit.

The LEO/GPS data set to be processed is selected in Panel LEOKIN 2 (Figure 7.17) together with the corresponding precise ephemerides file of the GPS satellites (first two input items in the panel). The program needs the GPS observations of the spaceborne GPS receivers in the so-called RINEX-format. The observation files are made available through the IGS in this format. The data sets (the LEO/GPS RINEX file and the precise ephemerides file) used below as examples were retrieved from the CDDIS (Crustal Dynamics Data Information System, internet address ftp://cddisa.gsfc.nasa.gov/pub/gps). The IGS precise ephemerides files contain the satellite positions *and* the corresponding satellite clock corrections at 15-minutes intervals. LEOKIN assumes that both, the positions (velocities, accelerations, etc.) and the clock corrections of the GPS satellites may be computed with cm-accuracy (3 cm corresponding to about 0.1 ns for the clock corrections) for any LEO/GPS observation epoch. Whereas it is no problem to interpolate the satellite ephemerides to any epoch covered by the SP3-file (internally, a polynomial of degree $q = 10$ is used for the purpose), the interpolation of the clock correction is problematic: The characteristics of the GPS quartz oscillators would require precise clock estimates at, let us say, 30 second intervals in order to justify neglecting the stochastic part of the clock correction terms. Such clock information will most likely become routinely available in the near future through the IGS and its Analysis Centers. One may of course use anyway the clock information contained in the precise ephemerides files – the drawback resides in a significant reduction of the achievable orbital accuracy (more information will be

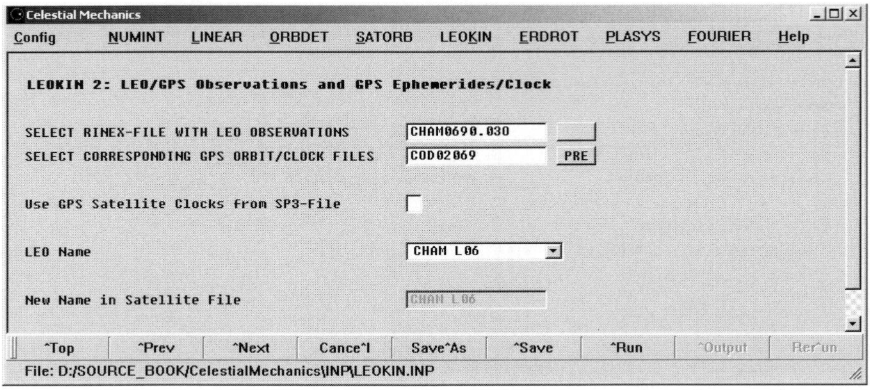

Fig. 7.17. Observations and LEO selection, GPS orbits and clocks

provided below). When using the satellite clocks from the SP3-files one has the advantage that no additional GPS clock information is needed. In this case the clock corrections found in the SP3-files are linearly interpolated to the observation epochs. If the decision is made not to use the SP3-clocks, a special clock correction file has to be supplied (see Panel LEOKIN 3, Figure 7.18).

The LEOKIN positions refer to the satellite's center of mass, the raw measurements refer, however, to the antenna phase center. This implies that, internally, the transformation between the two reference points has to be performed for each measurement epoch. This implies in turn that we have to know the *attitude* of the satellite in inertial space as a function of time *and* the antenna phase center coordinates w.r.t. the center of mass in a body-fixed coordinate system. The latter information is contained for some satellites (CHAMP, SAC-C, etc.) in the "SATELLIT" file. The three Cartesian coordinates are $DX = flight - direction$ (it is assumed that the DX-axis is always parallel to the velocity vector), $DZ = vector\ satellite \rightarrow geocenter$, and DY (completing the right-handed Cartesian system DX, DY, DZ). This definition already defines the satellite's attitude. If the satellite one wishes to process is not available in the list provided in Panel LEOKIN 2 (Figure 7.17), "OTHER" has to be selected in the input line "LEO Name". In this case, a name has to be provided in the line "New Name in Satellite File" and the file "SATELLIT" with the corresponding antenna coordinate information has to be updated. If one is satisfied with a modest accuracy or if the coordinates are not known, $DX = DY = DZ = 0$ may be entered in the line corresponding to the new LEO.

In Panel LEOKIN 3 (Figure 7.18) one has to provide a high-rate satellite clock file (only a better resolution than that of 15 minutes provided in the SP3-files makes sense) if the SP3 clock information is not used.

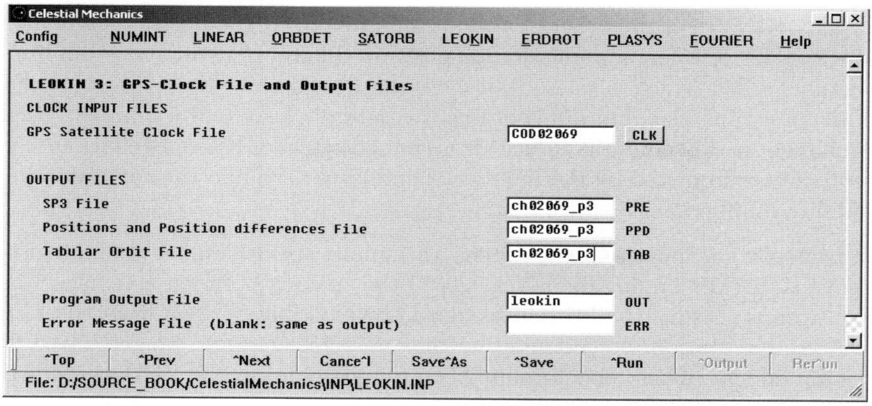

Fig. 7.18. Satellite clock input file and output files

Clock files with a 5 minutes time resolution are available at the address http://www.aiub.unibe.ch/download/BSWUSER/ORB/ under the name CODYYDDD.CLK, YY standing for the last two digits of the year, DDD for the day of the year.

Program LEOKIN generates a general output file and an error file (which are identical when the latter name is left blank) plus three files (with extension "PRE", "TAB" and "PPD", respectively) containing (in essence the same) orbit information in three different formats. The coordinates in the first file refer to the Earth-fixed system, the second and third contain the same information in the inertial system (derived from the Earth-fixed coordinates using the Earth rotation parameters in the corresponding file of Panel LEOKIN 1, Figure 7.16). The coordinates refer to the center of mass in the PRE- and TAB-files, to the antenna phase center in the PPD-file.

The PPD-file (Position and Position Differences) contain the "raw" position estimates based on the LEO/GPS code observations *and* the position differences between subsequent epochs as obtained by analyzing the GPS phase observations (see section I- 8.5.4 for details). The file should be viewed as some kind of a protocol of the position and position difference estimating process. The results of the first kind are referred to as measurement type 1, those of the second kind with 2 in column 1 of the PPD file. The measurement epoch is given in the second column of the file for position estimates, in columns 2 and 3 for position difference estimates. Observe that the reference epoch (MJD=...) in the file header has to be added to give the measurement epochs in the modified Julian date. The solution vectors (referring to the inertial system) are then provided in columns 4 to 6, the estimated rms-error of the observations in column 7. The (rather optimistic) rms-errors computed by LEOKIN are of the order of about half a meter for the position estimates, of the order of a few millimeters for position difference estimates. Column 8 contains a flag which is set to 0, if everything seemed to be satisfactory. For position estimates column 9 contains the sum of the satellite numbers used. A change in this columns indicates a change of the GPS constellation. The protocol is completed by the number of satellites used and the corresponding satellite numbers.

In principle the measurement epochs are equally spaced. In the example used above (CHAMP on day 69 of year 2002) the spacing between epochs is 10 s. For various reasons (bad satellites, data downloading periods from the LEO to a ground station, etc.) it may be that no solution could be generated for certain epochs. These epochs simply do not show up in the PPD- and TAB-files, a record $0, 0, 0, 999999.999999$ is written for the corresponding epoch in the PRE-file. The zeros correspond to the coordinates, the $999...$ to the LEO clock corrections.

Apart from the reference point (antenna phase center for the PPD-file and center of mass for the TAB-file) identical results are obtained in the PPD- and TAB-files *when processing only code, but no phase observations*. (In this case no estimates of type 2 are available in the PPD-file).

If satellite positions *and* position differences were estimated, the results in the TAB- and the PRE-file differ substantially from the raw results in the PPD-file: The positions in the former two files are obtained by a least-squares combination of all available positions and position differences (over all epochs) with the goal to obtain a series of positions. In the adjustment process the estimated position differences are assumed to be much more accurate (by a factor of 100) than the positions. The result consists, so-to-speak, of code-derived positions smoothed by the position differences emerging from the GPS phase analysis. The procedure is very efficient because the normal equation system associated with this combination is tri-diagonal and may be solved using efficient procedures described, e.g., in [88]. One LEOKIN run dealing with 10-s measurements of one entire day should take less than half a minute on a modern PC.

Panel LEOKIN 4 (Figure 7.19) contains general GPS-processing options and options used for the processing of the code observations. Based on the above assumptions concerning the attitude of the satellite, the LEO's GPS antenna used for precise orbit determination is pointing into the satellite's zenith direction (i.e., it lies in the line geocenter $\rightarrow$ satellite). It is not possible for a GPS-receiver situated on the surface of the Earth to gather GPS observations below the horizon. For spaceborne GPS receivers this is, however, possible. For LEOs, observations down to about $116°$ are possible. The GPS antennas are, on the other hand, not ideal for acquiring such low elevation observations. Two options are feasible to cope with this problem: One may define a cutoff angle for observations exceeding a certain zenith distance. This may be done

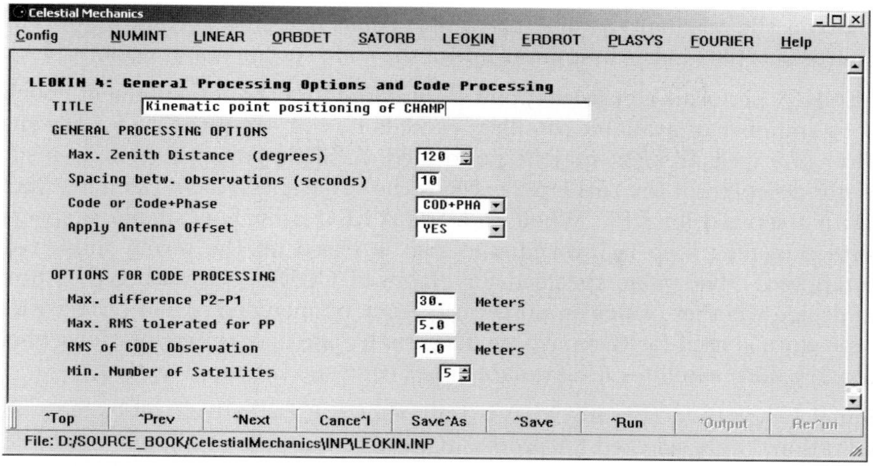

Fig. 7.19. General processing and code processing options

in the second input line of Panel LEOKIN 4 (Figure 7.19). One may, however, also follow the strategy to use all observations, but to impose a weighting scheme to the observations, which decreases the contribution of low-elevation observations in the adjustment process. In LEOKIN the observations are weighted proportional to $\cos^2 z$, when a cut-off zenith distance of $z_{max} \leq 90°$ was used proportional to $\cos^2(0.75 \cdot z)$ for maximum zenith distances of $z_{max} > 90°$.

The observations may be sampled in the next line: In the example processed above 10-s data were available and all were used. Specifying 30 s in the input line "Spacing between Observations" has the consequence that only every third observation is actually used. Only code observations or code and phase observation may be used (the latter, however, only if an a priori LEO orbit was made available in Panel LEOKIN 1, Figure 7.16). The next input line ("Apply ...") allows you to disregard the satellite-specific antenna phase center coordinates – LEO center of mass and antenna phase center are assumed to coincide, when selecting "NO" in this line. The option is only recommended for test purposes.

All spaceborne GPS receivers are (hopefully) dual-band receivers providing L_1- and L_2-observations. (This version of) LEOKIN assumes that dual-band observations are available *and* it automatically forms and processes the so-called ionosphere-free linear combination L_3 of the actual code and phase observations on L_1 and L_2. Residual ionospheric refraction effects are thus eliminated. LEOKIN makes a plausibility check of the code observations. It refuses to analyze a particular code observation if the difference (in m) between the original L_1- and L_2- observations exceed a certain limit. This limit is specified in the first input line of the code-specific input options block.

LEOKIN performs a classical point positioning for each measurement epoch. If the number of available satellites exceeds $n_s = 4$, an estimate for the rms error of a measurement may be performed. A single point positioning result is *not* accepted, if the rms error exceeds the value specified in the line "Max. Error tolerated for PP". When an a priori LEO orbit is available, a special data screening step is invoked; the step is based on the terms "observed-computed". The value specified for "RMS of CODE Observations" is used to decide whether or not an observation may be included. A majority voting procedure is used for this purpose. Eventually one may reject solutions when too few GPS satellites are available (last code-specific input option).

Panel LEOKIN 5 (Figure 7.20) is only shown if both, code and phase observations, were selected for processing. First of all, the minimum number of satellites available for position difference processing may be specified here. A minimum of four are (of course) required, because four unknowns (the three coordinates of the position difference vector and the the clock difference) have to be determined. The value of four may actually be recommended, because the tests performed prior to the actual adjustment are rather rigorous.

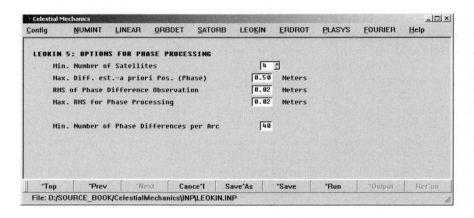

Fig. 7.20. Phase processing options

Analogously as in the case of code observations, the ionosphere-free linear combination L_3 of the phase difference observations is processed iby LEOKIN. A first plausibility check limits the plain difference (in meters) of the original L_1- and L_2- phase differences (the third input line of Panel LEOKIN 4, Figure 7.19). Values around 0.5 m are appropriate for 10 s data. The strongest a priori check is performed (as in the case of the code processing) on the terms "observed-computed (o-c)". If a good a priori orbit is available the terms "o-c" should be virtually free of errors due the satellite positions. One may therefore mark observations which are outside a $3 \cdot rms$ interval centered around the mean value of the observations considered as good. The rms used for this purpose is specified in input line four of Panel LEOKIN 5 (Figure 7.20). If more than four satellites are available (after having performed the above test) the estimated rms error of the phase difference solution may be used as a further, final quality indicator for the position difference estimation. If this rms error exceeds the $2 - 3$ cm level the result should be considered with caution.

The last parameter to be defined only concerns the combination of positions and of position differences in program LEOKIN. If only a short series of uninterrupted phase differences is available, the smoothing effect is insignificant and the result should not be used. The last input line in Panel LEOKIN 5 (Figure 7.20) limits the minimum number of contiguous phase differences per "arc". By reducing this number to below a value of, let us say, 20, one runs the risk that positions with a much lesser quality are contained in these files. The positions and position differences eliminated by this procedure are still available in the PPD-file.

7.3 Dynamic and Reduced Dynamics LEO Orbits Using Program SATORB

The content of the three output files *.TAB, *.SP3, and *.PPD, as generated by program LEOKIN, may be used as pseudo-observations in the program SATORB (where only one file type may be used at the time). The program will generate the best possible particular solution of the equations of motion using the parametrization already introduced in section 7.1 (more specifically in the subsection 7.1.3). It is thus only necessary to address the peculiarities of program SATORB when using LEO-data.

Panel SATORB 1 (Figure 7.1) is first presented to the user. The main option "ORBIT_DET (PSEUDO_OBS)" has to be selected for LEO orbit determination. The standard orbit file (last input line) must be generated, if one wishes to run program LEOKIN subsequently using an a priori orbit (last two input lines of Panel LEOKIN 1, Figure 7.16).

The residuals of the positions in the TAB-, PRE- or PPD-files w.r.t. the (reduced-) dynamic orbit established by SATORB are written into the file with extension "RES", the format of which was already given in section 7.1.3.

In Panel SATORB 2b (Figure 7.21) (compare with Panel SATORB 2b, Figure 7.11) one has to select the "LEO"-option in the first input line. The three file types (TAB, SP3, PPD, second input line) may be used. When analyzing TAB- or SP3-files, which were established by program LEOKIN using only

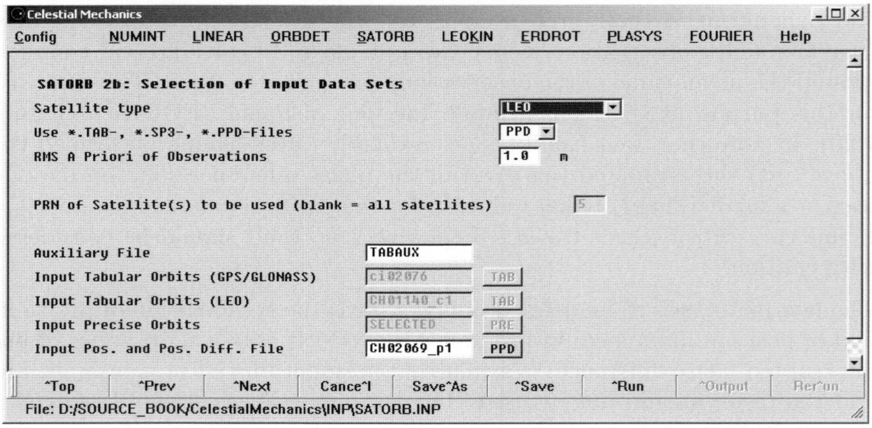

Fig. 7.21. Satellite type, input file type, and file names

GPS code observations, one should use an rms a priori of about 1 m; if the files established by program LEOKIN were based on code and phase observations, one should rather select a value of 0.10 m.

In the example of Panel SATORB 2b (Figure 7.21) the PPD file type was selected. In this case positions and position differences are available for orbit determination, which is why the additional Panel SATORB 3b(cont) (Figure 7.22) has to be considered (if alternatively either the SP3- or TAB-file was selected, this panel is skipped by the menu system). In the first input field

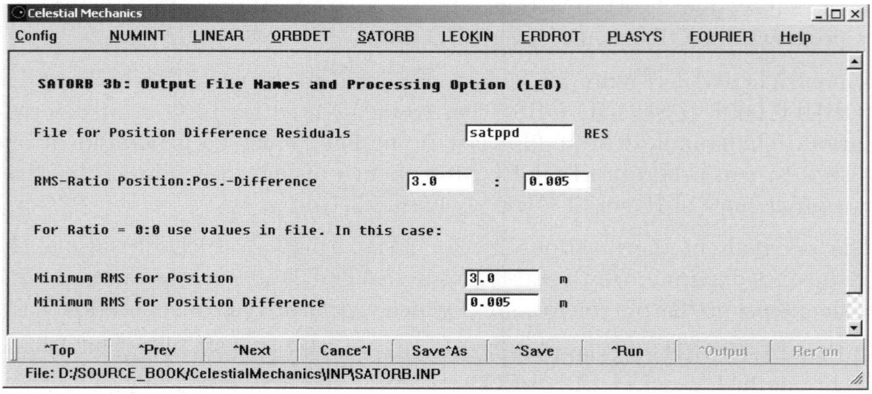

Fig. 7.22. Options for processing positions and position differences

the name of the file containing the residuals of the position differences has to be defined. The file format is identical with the residuals file of the satellite positions. The program user may then define the accuracy ratio of code and phase observations. If zero-values are specified, the accuracy ratio is derived from the rms values available in the PPD-file (i.e., the ratio is defined individually for each epoch). In order to avoid "unreasonable" values, minimum values for the two rms errors may be provided (if the rms values in the PPD-file are smaller then these values, the file values are replaced by the minimum values).

Independently on the input file type one may generate up to two more output files, a PRE-file and/or a TAB-file. The file names and the spacing between observations are defined in this panel. The characteristics of the standard orbit file, namely the degree q and the subinterval length h, are defined here. For orbits with a small eccentricity, the interval length should be of the order of 1/50 of the revolution period.

The panels following Panel SATORB 3b(cont) (Figure 7.23) presented to the user when processing LEO data are identical with those associated with the GPS/GLONASS processing. The force model naturally has to be adapted to the LEO case. It is in particular important to select a high degree and order gravity field in panel SATORB 4 (Figure 7.4) – degree and order 70, as provided in the JGM3 model may be considered as a minimum. One should use the relativistic version of the equations of motion, take the direct gravitational perturbations by Sun and Moon into account, and apply the tidal effects. The nonconservative forces (atmospheric drag and radiation pressure) are naturally of greatest importance to model the orbit of a LEO. It is in general not too difficult to develop an appropriate model for radiation pressure

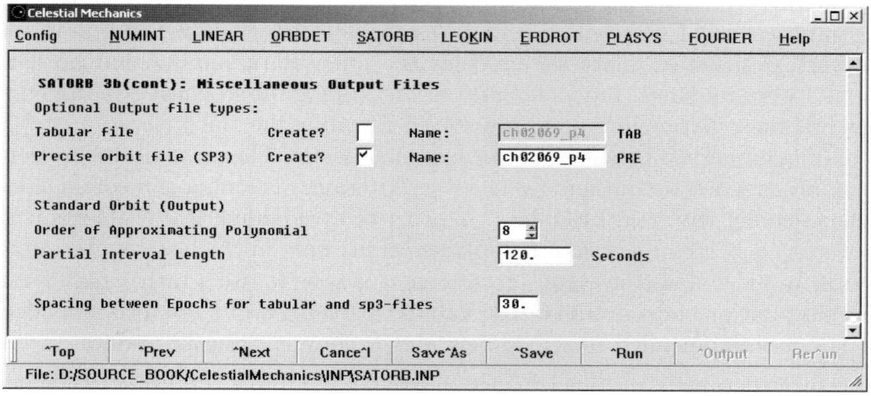

Fig. 7.23. Definition of up to three output orbit files

(based on the knowledge of the satellite surface and its reflective properties, its attitude, and the solar constant). It is, on the other hand, not trivial to specify good a priori models for atmospheric drag. One might, e.g., use the MSIS-models [51] and solve for scaling parameters associated with the two key input quantities (solar flux and planetary index). Such "exercises" undoubtedly make sense when dealing routinely (for a few years) with precise orbit determination for a particular satellite. If the only interest resides in an accurate representation of the orbit established via the spaceborne GPS receiver, these a priori models do not matter too much: modelling deficiencies will be (almost perfectly) absorbed or at least greatly reduced by the stochastic parameters, which have to be set up anyway (see remarks below). In the tests performed at the end of this section no a priori models were introduced for radiation pressure and for drag.

The definition of the integration characteristics for solving the equations of motion in Panel SATORB 5 (Figure 7.6) is not critical: An integration order of about $q = 10$ and an initial step size h of about one minute (about 1% of the revolution period) are appropriate. The integration procedure used in SATORB is the collocation method adjusting the stepsize automatically. The accuracy limit specified for the velocity promises a relative local accuracy of about 10^{-10} – which is amply sufficient for a total LEO arc length of about one day. The variational equations are dealt with separately. The characteristics for their solution are defined in Panel SATORB 6 (Figure 7.9). An integration order of $q \approx 20 - 30$ and a stepsize h between a quarter and half a revolution period $(30 - 45$ minute for LEOs) are appropriate.

The iterative orbit determination process is also specified in this panel. SATORB does not use a priori orbit information when analyzing LEO orbits. The a priori orbit of the first iteration step is obtained by solving a boundary value problem using those two position estimates at the beginning of the file, which are separated by (about) the selected stepsize h. If, e.g., an initial stepsize of 60 s was specified (as recommended above), the first and seventh position estimate are used for the purpose (assuming that no data points were missing). Independently of the number of stochastic parameters set up, these parameters are only solved for after the third iteration step – avoiding numerical problems to the extent possible. The pseudo-observations may be screened for outliers using a $3 \cdot \sigma$ criterion. Screening is recommended when dealing with real LEO data. In order to avoid numerical problems it is, however, also recommended to apply screening only in the last few iterations steps. In view of all these considerations, it is wise to use a rather high number of iteration steps in SATORB. Values between 6 and 8 are recommended. It is not recommended to set up new arcs when analyzing LEO data.

The orbit parametrization is defined in panel SATORB 7 (Figure 7.10). When analyzing LEO data the R, S, W decomposition is usually selected, and all nine deterministic parameters are set up in the deterministic part. In addition one usually has to introduce stochastic parameters in R, S, and W directions.

When processing tabular positions (obtained by LEOKIN) based only on code observations, it makes sense to set up one set of stochastic parameters (about) every 30 minutes. If the pseudo-observations are based on code and phase observations, one set of these parameters should be set up every 5 to 10 minutes. The velocity changes should be constrained to about 20 mm/s .

Results of orbit determination procedures for LEOs were presented in section I- 8.5.4. We complement these results using the GPS measurements of the CHAMP POD receiver of day 69 (10 March) of the year 2002. We focus in particular on the impact of the GPS clock information. The results are summarized in Table 7.3. Three kinds of GPS satellite clock corrections were used:

- The clock corrections contained in the CODE SP3-files (at 15 minute intervals).

- The clock corrections contained in the file COD02069.CLK. The clock corrections in this file, established routinely by the CODE Analysis Center, are provided at five minutes intervals.

- The clock corrections contained in the file COD02069x.CLK are given at 30 s intervals. They are produced for special research purposes.

It should be mentioned that the clock corrections contained in the three files are of the same accuracy. The only difference is the sampling rate.

The rms error of a particular orbit in Table 7.3 stems from a translation (three unknown parameters) between the tabular positions of this orbit and the

Table 7.3. LEO orbit quality using LEOKIN and SATORB with different GPS clock estimates

Solution	Code	Phase	Orbit	Clocks	Program	RMS [m]
C1	Y	N	N	30 s	LEOKIN	1.80
C2	Y	N	N	30 s	SATORB	0.40
P1	Y	Y	C2	30 s	LEOKIN	0.09
P2	Y	Y	C2	30 s	SATORB	0.08
PL	Y	Y	P2	30 s	LEOKIN	0.03
PS	Y	Y	P2	30 s	SATORB	—
C1a	Y	N	N	5 m	LEOKIN	1.80
C2a	Y	N	N	5 m	SATORB	0.41
P1a	Y	Y	C2a	5 m	LEOKIN	0.11
P2a	Y	Y	C2a	5 m	SATORB	0.11
C1b	Y	N	N	SP3	LEOKIN	3.69
C2b	Y	N	N	SP3	SATORB	1.56
P1b	Y	Y	C2b	SP3	LEOKIN	0.42
P2b	Y	Y	C2b	SP3	SATORB	0.41

tabular positions of the (probably) best possible orbit, labelled PS, that can be produced for this experiment using the programs LEOKIN and SATORB. The orbit PS was obtained by

- first using program LEOKIN to establish the kinematic orbit C1 with only code observations and no a priori orbit,

- then by using the C1 SP3 positions in program SATORB to establish the standard orbit C2 (parametrization: initial osculating elements, the nine deterministic parameters of Panel SATORB 7 (Figure 7.10) using the R, S, W decomposition, and three pseudo-stochastic pulses every 30 minutes in the same directions),

- then by using code and phase observations in program LEOKIN to establish the kinematic orbit P1 making use of the a priori orbit C2,

- then by using the P1 SP3 positions in program SATORB to establish the orbit P2 (parametrization: initial osculating elements, the nine deterministic parameters of panel SATORB 7 (Figure 7.10) using the R, S, W decomposition, and three pseudo-stochastic pulses every 10 minutes in the same directions),

- then by using code and phase observations in program LEOKIN to establish the kinematic orbit PL making use of the a priori orbit P2, and eventually

- by using the PL SP3 positions in program SATORB to establish the orbit PS (parametrization: initial osculating elements, the nine deterministic parameters of Panel SATORB 7 (Figure 7.10) using the R, S, W decomposition, three pseudo-stochastic pulses every 10 minutes in the same directions).

A comparison of the PS- and PL-orbits with a reduced-dynamics orbit established with the Bernese GPS software indicates that the PS-and PL-orbits are good to about 5 cm rms per coordinate.

The results in Table 7.3 speak a clear language. The better the time resolution of the GPS clock correction, the better the resulting orbit:

- The orbit that can be achieved using only code observations *without* screening the code observations with the help of an a priori orbit is of the order of one to two meters rms.

- Substantially better code-only orbits can be achieved if the C1 orbit is used in LEOKIN to screen the code observations. The quality of the orbit corresponding to C2 (but having used the a priori orbit C2 in LEOKIN) is of the order of 30 cm rms (not documented in Table 7.3).

- It is interesting to note that the accuracies of the P1- (a purely kinematic orbit) and the P2-orbits are of the same order of magnitude (of the order of 10 cm when compared to orbit PS). When compared to each other, the corresponding rms is about 3 cm, indicating that the parametrization used in SATORB to generate the orbits P2 and PS is adequate.

- The accuracy achieved when using five minute clock corrections (orbits C1a, C2a, P1a, and P2a) is surprisingly good. It is, in essence, possible to obtain sub-decimeter LEO orbits with the routinely available IGS clock and orbit information!

- The clock information contained in the SP3-files is (as mentioned) accurate, but obviously the linear interpolation between the 15 minute intervals seriously affects the results (see orbits C1b, C2b, P1a, P2b). Despite the fact that sub-meter accuracies is obtainable using this very convenient information, it cannot be recommended for "serious" LEO orbit determination.

Results of the kind achieved with the IGS standard products (precise SP3-orbits and five minute GPS clock corrections) are sufficient for many applications in practice. If even higher accuracy (of the order of one cm) or a higher productivity is required, it is unavoidable to use one of the well-established scientific GPS packages, which generate kinematic and/or reduced-dynamics orbits by the direct and correct use of all GPS observables.

8. The Computer-Program ORBDET

8.1 Introduction

Program ORBDET may be used to determine the orbits of minor bodies in the planetary system and the orbits of artificial Earth satellites and of space debris. ORBDET is the only interactive program of the program package. Its primary menu is reproduced in Panel ORBDET 1 (Figure 8.1). The program user first has to decide whether to determine the orbit of an object in the planetary system or in the Earth-near space. The algorithms used for the two purposes are identical. They are explained in detail in Chapter I-8.

General input and output files are defined after this primary selection. The file with constants (CONST), contained in the subdirectory GEN of the program system, is used throughout this program. The file DE200 contains the

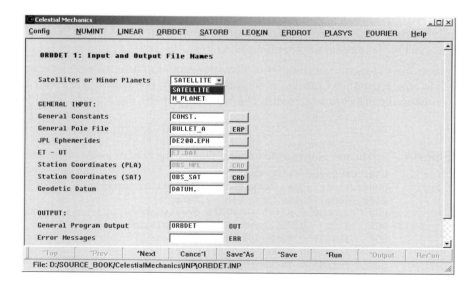

Fig. 8.1. Primary menu of program ORBDET

coordinates of Sun, Moon and planets for the time interval between 1981 and 2025 as computed from the JPL ephemerides [111]. Files for other time intervals might be generated as well (see [111] for more details). Depending on the application ("M_PLANET" or "SATELLITE" in the first input field) either the files "ET.DAT" (with the values "Ephemeris Time - Universal Time") and "OBS_MPL" or the files "OBS_SAT" and "DATUM" are required as input files.

The file containing the station coordinates of astronomical sites observing minor planets and comets is partly reproduced in Table 8.1. The file "obs_mpl.crd", containing the essential information for some of these observatories, is contained in the subdirectory ORBDET/STA of the directory CelestialMechanics. The format of the coordinate file is that defined and used by the MPC (60 Garden St., Cambridge MA 02138 USA) of the IAU. The ASCII-file may be easily edited to contain more or other observing sites.

The station code (three ASCII characters, first column) is used to identify the observatories (of minor planets etc.). Observations of more than one station may be analyzed in one and the same program run of ORBDET – provided the coordinates of the observing sites are contained in the file of Table 8.1. The geocentric coordinates of the stations are provided for each observing site in units of degrees for the geocentric longitude (LONG) and in units of equator radii for the Z-coordinate (column headed "sin") and for the length $\sqrt{X^2 + Y^2}$ of the projection of the geocentric station vector onto the equatorial plane (column headed "cos"). The file ET.DAT (not reproduced here), containing the final and predicted values for "Ephemeris Time - Universal Time" for the time interval 1965 - 2008, is contained as well in the subdirectory ORBDET/STA of the program system. This file must be updated from time to time (information in the Astronomical Ephemerides and Nautical Almanachs). It may be updated easily, as well.

Table 8.2 contains the coordinates of sites observing Earth satellites and space debris (the reproduced format is slightly reduced). The coordinates refer to a geodetic datum characterized in the third line of the file 8.2 (the ITRF97

Table 8.1. Coordinates of astronomical sites monitoring minor planets and comets

```
List of Observatory Codes
Observatory code, longitude and parallax constants
Code   Long.    cos      sin       Name
000    0.0000  0.62411  +0.77873  Greenwich
003    3.90    0.725    +0.687    Montpellier
004    1.4625  0.72520  +0.68627  Toulouse
009    7.4417  0.6838   +0.7272   Berne-Uecht
026    7.4648  0.68489  +0.72640  Berne-Zimmerwald

...    ......  .......  ........  ................
```

is used in the example of Table 8.2). The geodetic datum is defined by the reference ellipsoid (parameters: equatorial radius AE, flattening F, scale SC, the offsets w.r.t. the geocenter, and the rotations w.r.t. the "best possible" geocentric, Earth-fixed system). The file "DATUM", containing a wide selection of (local) geodetic datums, is provided in the GEN-subdirectory of the program system; the file is partly reproduced in Table 8.3 in the format of the Bernese GPS software [58]. Only the offsets and the rotations are used by the program ORBDET. If a satellite orbit shall be analyzed, a file containing the so-called Earth rotation parameters (see Chapter 2) is required as well. The file "rap_all.erp" (not reproduced here) is contained in the GEN subdirectory of the program system. The most recent version for this file may be retrieved from the CODE Analysis Center of the IGS (www.aiub.unibe.ch).

Table 8.2. Coordinates of astronomical sites monitoring satellites

```
ITRF93 EPOCH 1993.0: ITRF97                                23-DEC-94
-------------------------------------------------------------------
LOCAL GEODETIC DATUM: ITRF97           EPOCH: 1996-06-15  0:00:00

NUM  STATION NAME          X (M)           Y (M)            Z (M)

000  Geocenter              0.0000          0.0000           0.0000
001  Zimmerwald          4331283.5610     567549.6610     4633140.0410
002  Herstmonceux        4033463.7650      23662.4200     4924305.1230
003  FGAN/TIRA           4023953.5500     503304.4900     4906853.3200
004  Grasse              4581691.7070     556159.4690     4389359.4530
005  Graz                4194426.5920    1162693.9740     4647246.6000
009  Dresden/Guensber    3900310.0000     963194.6000     4936741.5000
010  Tenerife/Teide      5390281.0000   -1597891.0000     3007078.0000
011  Gibraltar           5136160.9640    -480717.0460     3738636.3880
580  Graz BMK            4194439.6000    1162718.4000     4647224.5000
754  Matera              4641990.2430    1393042.3920     4133231.9160
759  Wetzell             4075582.5020     931837.2700     4801559.9090
999  Herstmonceux2       4033464.336       23671.341      4924297.653
```

The names of the general output file and of the file with error messages have to be provided in the last two input fields of the primary menu panel ORBDET 1 (Figure 8.1). The general output file contains the results (the orbital elements), but also a summary of the interactive orbit determination session.

Program ORBDET does not require the knowledge and makes no use of approximate orbital elements that might be known a priori. This is why a *first orbit* has to be determined initially. The method has to be specified as the first input item in Panel ORBDET 2 (Figure 8.2) of program ORBDET. Two methods, namely the *determination of a circular orbit* and the determination of a general two-body orbit, based on the formulation as a *boundary value*

Table 8.3. Definition of Geodetic Datum

```
LOCAL GEODETIC DATA FOR BERNESE GPS SOFTWARE VERSION 4.2
----------------------------------------------------------
```

DATUM	ELLIPSOID	SHIFTS (M)		ROTATIONS (")	
ITRF97	AE = 6378137.000	DX =	0.0000	RX =	0.0000
	1/F= 298.2572221	DY =	0.0000	RY =	0.0000
	SC = 0.00000D+00	DZ =	0.0000	RZ =	0.0000
....	....	....		....	
CH - 1903	AE = 6377397.200	DX =	679.0000	RX =	0.0000
	1/F= 299.1528000	DY =	-2.0000	RY =	0.0000
	SC = 0.00000D+00	DZ =	404.0000	RZ =	0.0000
PZ - 90	AE = 6378137.000	DX =	0.0000	RX =	0.0000
	1/F= 298.2572236	DY =	0.0000	RY =	0.0000
	SC = 0.00000D+00	DZ =	0.0000	RZ =	-0.3345

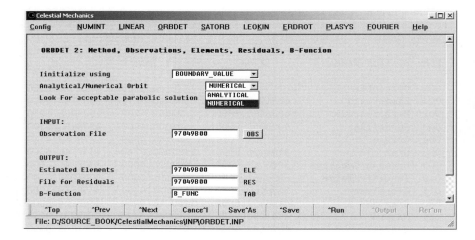

Fig. 8.2. Selection of object and orbit determination method

problem, may be selected. The theory for the determination of a circular orbit was presented in section I-8.3.1, the theory of the orbit determination as a boundary value problem was developed in sections I-8.3.2 and I-8.3.3. In the next input option ("Analytical/Numerical Orbit") one may decide to represent the orbit either analytically (using the procedure outlined in section I-8.3.2) or numerically as the solution of a local boundary value problem (see I-7.5.2). Both methods work equally well. They are of course not active, if a circular orbit shall be determined. Under the same circumstances (boundary value problem) you may also look for acceptable parabolic orbits. For some applications you may not be interested in parabolic orbits, which is why the option may be selected or switched off.

The name of the file containing the astrometric positions has to be specified as the fourth input item in Panel ORBDET 2 (Figure 8.2). In the example of Panel ORBDET 2 a file named gtoobj_7 containing 23 observations of a (non-catalogued) object in a geostationary transfer orbit was selected. The observations were made in September 2002 with the ESA 1-m telescope in Teneriffe. The search strategy and the observatory are described by [100].

The first three lines of the observation file "gtoobj_7.obs" are reproduced in Table 8.4. The observation files have to be located in the subdirectory ORBDET/OBS of the program system. Each line may, e.g., be read by the FORTRAN statement

```
          READ(LFN001,5010,END=1010)STNAME,OBJCHAR,XMJD,RA,DE
   5010   FORMAT(1X,A16,1X,A35,F17.9,F13.9,F13.8),
```

where STNAME and OBJCHAR, the name of the observing station and the characterization of the object, are character variables. XMJD, the Modified Julian Date (MJD), RA (right ascension) and DE (declination) of the object are REAL*8 variables. The unit for the modified Julian date MJD is days. The right ascension is specified in the form "hh.mmssxxxxx", where "hh" stands for hours, "mm" for minutes, "ss" for seconds, and "xx..." for fractions of seconds. The declination has to be provided in the form "dd.mmssxxx", where "dd" stands for degrees ($°$), "mm" for arcminutes ($'$) and "ss" for arcseconds ($''$), "xx..." are fractions of arcseconds. It is assumed that the observations refer to the system $J2000.0$.

Table 8.4. Observation file of a space debris object

```
Tenerife/Teide   geo-0004:sv1-20020908-S-6-1B   52525.960917109   0.43026812   -11.5626736
Tenerife/Teide   geo-0004:sv1-20020908-S-6-1B   52525.961657776   0.43372378   -11.5557413
Tenerife/Teide   geo-0004:sv1-20020908-S-6-1B   52525.962398748   0.44117492   -11.5527736
...
```

The observations of a minor planet or comet may be provided in two different formats. The first is illustrated in Table 8.5. The format is (more or less) self-explanatory. Observe that the observations may be provided in the system $J2000.0$ or in the system of its predecessor, the Bessel system $B1950.0$. The system is defined by the index in the column 30 (with title "X"). If the index is either blank or "1", the observations are assumed to refer to the system $B1950.0$, if it is "2", to the system $J2000.0$. Different observations in the same observation file may refer to different systems. This flexibility is required if "old" and more recent observations shall be combined. Before determining an orbit, all observations are transformed into the system $J2000.0$ and the resulting orbital elements refer to this latter system, as well. All observations to be used in one and the same orbit determination process have to reside in one file.

Table 8.5. Observation file of Comet Panther

NAME: PANTHER   REF.:

		UT		X	HH	MM	SS.SSS	VDD	MM	SS.SS	STA
JJJJ	MM	DD.DDDDD						EQUINOX: 1950.0			
1980	12	27.76076		1	18	47	55.60	39	22	27.5	026
1980	12	28.40625		1	18	48	14.47	39	32	01.3	026
1980	12	28.72986		1	18	48	24.52	39	36	55.2	026
1980	12	29.42986		1	18	48	45.24	39	47	40.5	026
1980	12	30.95234		1	18	49	32.25	40	11	53.0	026
1980	12	31.06597		1	18	49	35.87	40	13	52.4	026
1980	12	31.74792		1	18	49	57.37	40	24	53.0	026
1981	01	8.22194		1	18	54	10.61	42	44	30.3	026
1981	01	23.77500		1	19	04	20.58	49	25	27.3	026
1981	01	28.77222		1	19	07	59.99	52	11	39.3	026
1981	04	9.06250		1	07	59	31.09	59	29	06.4	026

The observations may also be provided in the format defined and used by the MPC. The format is not reproduced here. An example is provided in the help panel of the program.

The next three file names to be specified in Panel ORBDET 2 (Figure 8.2) refer to specific output files. The "*.ele"-file contains the estimated elements. If a satellite orbit was estimated, the resulting elements file may be used as an input file by the program SATORB, where the information is used to initialize an orbit improvement process with a sophisticated force field (much more complex than the force model available in program ORBDET; for details see Chapter 7).

The "*.res file" contains the observation number in the first column, the time (MJD in the case of minor planets and comets, seconds relative to the first observation epoch in the case of satellite observations) in the second column, the residuals in $\alpha \cdot \cos \delta$ and in δ (in units of arcseconds) in the third and fourth column. The fifth column is left blank, except for observations which were marked during the orbit determination process. A mark in this column indicates that the corresponding observation was not used to determine the orbit relative to which the residuals have been calculated. The last three columns contain the right ascension and declination (both in degrees) and the topocentric distance of the object at observation time (in AU for minor planets and comets, in meters for artificial Earth satellites). The residuals refer to the last orbit improvement step, performed in program ORBDET. The "*.res" may be visualized by the menu function "display results".

The output file "*.tab" specified in Panel ORBDET 2 (Figure 8.2) contains the essential information related to the *first* orbit determination step. If a *circular* orbit was determined initially, the file contains only three relevant

columns, where the record number is contained in the first, the assumed value
a for the semi-major axis in the second, and the difference in the argument
of latitude $B(a) \overset{\text{def}}{=} \Delta u_g(a) - \Delta u_d(a)$ for the two observations selected in the
third column. If the solution was sought in the form of a boundary value prob-
lem, the first column contains the record number, the second the topocentric
distance Δ_{b1} corresponding to the first boundary epoch, and the third contains the value $\log(\sigma(\Delta_{b1}))$, the logarithm of the estimated rms a posteriori.
In addition, the semi-latus rectum p and the eccentricity e of the conic section
associated with the assumed value Δ_{b1} are provided in the columns four and
five.

The third and last panel shown to the program user depends on whether the
orbit of a minor planet (or comet) or of a satellite (or a space debris) has to be
determined. In the case of a satellite orbit, the force model for the concluding
orbit improvement step has to be defined in panel ORBDET 3b (Figure 8.3),
in the case of a minor planet, the force model for the orbit improvement step
has to be specified in the Panel ORBDET 3a (Figure 8.4).

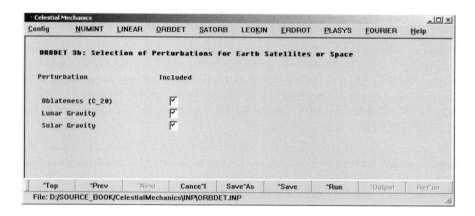

Fig. 8.3. Orbit model for satellite orbits

The force models for the final orbit improvement step thus are comparatively
simple in program ORBDET. Only the Earth's oblateness and the gravitational attraction exercised by Sun and Moon may be taken into account for
satellite orbits, whereas the gravitational attractions exercised by the planets
(their positions are approximated by the formulae provided by Meeus [72])
may be taken into account for orbits in the planetary system.

In the following two sections the two orbit determination procedures (boundary value problem and circular orbit) are dealt with separately using the
observations in the file "gtoobj_7.obs".

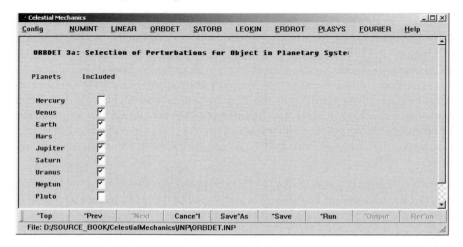

Fig. 8.4. Orbit model for planetary orbits

8.2 Orbit Determination as a Boundary Value Problem

Subsequently we will establish an orbit for a satellite-like object (a space debris particle). The main switch in Panel ORBDET 1 (Figure 8.1) was thus set to "SATELLITE". The force field to be used for the concluding orbit improvement process was defined as shown in Panel ORBDET 3b (Figure 8.3). "BOUNDARY_VALUE" was selected as the orbit determination method in Panel ORBDET 2 (Figure 8.2).

The program user then obtains the list of observations as shown in Figure 8.5. Based on this list the decision has to be taken, which observations to include into the first orbit determination process. It is recommended to include only observations stemming from a relatively short time interval (shorter than the expected revolution period). Observations from different observatories may be used here. In the example of Figure 8.5 all 23 observations stemming from the ESA 1-m telescope on Teneriffe were selected for the first orbit determination process. Observe that observations which are not selected here, may be included later on, in the orbit improvement step. The program user then has to specify the two boundary epochs. In the example of Figure 8.5 the epochs 1 and 23 were chosen. It is recommended to select two boundary epochs which are relatively close together. One can, e.g., easily verify that in the above example almost any other selection (e.g., 1 and 5) would work, as well.

The program then proposes to perform a search over the topocentric distance Δ_{b1} associated with the first boundary epoch. The initial (250 km) and the final (70000 km) search value for Δ_{b1} are proposed together with the mesh size (250 km). The proposed search pattern should be adequate for most cases

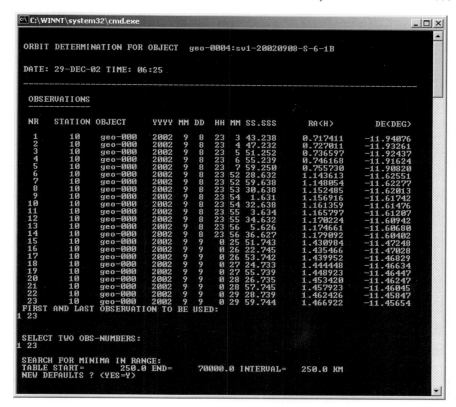

Fig. 8.5. List of observations, selection of observations and of boundary epochs

(except perhaps for the determination of very low orbits, where one should reduce the mesh size). If the program user agrees with the specified search pattern (or after the redefinition of this pattern), Figure 8.6 is generated (where only the last few lines of this table are included in this figure).

A similar search table is proposed when determining the orbits of minor planets, comets, etc. The provided search table is, however, optimized for "ordinary" celestial bodies. When dealing, e.g., with objects in the Edgeworth-Kuiper belt, this table should be adapted to include large topocentric distances. This may be done, when no satisfactory solution was found within the range of the proposed search table. When dealing with NEAs, it is sometimes necessary to reduce the search range and to refine the table spacing.

The table at the top of Figure 8.6 gives the estimated rms error $\sigma(\Delta_{b1})$, the semi-latus rectum p (in meters), and the eccentricity e associated with each topocentric distance Δ_{b1} (and the corresponding best possible value for the topocentric distance Δ_{b2} resulting from a parameter estimation process using

```
C:\WINNT\system32\cmd.exe                                              _ □ ×
R=      69000.0 KM    RMS=    154.55"  P=   104588958.132  E=   0.4551276   ▲
R=      69250.0 KM    RMS=    154.73"  P=   105990964.772  E=   0.4695626
R=      69500.0 KM    RMS=    154.90"  P=   107407274.732  E=   0.4840976
R=      69750.0 KM    RMS=    155.06"  P=   108837985.003  E=   0.4987329
ROOT # 1 TOP. DIST. 1,2   36216.701    36944.803 KM
SELECT ROOT:
1
ORBIT IMPROVEMENT WITH PERTURBATIONS (YES="y")
y
RESET ALL MARKS (YES="Y")
y

RESIDUALS IN RIGHT ASCENSION AND DECLINATION
----------------------------------------------------
    #       TIME          RA*COS(DE)       DE     MARK
                             (")          (")
    1       0.0000000        0.09        0.70
    2      63.9936292       -0.14        0.02
    3     128.0136099       -0.28       -0.05
    4     192.0003265       -0.08       -0.23
    5     256.0115810        0.44       -0.53
    6    2925.3934943       -0.08        0.09
    7    2956.3998628        0.12        0.16
    8    2987.4001825       -0.12       -0.07
    9    3018.3924675       -0.31       -0.01
   10    3049.4001316        0.10       -0.12
   11    3080.3962180        0.27       -0.04
   12    3111.3935139       -0.11       -0.07
   13    3142.3881312        0.12       -0.18
   14    3173.3891424       -0.06        0.33
   15    4928.5046596        0.31       -0.19
   16    4959.5071393        0.10        0.46
   17    4990.5039172       -0.01        0.33
   18    5021.4943873        0.42        0.16
   19    5052.5012738       -0.54       -0.30
   20    5083.4971012       -0.37       -0.25
   21    5114.5068386       -0.05       -0.08
   22    5145.5009378        0.22       -0.03
   23    5176.5061823       -0.04       -0.11
MARK OBSERVATION NUMBER (STOP=0; FROM TO= n 0 n )
0                                                                       ▼
```

Fig. 8.6. RMS values, semi-latus rectum, and eccentricity as a function of geocentric distance; selection of solution

the observations specified in the first interactive input selection in Figure 8.5). If acceptable (relative) minima were found, a table with these minima is provided and the program user may select one of them to initialize the subsequent orbit improvement process. In the example of Figure 8.6 only one minimum was found; obviously this minimum was selected. Subsequently, an orbit improvement process with the inclusion of perturbations (based on the force model of Panel ORBDET 3b 8.3) was invoked. It is now possible to include all observations by using the option "RESET ALL MARKS". Use was made of this option – with no effect because in this example all available observations were already used for the initial orbit determination. In general it is recommended *not* to use other observations than those already used for initial orbit determination at this stage – the orbit improvement process may be repeated later on with more observations.

The program then displays the list of residuals (for all observations, those used and those marked). Based on this list the program user may decide to repeat the orbit determination process after having marked a few observations. Individual observations may be marked by typing the observation number and pressing <enter>, a sequence of observations from n_1 to n_2 by typing n_1 0 n_2 and then pressing <enter>. The orbit improvement process is repeated until no further changes are detected by the program, in which case the program is terminated.

The program user may then inspect and review the essential steps of the orbit determination session in the general output file. The tables displayed in real time and contained in Figures 8.5 and 8.6 may be reviewed in this file. Moreover, the detailed results (orbital elements and residuals) are contained in this file for each orbit determination and orbit improvement step. For each orbit improvement step the estimated elements *and* the corresponding mean errors are provided. Table 8.6 is a small extraction from the general output file "orbdet.out" containing the final result. More examples may be found in Chapter I-8. The program system allows it to display the function $\log(B(\Delta_{b1}))$ by using the option "display results" associated with the menu-button "ORBDET". Figure 8.7 shows the result after having decided to display (the logarithm of) the function "B-values". The figure is drawn with the use of the file "*.tab" produced by the program ORBDET. The orbit determination process obviously had exactly one solution. Instead of producing a graph of the function $\log(B(\Delta_{b1})) \overset{\text{def}}{=} \log(\sigma(\Delta_{b1}))$ it is also possible to visualize the semi-latus rectum p or the eccentricity e associated with the topocentric distances Δ_{b1} (see Figures I-8.7 and I-8.8).

The option "display results" provided for program ORBDET also allows it to inspect the residuals as a function of time. Figure 8.8 illustrates a typical result.

The files "*.tab" and "*.res" produced by the program ORBDET may of course also be used by any plot-package. Figure 8.9 shows, as an example, two curves $\log(\sigma(\Delta_{b1}))$ for the example used here, once including all observations (and using the first and last observation epochs as boundary epochs),

Table 8.6. Orbit elements for the previously unknown object geo-004

```
ORBIT DETERMINATION FOR OBJECT  gtoobj_07:sv1-20020908-S-6-1B
-----------------------------------------------------------------
# OBS         =   23
   RMS        =   0.27 "
TIME INTERVAL =   5176.506 SEC
P    =      11656870.286381 +/-         3303.218992 M
A    =      23819403.495614 +/-         2604.819940 M
E    =          0.7145729612 +/-      0.0000595942
I    =            6.6971388 +/-         0.0005283 (DEG)
NODE =           82.3293036 +/-         0.0032306 (DEG)
PER  =         -248.1323982 +/-         0.0024953 (DEG)
TPER =       -14061.4252978 +/-         3.0547606 SEC
```

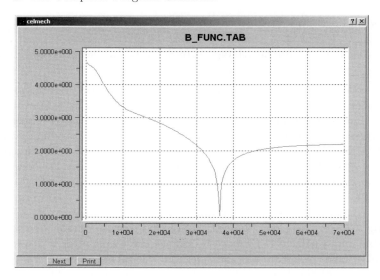

Fig. 8.7. Logarithm of RMS values $\sigma(\Delta_{b1})$ as a function of geocentric distances Δ_{b1}

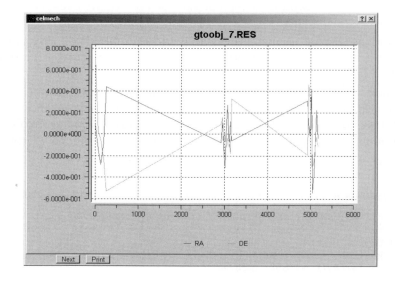

Fig. 8.8. Logarithm of RMS values $\log(\sigma(\Delta_{b1}))$ as a function of geocentric distances Δ_{b1}

once using only observations 1 to 5 with the observation epochs 1 and 5 as boundary epochs. The example clearly shows that the length of the observation interval (and, to a lesser extent, the number of observations) plays an essential role for determining a robust orbit.

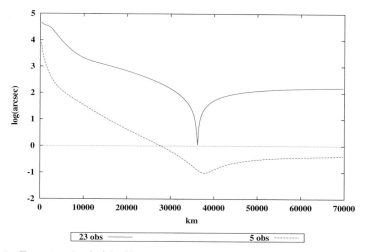

Fig. 8.9. Function $\log(\sigma(\Delta_{b1}))$ using 23 astrometric positions of a GTO-object within $1^{\mathrm{h}}26^{\mathrm{m}}$ (solid line) or 5 positions within 4^{m} (dotted line)

8.3 Determination of a Circular Orbit

It is not really a wise decision to determine a circular orbit if more than two observations are available (except, perhaps, if the observations are contained in a very short time interval). The statement is particularly true for the example of the object "gtoobj_7", which moves around the Earth in an orbit with an eccentricity $e \approx 0.7$ (see Table 8.6).

Let us make the attempt, anyway. It does not make sense to include the initial dialogue for the determination of a circular orbit, because the pattern of Figure 8.5 is followed more or less when determining a circular orbit: The observation epochs to be used for the determination of a circular orbit have to be selected instead of the boundary epoch and the search pattern is proposed for the semi-major axis a and not for the topocentric distance. Let us assume that we made the decision to determine the orbit with observations 1 and 5.

The determination of a circular orbit is a rather risky business. It may well be that no suitable orbit is found – in particular if the eccentricity e of the true orbit is rather high (as in our case).

Despite these negative remarks a circular orbit actually was found, the result of which is reproduced in Table 8.7. The table should be compared to Table 8.6 containing the result of the orbit improvement step. The residuals for the first and fifth observation are zero (as they should). Also, the residuals of observations 2, 3, and 4 are reasonably small. If only observations 1 to 5 were available, one could believe that the circular orbit is close to reality. Table 8.7 illustrates the difficulty to determine a good orbit from observations contained in a very short time-span. That the circular orbit is no longer an acceptable approximation if observations from a substantial part of the object's revolution period are available, is illustrated by Table 8.8, where the first and last observation of the file "gtoobj_7.obs" were used to determine a circular orbit. The insufficiency of the orbit model is now apparent.

Let us return to the discussion of the results in Table 8.7. The comparison with Table 8.6 containing the "true" results show that the determination of the orbital plane was more or less successful (differences of only a few degrees w.r.t. the truth). The semi-major axis a seems to be, however, unacceptably wrong: instead of a value of $a \approx 23819$ km (see Table 8.6) a value of about $a \approx 62000$ km is obtained. Observe that the estimate does not improve by

Table 8.7. Circular orbit for the previously unknown object gtoobj_07 when using observations 1 and 5

```
ORBIT DETERMINATION FOR OBJECT  gtoobj_07:sv1-20020908-S-6-1B
-----------------------------------------------------------------
    RESIDUALS IN ARCSECONDS                      ELEMENTS
    -----------------------                      --------

    I      RA         DE
    1     0.00       0.00      A    =     62210896.6
    2     0.94       0.38      E    =      0.000000
    3     1.12       0.87      I    =      9.287203
    4     0.81       0.75      NODE =     73.566100
    5     0.00       0.00      PER  =      0.000000
    6  -519.65    -429.57  *   TPER =     28864.319
    7  -528.07    -438.91  *
    8  -536.92    -448.65  *
   ..
   15  -989.69   -1229.10  *
   16  -994.37   -1243.79  *
   ..
   22 -1018.25   -1338.20  *
   23 -1022.02   -1354.25  *

 *** SR BNIMPP: BIG CORRECTIONS IN ITER-STEP  1
```

Table 8.8. Circular orbit for the previously unknown object gtoobj_07 when using observations 1 and 5

```
ORBIT DETERMINATION FOR OBJECT  gtoobj_07:sv1-20020908-S-6-1B
--------------------------------------------------------------
    RESIDUALS IN ARCSECONDS                      ELEMENTS
    -----------------------                      --------

    I      RA        DE
    1     0.00      0.00     A    =    63300002.4
    2    13.89     16.72     E    =     0.000000
    3    27.01     33.56     I    =     8.849607
    4    39.63     49.79     NODE =    83.445883
    5    51.76     65.43     PER  =     0.000000
    6    63.93    329.14     TPER =    33891.208
    7    61.62    327.95
    8    58.87    326.36
    9    56.16    324.97
   10    54.05    323.32
   11    51.71    321.77
   12    48.83    320.00
   13    46.56    318.07
   14    43.90    316.66
   15   -15.53     59.26
   16   -14.23     52.80
   17   -12.69     45.47
   18   -10.48     38.02
   19    -9.53     30.19
   20    -7.31     22.77
   21    -4.81     15.39
   22    -2.21      7.81
   23     0.00      0.00


  *** SR BNIMPP: BIG CORRECTIONS IN ITER-STEP  1
```

increasing the length of the time interval between the two observation epochs used for the determination of a circular epoch.

It is close to a miracle that the orbit improvement process converges with an a priori orbit of such a bad quality. The last line of Tables 8.7 and 8.8 documents that this is also the opinion of program ORBDET: Unusually big increments of the estimated parameters were encountered in the first step of the orbit improvement (they were, as a matter of fact, automatically reduced to a "reasonable" size by the program).

The result $a \approx 63000$ km for geostationary transfer orbits when determining a circular orbit with observations near the apogee may be explained easily by taking into account that the actual velocity v of an object with eccentricity e near the apogee is defined by (compare eqn. (I- 4.20):

$$v^2 = GM \left\{ \frac{2}{a\,(1+e)} - \frac{1}{a} \right\} = \frac{GM}{a} \frac{1-e}{1+e} \, , \tag{8.1}$$

and that the apogee is in the geostationary belt at $a(1+e) \approx 41000$ km . The proof that $\Delta u_g - \Delta u_d \stackrel{\text{def}}{=} 0$ leads to $a \approx 63000$ km under such circumstances may be left to the reader!

9. The Computer-Program ERDROT

As indicated by the primary panel ERDROT 1 (Figure 9.1) the program ERDROT may be used for the following four tasks:

1. Solve Euler's equations (I- 3.124) of Earth rotation, assuming that the geocentric orbits of Moon and Sun are known (given by the JPL ephemerides or by approximate formulas).

2. Solve Euler's equations (I- 3.124) of lunar rotation, assuming that the geocentric orbits of Moon and Sun are known (given by the JPL ephemerides or by approximate formulas).

3. Solve the three-body problem Earth-Moon-Sun according to the theory developed in section I- 3.3 assuming that Earth and Moon are rigid bodies.

4. Study the correlation between Earth rotation parameters as established by space geodetic methods (e.g., IGS, IERS) and atmospheric angular momenta (AAM), as established by the IERS Special Bureau for the Atmosphere (SBA).

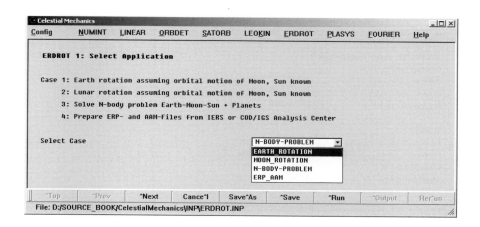

Fig. 9.1. Primary menu of program ERDROT

In the first case (Earth rotation), the Earth may be assumed to be either rigid or elastic (see eqns. (2.120) in section 2.3), or to consist of a rigid mantle and a liquid core (Poincaré model, see eqns. (2.175) in section 2.3.4).

Option 4 (correlation) does not involve simulations, but the analysis of real data. The theory behind this application is contained in section 2.3.3. The four program options will now be dealt with separately.

9.1 Earth Rotation

Panel ERDROT 2 (Figure 9.2) shows that up to three input files are required to run the program. The first file is the general file with program constants (contained in the GEN-subdirectory of the program system and used for practically all programs of the system). The second input file contains the Earth rotation parameters as retrieved from the address of the CODE analysis center http://www.aiub.unibe.ch/download/BSWUSER/GEN/ under the file name BULLET_A.ERP. The file is only required if the third program option (N_BODY_PROBLEM) is activated. The third input file contains the JPL ephemerides in a binary version. If this input file is left blank, the program uses approximate ephemerides for Sun, Moon and planets (where required). When using the file DE200.EPH for pure Earth or Moon rotation studies (first two options)) the applications have to lie within the time interval between 1981 and 2025. When using the N-body option, the file DE200.EPH is only used to define the initial values, implying that the initial epoch has to lie within the mentioned time interval.

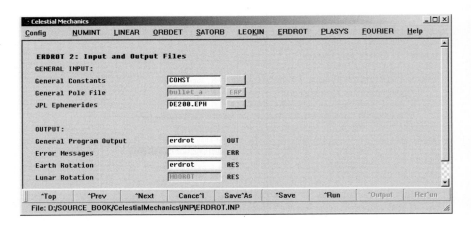

Fig. 9.2. Output files of program ERDROT (option Earth rotation)

There are at maximum three output files to be specified in Panel ERDROT 2 (Figure 9.2). The general output file echoes the general program characteristics and contains in addition the statistical information like the CPU time required for running the program. The error messages are written into the same file, unless the second output file name is not left blank.

The third output file contains the Earth rotation parameters. The columns contain the following information:

1. Time in the format to be selected in Panel ERDROT 6 (Figure 9.6).

2. x-coordinate $x \overset{\text{def}}{=} \frac{\omega_1}{\omega_3} \cdot 206264.8$ of the pole.

3. -y-coordinate $-y \overset{\text{def}}{=} \frac{\omega_2}{\omega_3} \cdot 206264.8$ of the pole.

4. Excessive day length in milliseconds.

5. Precession $\Psi(t)$ in longitude.

6. Nutation in longitude $(\Delta\Psi(t))$ reduced by a precession with a rate of $-50.387784''$ per year.

7. Nutation $\Delta\varepsilon$ in obliquity (minus mean obliquity $J2000.0$).

8. Sidereal time at Greenwich $\Theta(t)$ reduced by initial value and the rate specified in Panel ERDROT 4a (Figure 9.4).

9. $\chi_1(t)$, first component of vector $\boldsymbol{\chi}(t)$ characterizing the motion of the liquid core w.r.t. the rigid mantle.

10. $\chi_2(t)$, second component of vector $\boldsymbol{\chi}(t)$ characterizing the motion of the liquid core w.r.t. the rigid mantle.

11. $\chi_3(t)$, third component of vector $\boldsymbol{\chi}(t)$ characterizing the motion of the liquid core w.r.t. the rigid mantle.

The information was written using the following (FORTRAN) format:

FORMAT(F15.8,2F12.6,3F12.4,F15.3,5F15.6),

Each output file *.res contains a header consisting of eight lines, including the date and the time of its generation (for identification), and a title characterizing each column. The last three columns are only written, if a liquid core was included.

Panel ERDROT 3 (Figure 9.3) shows the value of the constant of gravitation used in the program (without giving you the chance to change it) and it defines the mass ratios $m_{\delta} : m_{\mathbb{C}}$ and $m_{\odot} : m_{\delta}$. The values provided on the CD are "the true values", but one may change them within reasonable limits. When producing so-called "free solutions" (the two options at the bottom of Panel ERDROT 3, Figure 9.3), the corresponding mass ratios are set to zero (and the values masked with shades in this panel, retaining, however, the values for later use).

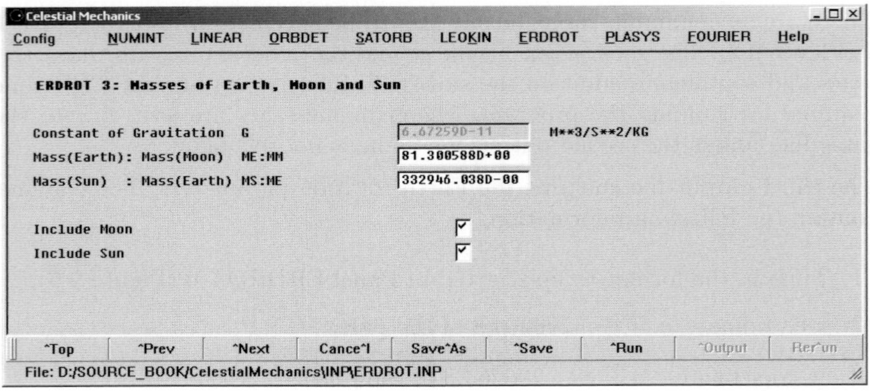

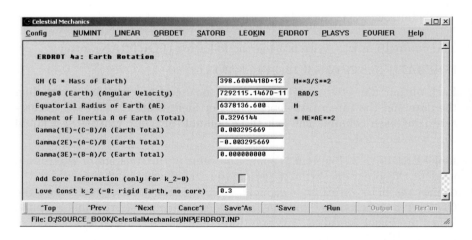

Fig. 9.3. Definition of parameters (part 1) in program ERDROT (option Earth rotation)

Panel ERDROT 4 (Figure 9.4) offers even more modelling options. The mass of the Earth may, e.g., be redefined (via the GM-value and the fixed value of G in the previous panel). Then, the angular velocity referring to the initial epoch has to be specified (the value is given in radian per second). The next three input lines are reserved for the definition of the values γ_{δ_i} as defined by eqns. (I- 3.123). The values provided on the CD are those corresponding to a realistic, rotationally symmetric Earth. Other experiments with other values or models are easily set up. Eventually, one may decide to introduce a

Fig. 9.4. Definition of parameters (part 2) in program ERDROT (option Earth rotation)

liquid core (in which case the Panel ERDROT 4a(cont) (Figure 9.7), discussed towards the end of this section, will allow it to refine the model even further). Observe that as soon as a liquid core is introduced, the value for k_2 will be automatically set to $k_2 = 0$ (and the corresponding value shaded in the panel).

The value $0 \leq k_2 \leq 1$ allows it to continuously switch from a completely rigid ($k_2 = 0$) to an ideally elastic Earth ($k_2 = 1$).

In Panel 5 (Figure 9.5) the numerical integration is set up (a collocation procedure of fixed stepsize is used), the integration interval, the initial conditions for polar motion, and the output interval are defined. Please observe that the file "*.res" may become very long if the integration interval is long and the output data rate high.

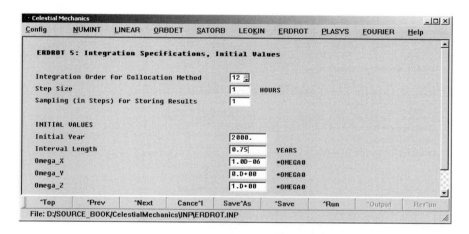

Fig. 9.5. Simulation and integration characteristics in program ERDROT (option Earth rotation)

The last panel allows it to define the time scale in column 1 of the file *.erp. The time may be provided either in days (relative to the first epoch), in the modified Julian day (MJD) or in years (and fractions thereof). One may decide to either include or exclude the quasi-daily terms (see eqns. (2.25)) in nutation and sidereal time.

Panel 4a(cont) (Figur 9.7) is only activated if a Poincaré-type model for the Earth was activated in Panel ERDROT 4a (Figure 9.4). In this case the moments of inertia associated with the liquid core (in units of the Earth's total first moment of inertia A), the elipticity of the core-mantle boundary, and the initial values for vector χ must be specified. The values proposed on the CD are the realistic values of eqns. (2.151).

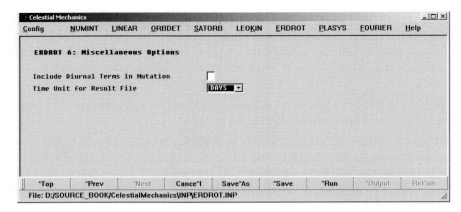

Fig. 9.6. Output characteristics in program ERDROT (option Earth rotation)

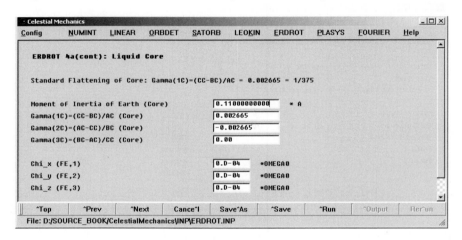

Fig. 9.7. Parameters of a Poincaé model in program ERDROT (option Earth rotation)

9.2 Rotation of the Moon

The rotation of the Moon may be studied by selecting the second option (MOON_ROTATION) in Panel ERDROT 1 (Figure 9.1). The next panel is very similar to that presented in the case of Earth rotation: In Panel ERDROT 2 (Figure 9.2) one has to select the name for the file with the lunar rotation results (and the corresponding file for Earth rotation is shaded). The file format and content will be provided at the end of this section. Exactly like in the case of Earth rotation one has to select the mass ratios in Panel ERDROT 3 (Figure 9.3). For obvious reasons one does not have the option,

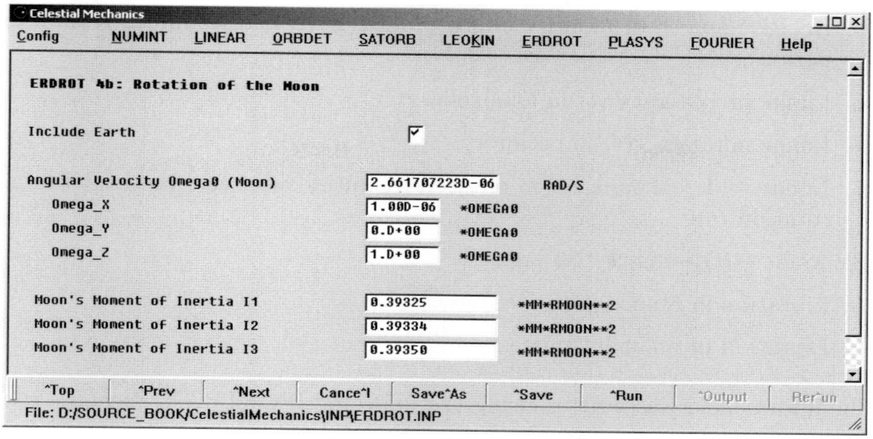

Fig. 9.8. Definition of parameters in program ERDROT (option lunar rotation)

however, to exclude the Moon! The key characteristics of lunar rotation are now specified in Panel ERDROT 4b (Figure 9.8). Here, the decision is made whether or not to include the Earth as a torque-generating body. If the Sun was excluded in the previous panel and the Earth here, studies of the free rotation of the Moon are invoked. Observe that in the case of lunar rotation only the rigid body model may be used.

The initial angular velocity of the Moon may be defined subsequently. The value provided on the CD corresponds to that contained in Table 2.1 (in radians/second), leading to the well-known coupling of the Moon's periods of rotation and revolution. The initial position of the Moon's rotation axis is specified in the Moon's PAI-system. The last three input lines are reserved for the definition of the Moon's three principal moments of inertia. Again, the values of Table 2.1 are provided on the CD.

Panel ERDROT 5 (Figure 9.5) is almost identical as in the case of Earth rotation, the exception being the last three lines of the input panel which are blocked (reserved for Earth rotation). When defining the integration characteristics, much longer stepsize are possible than in the case of Earth rotation. $h = 24$ hours is about appropriate when using a collocation method of order $q \approx 12$. The time scale may be defined as in the case of Earth rotation (see Panel ERDROT 6, Figure 9.6).

The file "*.res" of lunar rotation, defined in Panel ERDROT 2 (Figure 9.2), contains the following parameters:

1. Time in the format selected in Panel ERDROT 6 (Figure 9.6).

2. x-coordinate $x \stackrel{\text{def}}{=} \frac{\omega_1}{\omega_3} \cdot 206264.8$ of pole in the Moon's PAI-system.

3. -y-coordinate $-y \stackrel{\text{def}}{=} \frac{\omega_2}{\omega_3} \cdot 206264.8$ of pole in the Moon's PAI-system.

4. Relative length of month (length of month divided by initial revolution period).

5. Lunar precession $\Psi(t)$ in longitude.

6. Lunar nutation $\varepsilon(t)$ in obliquity.

7. Lunar "sidereal time" $\Theta(t)$ reduced by initial value and the initial lunar rotation rate.

8. Angle $\Theta(t) - n_0\,(t - t_0) - \Psi(t)$.

9. Libration in lunar longitude.

10. Libration in lunar latitude.

The information was written using the following (FORTRAN) format:

FORMAT(F15.8,5F16.5,F15.5,5F15.5)

Each output file *.res contains a header consisting of eight lines, including the date and the time of its generation (for identification) and a title characterizing each column.

9.3 The N-Body Problem Earth-Moon-Sun-Planets

When the option "N-BODY-PROBLEM" (Panel ERDROT 1, Figure 9.1) is selected, the three-body problem Earth-Moon-Sun is numerically integrated together with a user-defined number of planets. The planets may be selected in Panel ERDROT 4b(cont) (Figure 9.9). If no planet is selected in this panel, the pure three-body problem, as discussed in section I-3.3, is treated. Earth and Moon are considered as bodies of finite size, whereas all other planets assumed to be point masses. The system is a coupled second-order differential equation system with $d = 3 \cdot (n_p + 1) + 2 \cdot 3$ (n_p being the number of planets) equations. Three equations have to be set up for each center of mass of the planets and the Moon, and there are three second-order equations for each set of Euler angles for Earth and Moon (as introduced in section I-3.3.6).

In this mode the program ERDROT may be viewed as a generalization of program PLASYS, because it includes the orbital motion of the Earth's only natural satellite, and because it takes the rotational motion of Earth and Moon into account. As opposed to the treatment in program PLASYS, relativistic effects are not considered here – except for the Moon. The Earth's oblateness is also taken into account for the center of mass motion of the Moon. This option of ERDROT may, however, also be viewed as a generalization of the options "EARTH_ROTATION" and "MOON_ROTATION" (see Panel ERDROT 1, Figure 9.1) in the same program – except for the fact that Earth and Moon are assumed to be rigid in this mode.

The multistep integration method (see sections I- 7.4.2 and I- 7.5.6) is used to solve the generalized *N*-body problem. This is undoubtedly the best choice under the circumstances given (no highly eccentric orbits, no big variations in the rotational motions).

After having selected the *N*-body mode in Panel ERDROT 1 (Figure 9.1) the output names for both, the files containing the information concerning Earth and Moon rotation, have to be defined in Panel ERDROT 2 (Figure 9.2). Note that the file DE200.EPH *must* be provided for the definition of the initial state vector of the centers of mass of Sun, Moon and planets. The rigid-body characteristics of Earth and Moon are then defined in Panels ERDROT 4a (Figure 9.4) and ERDROT 4b (Figure 9.8). The planet list (including for obvious reasons the Earth) may then be completed in Panel ERDROT 4b(cont) (Figure 9.9). Note that only the perturbations of the planets in this list acting on the centers of mass (but not on the torques influencing the rotational motion) are taken into account. In the same panel one also must specify the name of the output file containing the orbital elements of the Moon. It is possible to store the osculating elements (specify 0 in the second input field) or mean elements, averaged over an integer number of lunar revolutions. The columns of the file with the lunar elements contain the following information:

1. Time in days elapsed since the starting epoch.

2. Semi-major axis *a* (in meters, relative to a value specified in the header) of the lunar orbit.

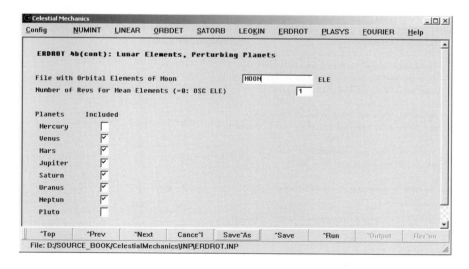

Fig. 9.9. Selection of planets for *N*-body option in program ERDROT

3. Numerical eccentricity e of the lunar orbit.

4. Inclination i (in degrees) of the lunar orbit w.r.t. the ecliptic $J2000.0$.

5. Longitude Ω (in degrees) of the ascending node of the lunar orbit w.r.t. the ecliptic $J2000.0$.

6. Distance ω (in degrees) of lunar perigee from the ascending node (according to the above definition).

7. Angle $\Omega + \omega + v - \dot{\Theta}_m\left(t - t_0\right)$.

The initial epoch, the length of the integration interval, as well as the initial position of the rotation pole in the Earth-fixed system is selected in Panel ERDROT 5 (Figure 9.5). Note that in this program option the integration order, the step size and the sampling of the output data may not be defined in this panel. This is actually performed in Panel ERDROT 5(cont) (Figure 9.10). The panel shows that the generalized N-body problem may be solved either in a straight forward, correct way, or in an approximative, much more rapid way. The straight forward way must be rather inefficient, because the constant stepsize h is dictated by the shortest period in the system, which is the period of Earth rotation, namely one sidereal day. With the experiences gained in Chapter I-7 one would guess that the stepsize for a multistep procedure of the order $8 \leq q \leq 14$ should be of the order of 0.001 days, i.e., of the order 15 minutes. It is a pity to integrate the equations of motion for

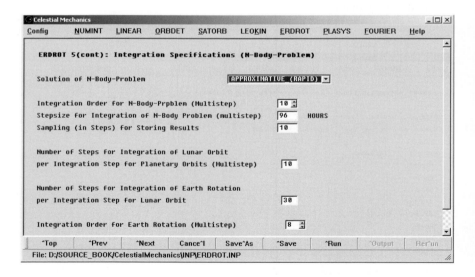

Fig. 9.10. Integration specification in program ERDROT (option N-BODY-PROBLEM)

the planets, where stepsizes of the order of several days would be appropriate, with this ridiculously short stepsize, as well. The rapid solution version of the three-body therefore problem therefore solves the integrations of motion in three steps:

Step 1: The planetary N-body problem including the point masses of the Sun, the center of mass Earth-Moon, and the centers of mass of the planets selected in Panel ERDROT 4b(cont) (Figure 9.9) are solved using the initial conditions taken from the JPL Development Ephemeris DE-200 [111]. A stepsize $h_1 \approx 3$ days and a conventional multistep procedure of a user-defined order ($q = 12$ may be recommended for the purpose, if Venus is the innermost planet included).

Step 2: Each of the above integration steps is covered with an integer number n_1 of integration steps for solving the equations of motion of the Moon (and the Eulerian equations for the rotation of the Moon) using a multistep procedure of the same order as in *Step 1*. As the Moon's revolution period is about ten times shorter than that of Venus, $n_1 \approx 10$ should be bout appropriate (see Panel ERDROT 5(cont), Figure 9.10). The initial conditions for the orbital motion were taken from [111]. "Plausible values" are taken for the initial values for the rotation of the Moon: It is assumed that the axis of the Moon's axis of minimum inertia is pointing precisely from the seleno- to the geocenter at the initial epoch. The geocentric positions of the Sun and the planets are taken from the integration in *Step 1*, the correction from the Earth-Moon barycenter to the geocenter is performed using the results of the integration in *Step 2*. A linear approximation for the three Eulerian angles Ψ_δ, ε_δ, and Θ_δ (as they emerge from the solution of the Eulerian equations for the rotation of the Earth in *Step 3*) is used to compute the orientation of the Earth in inertial space.

Step 3: After each integration step of type *Step 2* the differential equations for the rotation of the Earth are solved in a user-defined number n_2 of steps per *Step 2* using a multistep procedure of a user specified order q_2. As the rotation period of the Earth is about 30 smaller than the revolution period of the Moon, a value of about $n_2 \approx 30$ seems appropriate.

The above approximations imply that the relative motion of Earth and Moon about their barycenter are not taken into account for the integration in *Step 1*, and that only linear predictions (over a very short time interval) of the integration in *Step 3* are used in the integration in *Step 3*. In view of the weak coupling between the respective equations, this approximation is perfectly justified.

The advantage of using the approximative instead of the correct solution of the entire system (which is also implemented in program ERDROT) is a significant gain in integration velocity. The integration over an interval of one thousand years takes less than 9.4 minutes using a 1.2 GHz PC. In this rapid integration mode, whereas more than 26 minutes would be required, when integrating the equations rigorously with a stepsize of 15 minutes and an integration order of $q = 8$. By using the rapid option, one therefore gains a factor of about 6 in computation time.

The content of the files with the results related to the rotation of Earth and Moon is slightly different from that documented, when using the options "EARTH_ROTATION" and "MOON_ROTATION" in Panel ERDROT 1 (Figure 9.1). The first eight columns in the file containing the Earth rotation data are identical with those already mentioned under the option "EARTH_ROTATION". The last three columns are replaced by:

9. True obliquity $\varepsilon(t)$ of the equatorial plane.

10. Node Ω_δ of the Earth's orbit in the true equatorial system at time t.

11. Inclination $i_\delta(t)$ of the "true" Earth's orbit w.r.t. the ecliptic of system $J2000.0$.

The first ten columns in the file containing the lunar rotation data are identical with those documented in the preceding section. Column 11 contains in addition the inclination $i_{\delta_{eq}}$ of the lunar orbit w.r.t. the equatorial plane. This option was used to produce Figure 2.12.

9.4 Space Geodetic and Atmospheric Aspects of Earth Rotation

Panel ERDROT 2(cont) (Figure 9.11) shows the principal panel when using the option "ERP_AAM" of program ERDROT in Panel ERDROT 1 (Figure 9.1). As opposed to the other three options, which serve to solve the equations of motion of Earth and/or lunar rotation (possibly together with the equations of motion for the centers of mass of these bodies), real data are analyzed when activating option "ERP_AAM". Earth rotation parameters (in particular x-, y-coordinates of the pole, LOD produced by the IERS or by the CODE Analysis Center of the IGS (both input formats are supported) are analyzed. The format of the ERP input file (IERS or CODE) is selected at the top of Panel ERDROT 2 (cont) (Figure 9.11). If no AAM-files are included, the program generates only one output file (apart from the general output file), namely that containing the Earth rotation parameters in a format which may be easily analyzed by program FOURIER to generate spectra. This option is in particular useful, when using the IERS input format, because the presence of non-numerical characters prevents the direct

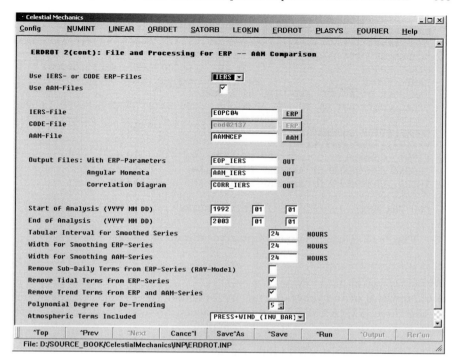

Fig. 9.11. Options of program ERDROT (option ERP_AAM)

use of this file by program FOURIER. The output file (in the example of
Panel ERDROT 2(cont), Figure 9.11, the file "EOP_IERS.OUT") contains
in essence the same information as the selected input file, except for the fact
that the tidal terms up to periods of one year may be eliminated from the
LOD-data. It is also possible to remove the a priori known sub-daily terms
(using the Ray-model) from the original polar motion results. This option
only makes sense, however, if polar motion of a higher than daily time resolution is analyzed.

From the Earth rotation parameters one may calculate the three components
of the Earth's angular momentum vector. This is only done when selecting the
corresponding program option in the second input field (Use AAM-Files). The
program calculates the so-called angular momentum (AM) functions from
the Earth rotation parameters, then it correlates these AM functions with
the atmospheric angular momentum (AAM) functions provided by the IERS
Special Bureau for the Atmosphere (SBA). The general output file contains
the correlation coefficients (defined according to formula 2.146) for the three
momentum functions. Figure 9.12 shows an example of the output generated
by program ERDROT, when using the option "ERP_AAM". The essential

```
ANALYSIS OF ERP FILES AND AAM-FILES
DATE: 02-FEB-03    TIME: 06:04
********************************************************************
NUMBER OF DATA POINTS   :              3836
FIRST EPOCH             :   1991 12  1.00000 (MJD= 48591.00000)
LAST  EPOCH             :   2002  6  1.00000 (MJD= 52426.00000)
PRESSURE DATA (IB) AND WIND DATA USED
TIDAL CONTRIBUTION TO LOD (PERIODS < 366 DAYS) WERE REMOVED IN ERPS

CORRELATION COEFFICIENTS
************************
FIRST  COMPONENT OF AAM/ERP DERIVED AM:       0.596
SECOND COMPONENT OF AAM/ERP DERIVED AM:       0.733
THIRD  COMPONENT OF AAM/ERP DERIVED AM:       0.983

CPU FOR RUN:              2.374 MINUTES
```

Fig. 9.12. General output of program ERDROT (option ERP_AAM)

characteristics of the program run, in particular the correlation coefficients for the three AM/AAM function components, are contained in the output file.

Apart from the reformatted (and, depending on the option, slightly reduced) ERP-file the program generates two other output files, when using the AAM-option, namely (1) the file containing the angular momentum functions (from space geodesy and from IERS Special Bureau for the Atmosphere (SBA)) and (2) the file allowing it to draw the correlation diagrams of the three components. The former file contains the time in two different formats in columns one and two (modified Julian date in the first, year in the second column). Columns 3-5 contain the three components of the space geodesy derived AM functions, columns 6-8 the three components of the AAM functions. This file may be easily used by program FOURIER to generate spectra of the six AM-functions. The latter file contains the the time in the same two formats as the former file in columns two. Columns 3 and 4, 5 and 6, 7 and 8 contain the normalized "residuals" $\tilde{x}_i$, $\tilde{y}_i$ (see eqn. (2.146)) of the correlation process for the first, second, and third components of the AM/AAM functions, respectively.

Several program options may be defined in Panel ERDROT 2(cont) (Figure 9.11): A time window may be set by specifying the start and end time of the series. If these borders are outside the data interval covered by the series, they are automatically adjusted. The actual limits are provided in the output file (see Figure 9.12). The tabular interval of the ERP- and the AAM-files usually do not agree. (In the above example daily values of polar motion data were analyzed, whereas four AAM-values are available per day). The two series thus must be synchronized by specifying the tabular interval. Moreover, it makes sense to smooth the series using the values within a window around the selected tabular times. It was already mentioned that the tidal and the sub-daily polar motion terms may be removed from the LOD-values using the Ray-model contained in the IERS conventions [71]. In addition it makes often sense to remove the low frequency part of the spectrum by subtracting the best-fitting polynomial from the time series. This was actually done in the example of Panel 2(cont) (Figure 9.11) by removing polynomials of degree five in the angular momentum series. Last, but not least, it is possible to select "only" particular constituents of the contributions to the AAM series. One can either use only the pressure- or the wind-terms, the combination of both, and the combination of both, assuming that the ocean surface reacts like an "inverted barometer" on atmospheric pressure. We refer to [96] or to the IERS homepage for more information concerning these options.

10. The Computer-Program PLASYS

Program PLASYS allows it to repeat and considerably extend all investigations presented in Chapter 4. The program was developed as a demonstration tool for lectures in classical Celestial Mechanics, but it may also be used for many research tasks.

PLASYS is based on the integration techniques developed in Chapter I-7. Two integration methods, namely collocation and multistep, may be used. The planetary system is configured by the program user, where all nine planets (or a subset thereof) and one minor planet (object of negligible mass) may be included. The starting conditions are either taken from the JPL's planetary ephemerides DE-200 (see [107] and [111]) or from approximate orbital elements as published by Jan Meeus [72].

Panel PLASYS 1 (Figure 10.1) shows the primary panel of Program PLASYS. Two files, an ASCII file containing constants and one containing the planets' ephemerides, are required to run the program. The constants file is shared by all programs of the package. It has to reside in the subdirectory "GEN" of the directory "CelestialMechanics". The file may be browsed but should not be altered by the program user (unless he/she actually wants to use different constants). The JPL development ephemerides DE200 are read from the binary file "de200.eph" in the same subdirectory. The file covers the time interval between 1981 and 2025. Other DE-files might be attached to the program to cover other time periods. We refer to [111] for more information. The program PLASYS uses the JPL ephemerides uniquely to define the initial position and velocity vectors of the planets selected.

The names of four general output files are defined in the panel shown in Panel PLASYS 1 (Figure 10.1). The first file is the general output file, reflecting essential characteristics and statistical information of the program run. Figure 4.3 is an example related to the generation of the outer planetary system. The file has the extension "OUT" and is automatically stored into the subdirectory "PLASYS/OUT" of the directory "CelestialMechanics". The file with the extension "ERR" contains the error messages and warnings. If the field for the file name is left blank (as in Panel PLASYS 1, Figure 10.1) the error messages and warnings (if any) are written into the log-file "*.out".

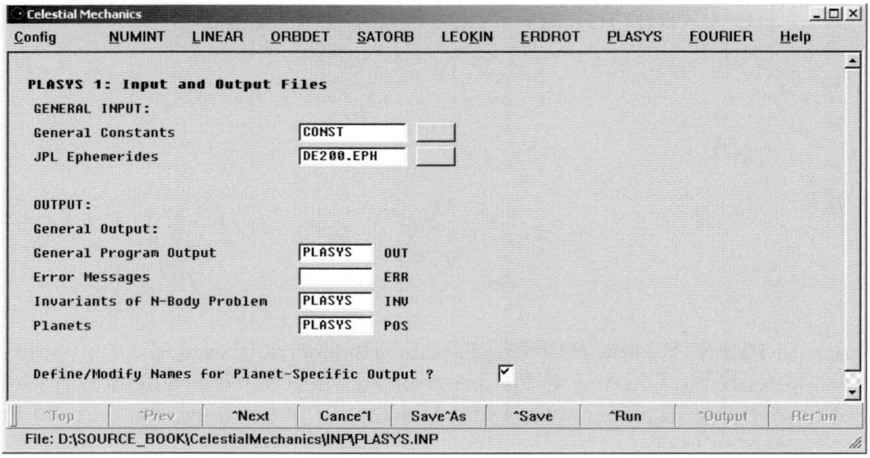

Fig. 10.1. Primary menu of program PLASYS

The file with extension "INV" contains the errors of the invariants of the N-body problem and the stepsizes $h(t)$. Seven columns are stored in the file. The first column lists the time in years, columns 2-4 contain the relative errors of the total angular momentum components (in units of 10^{-9}, parts per billion). The relative error in parts per billion of the energy E may be found in column 5. The initial value of the energy and of the three components of the total angular momentum vector are given in the general output file (see, e.g., Figure 4.3).

According to the developments in section I- 3.2.2 of Chapter I- 3 (see eqns. (I- 3.40) and (I- 3.47)) the invariants are first integrals, and their variation in the course of the integration should be zero. In particular, no deterministic trend should be visible, when numerically integrating the N-body problem. It is therefore recommended to check the invariant file after long integrations. Column 6 contains the time development of the polar moment of inertia (I- 3.48) for the N-body problem considered. The polar moment of inertia is conserved only in a statistical sense, as an average over long time spans. When inspecting this column, no trend should be visible, but rather strong (quasi-)periodic variations have to be expected. Figure 4.5 illustrates the development of the "true" invariants of the N-body problem, Figure 4.6 that of the polar moment of inertia when integrating the outer planetary system. Column 7 of the "*.INV"-file contains the stepsize (this information is only of interest, if the integration was performed with automatic stepsize control).

The planetary positions for all planets included in the integration are written into the file "*.POS". The time argument is contained in the first column, the components of the planets (in the system $J2000.0$) in columns 2,3, and 4. This file may be used to visualize the geometry of the N-body problem integrated (see, e.g., Figure 4.4). The last entry in Panel PLASYS 1 (Figure 10.1) allows it to define or modify the names of the planet-specific output. If this field is left blank, the panel in Panel PLASYS 3(cont) (Figure 10.2) is not presented and previously defined files are overwritten. Panel PLASYS 3(cont) (Figure 10.2) shows that program PLASYS will in the maximum integrate all nine planets and one minor planet. Only the files corresponding to the celestial bodies included in the integration may be altered in Panel PLASYS 3(cont) (Figure 10.2) corresponding to the integration of the outer planetary system.

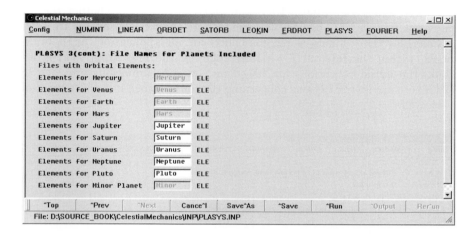

Fig. 10.2. Definition of planet-specific output files in program PLASYS

The structure of the planet-specific files was already explained in Chapter I-7 for tests based on the two-body problem: the differences of the osculating elements referring to t and t_0 in columns 2-7 are listed as functions of the time t; in addition the difference of the arguments of latitude δu of the numerically integrated w.r.t. the initial orbit is given in column 8.

In all other cases, these files contain eleven columns, where

- column 1 contains the time argument in years,

- columns 2 to 6 the orbital elements a, e, i, Ω, ω (either osculating or mean elements, see section I-4.3); The semi-major axis a is given in AU, the eccentricity is dimensionless, the units for all other elements are degrees,

- column 7 contains $\Delta\sigma(t) \stackrel{\text{def}}{=} \left(n(t) - n(t_0)\right)(t - t_0) + \sigma_0(t)$,

- columns 8 and 9 contain the first two components (in the plane of the ecliptic J2000.0) of the unit vector $\boldsymbol{e}_h$ normal to the orbital plane, and

- columns 10 and 11 the first two components (in the plane of the ecliptic $J2000.0$) of the Laplacian vector (of length e), pointing to the pericenter.

All output files may be graphically visualized in the primary menu. This way of inspecting the output files is, however, in general based on a heavy sampling of the files' content (due to storage and CPU limitations). The other way of vizualising these files is to use the gnu-plot graphics package (see http://www.gnuplot.info for more information). No (heavy) restrictions concerning the number of data points exist for this way of visualizing the results.

In Panel PLASYS 2 (Figure 10.3) one first has to decide whether the initial state vectors for the specific application shall be based on the JPL ephemerides [111] – this is recommended for all applications with initial epochs within the interval $I = [1981, 2025]$. If the first input field is left blank, the initial conditions are based on the approximations provided by [72]. There are no restrictions concerning the initial epoch if the latter option is selected.

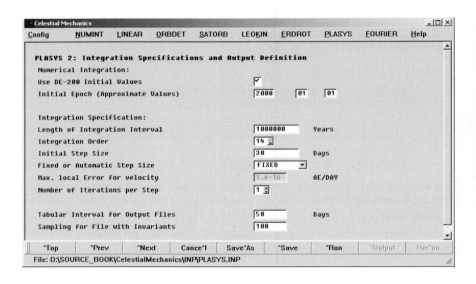

Fig. 10.3. Defining initial epoch and integration-specific input in PLASYS

The length of the integration interval has to be specified in years. If a negative value is supplied here, the integration is performed in reverse time direction (backward integration). The options referring to the numerical integration technique already were explained in Chapter I- 7, which is why only the last two input items have to be mentioned. The input line "tabular interval for output files" defines the epochs t_i, for which the orbital elements, the planets' positions, and the invariants are calculated:

$$t_i \overset{\text{def}}{=} t_0 \pm (i-1)\,\Delta t \,, \; i = 1, 2, \dots \,, \tag{10.1}$$

where the negative sign holds for backward integration. Δt must be a positive number.

In Panel PLASYS 3 (Figure 10.4) the N-body problem is configured. The integration may be based on the Newton-Euler equations developed in Chapter I- 3 (option "NEWTON-EULER"), on the relativistic counterpart represented by eqns. (I- 3.186) (option "RELATIVISTIC"), or on the approximate relativistic equations (option "RELATIVISTIC(APPROX)") based on the equations (I- 3.190). The integration of official ephemerides is performed today in the correct ppn-approximation (i.e., using option "RELATIVISTIC"). One should be aware of the fact that the integration is considerably (by a factor of 3-4) slowed down if the correct relativistic version of the equations of motion is used. Most investigations are therefore based on the Newton-Euler formulation or on the approximate relativistic version. If the option "RELATIVISTIC(APPROX))" is used, the integration efficiency suffers only marginally. Some results based on the correct ppn-formulation may be found in section I- 3.5.

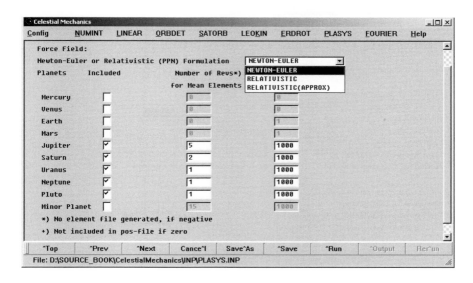

Fig. 10.4. Configuration of the N-body problem in program PLASYS

The N-body problem is configured by selecting the first input fields ("click" on field behind planets' names). In Panel PLASYS 3 (Figure 10.4) the planets Jupiter, Saturn, Uranus, Neptune, and Pluto were included. As already mentioned, one planet-specific file is generated for each of the selected planets.

In the column "number of revolutions for mean elements" one may enforce to store the osculating elements (by selecting "0" for the number of revolutions), or mean elements as mean values of all osculating elements over an entire (positive) number of sidereal (osculating) revolutions. More about the issue osculating vs. mean elements may be found in section I- 4.3. In the third (and last) planet-specific input field the sampling rate for storing the planet's positions in the file "*.POS" is defined. A sampling rate of 1000 was specified in Panel PLASYS 3 (Figure 10.4). As the stepsize of the integration was 30 days, one position was stored every 30000 day (i.e., about one position was stored per 82 years). Defining a high sampling rate reduces the size of the file with planetary positions.

If a minor planet is included in the integration, its elements have to be defined in Panel PLASYS 5 (Figure 10.5). In order to ease the study of resonances, each of the planets may be chosen as the *reference planet* w.r.t. which the orbital elements of the minor planet are referred. The semi-major axis a is defined via the revolution period of the reference planet, the perihelion and time of perihelion passage refer to the corresponding quantities of the reference planet, as well.

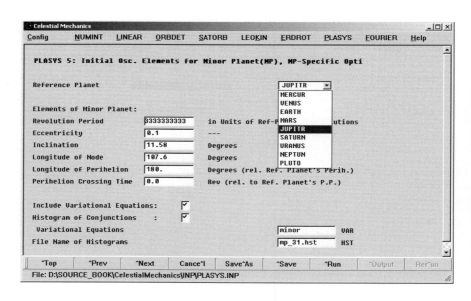

Fig. 10.5. Orbital elements for minor planet

The program user may generate a histogram of conjunctions of the minor planet with the reference planet by selecting the corresponding option (Histogram of Conjunctions). The file with the histogram contains three columns and 360 lines. The first column just contains the numbers (angles)

$i - 0.5$, $i = 1, 2, \ldots, 360$. The second column contains in row i the number of conjunctions, which took place in the interval $(i - 1)[\,^\circ\,] \leq v\,i\,[\,^\circ\,]$ of the mean anomaly v, the third the number of conjunctions taking place at a longitude difference $(i - 1)[\,^\circ\,] < \Delta w\,i\,[\,^\circ\,]$ of the minor planet w.r.t. Jupiter's perihelion. Figure 4.83 was generated using these histogram files.

If a minor planet is included in the integration, a selection of the variational equations (referring to the osculating elements at t_0) may be integrated simultaneously with the equations of motion. If the option "Variational Equations" is selected in Panel PLASYS 5 (Figure 10.5), Panel PLASYS 6 (Figure 10.6) is presented to the program user. One may select a sampling rate indicating that the solution of variational equations is only stored for every n-th output time interval defined in Panel PLASYS 3(cont) (Figure 10.2). One may include between one and six variational equations. The output file containing the solutions of the variational equations contains $1 + 5\,n_v$ columns, where n_v is the number of variational equations included. The time argument is in the first column. The first of the five columns referring to one particular variational equation refer to the element number (sequence as given in Panel PLASYS 6, Figure 10.6), the second to fourth of these columns are the mantissa of the three Cartesian components of $\frac{\partial \boldsymbol{r}}{\partial I}(t)$, the fifth is the exponent (base e).

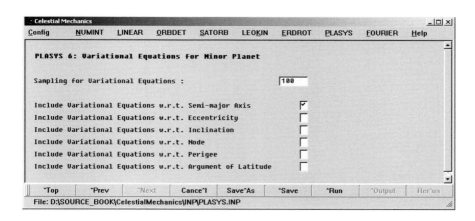

Fig. 10.6. Integration of variational equations

11. Elements of Spectral Analysis and the Computer-Program FOURIER

When analyzing the orbital and/or rotational motions of planets and satellites one usually has to deal with (quasi-)periodic processes. Examples are the development of osculating or mean elements of planets, minor planets, satellites, etc. over long time periods or long time series of empirically determined polar motion of the Earth. It is therefore often important to represent a function as a (finite) series of periodic functions. If sine and cosine functions of a basic frequency (and of multiples thereof) are used for this purpose, the resulting development is also referred to as Fourier series. The expressions "harmonic analysis", "Fourier analysis", "spectral analysis" are used as synonyms to characterize the methods.

If a function is given by a finite time series of discrete function values, the attribute "discrete" is used to characterize the analysis. Naturally, we focus uniquely on discrete Fourier analysis. The primary goal of discrete Fourier analysis may be defined as the determination of the coefficients of the Fourier series. The set of these coefficients may be called the Fourier transformed of the original series. Therefore, the term FT (Fourier Transformation) is also used as a synonym for Fourier analysis.

Fourier analysis is developed here as a topic in parameter estimation theory, starting from LSQ (Least Squares) adjustment. The reader is assumed with the basic properties of LSQ (some facts from this theory are reviewed, but not derived, in section 11.2).

Usually, discrete Fourier analysis is based on data series with equal spacing (of the independent argument) between subsequent data points. Moreover, all data points are assumed to be of the same accuracy. The two assumptions let the normal equation matrix become diagonal and the result of the LSQ adjustment is reduced to the simple formulas attributed to Fourier.

Today, FFT (Fast Fourier Transformation) is used almost as a synonym for Fourier analysis. FFT is nothing but a very efficient method to calculate the coefficients of the series of harmonic functions, i.e., to perform a FT. FFT became an openly available technique in applied science with the availability of fast computers in the second half of the twentieth century (based on fundamental contributions by J. W. Cooley and J. W. Tukey, G. C. Danielson, and C. Lanczos). Some of the underlying ideas are much older, however. For more information concerning the development of FFT the reader is referred to [88].

The three techniques for discrete spectral analysis considered in this chapter and implemented in program FOURIER are

- LSQ analysis for general discrete time series,
- the classical FT for time series with equal spacing between data points, and
- the FFT with user-defined optimization levels for an efficient treatment of long time series.

The program FOURIER was developed as a demonstration tool for introductory courses into applied spectral analysis.

After the statement of the problem in section 11.1 the three techniques are developed in the sections 11.2, 11.3, and 11.4 in the order of decreasing generality and mathematical purity and in the order of increasing efficiency.

In section 11.5 we consider the important special case of two Cartesian components $x(t)$ and $y(t)$ of a two-dimensional vector. Section 11.6 eventually gives an overview of the capabilities of program FOURIER.

11.1 Statement of the Problem

Let us assume that the real function $f(t)$ is given by the discrete values

$$f'_k \stackrel{\text{def}}{=} f(t_k), \quad k = 1, 2, \ldots, N \tag{11.1}$$

at N different epochs t_k .

Let us define the fundamental period P as

$$P \stackrel{\text{def}}{=} |t_N - t_1| \tag{11.2}$$

and the corresponding fundamental angular frequency ω by

$$\omega \stackrel{\text{def}}{=} \frac{2\pi}{P} . \tag{11.3}$$

The function $f(t)$ is then approximated in the interval $I \stackrel{\text{def}}{=} [t_1, t_N]$ by a truncated Fourier series

$$f(t) = a_0 + \sum_{i=1}^{m} (a_i \cos(i\omega t) + b_i \sin(i\omega t)) , \tag{11.4}$$

where it is assumed that

$$2\,m + 1 \le N \ . \tag{11.5}$$

The coefficients a_i, b_i are the Fourier coefficients of the representation (11.4).

If the number of parameters and the number of data points are equal,

$$2\,m + 1 = N \ , \tag{11.6}$$

the approximating function (11.4) must assume the function values f'_k at the corresponding epochs t_k and we may write

$$f'_k = a_0 + \sum_{i=1}^{m}\left(a_i\,\cos(i\omega t_k) + b_i\,\sin(i\omega t_k)\right) \ , \quad k = 1,2,\ldots,2\,m+1 \ . \tag{11.7}$$

The main problem of discrete Fourier analysis "only" consists of determining the coefficients a_i, b_i as efficiently as possible. If assumption (11.5) holds, it must be possible to determine the coefficients a_i and b_i with a least squares adjustment, using the function values (11.1) as pseudo-observations.

As a side issue (in our context (!)) we mention that under quite general conditions (Dirichlet conditions) each periodic function $f(t)$ with period P may be represented by a converging infinite series, the Fourier series

$$f(t) = a_0 + \sum_{i=1}^{\infty}\left(a_i\,\cos(i\omega t) + b_i\,\sin(i\omega t)\right) \ . \tag{11.8}$$

Arbitrary, non-periodic functions may be analyzed as well. One should, however, keep in mind that in this case it is implicitly assumed that $f(t+jP) = f(t)$ for any integer j. The approximation of $f(t)$ by the right-hand side of eqns. (11.4) or (11.8) therefore cannot be used outside the interval I in the case of non-periodic functions.

11.2 Harmonic Analysis Using Least Squares Techniques

In the spirit of least-squares adjustment, N observation equations for the unknown coefficients a_i and b_i are set up by substituting in eqn. (11.1) the function values $f(t_k)$ by eqn. (11.4) and by introducing the residual v_k to account for the fact the function value f'_k and its approximation by eqn. (11.4) may be different:

$$a_0 + \sum_{i=1}^{m}\left(a_i\,\cos(i\omega t_k) + b_i\,\sin(i\omega t_k)\right) - f'_k = v_k \ , \quad k = 1,2,\ldots,N \ . \tag{11.9}$$

By defining the row matrices $\boldsymbol{p}$, $\boldsymbol{f'}$, and $\boldsymbol{v}$ as

$$\begin{aligned}
\boldsymbol{p}^T &= (a_0, a_1, \ldots, a_m, b_1, b_2, \ldots, b_m) \\
\boldsymbol{f}'^T &= (f'_1, f'_2, \ldots, f'_N) \\
\boldsymbol{v}^T &= (v_1, v_2, \ldots, v_N) ,
\end{aligned} \tag{11.10}$$

the system (11.9) of observation equations may be written in matrix form

$$\mathbf{A}\,\boldsymbol{p} - \boldsymbol{f}' = \boldsymbol{v} , \tag{11.11}$$

where the elements of the first design matrix are defined by

$$\begin{aligned}
A_{k,1} &= 1 \\
A_{k,i+1} &= \cos(i\omega t_k) , \quad i = 1, 2, \ldots, m \\
A_{k,m+1+i} &= \sin(i\omega t_k) , \quad i = 1, 2, \ldots, m \\
k &= 1, 2, \ldots, N \geq 2\,m + 1 .
\end{aligned} \tag{11.12}$$

The method of least-squares asks that the sum of the squares of the residuals becomes a minimum:

$$\sum_{k=1}^{N} v_k^2 = \min . \tag{11.13}$$

If the function values f'_k are the values of a function generated, e.g., by numerical integration, it is fair to assume that all function values are of equal accuracy.

If, on the other hand, the function values f'_k stem from a parameter estimation process, the values f'_k may be of different accuracy. Let us assume that estimates σ_k, $k = 1, 2, \ldots, N$, are available for the mean errors of the values f'_k. In this case one should replace the minimum principle (11.13) by

$$\sum_{k=1}^{N} \left(\frac{v_k}{\sigma_k} \right)^2 \stackrel{\text{def}}{=} \sum_{k=1}^{N} w_k\, v_k^2 = \min , \tag{11.14}$$

where obviously $w_k \stackrel{\text{def}}{=} 1/\sigma_k^2$. In LSQ adjustment theory w_k is called the *weight* of observation k.

The principle (11.13) or (11.14) leads to one condition equation per parameter by demanding the partial derivatives of the function on the left-hand side of these equations w.r.t. this parameter to be zero. As the sum (11.13) or (11.14) are quadratic functions of the $2\,m + 1$ parameters p_i, $i = 1, 2, \ldots, 2\,m + 1$, the minimum principle leads to $2\,m + 1$ linear equations for the determination of the parameters. These linear equations are called *normal equations*. In the case of the minimum principle (11.13) the normal eqution system assumes the form

$$\mathbf{A}^T \mathbf{A} \, \boldsymbol{p} = \mathbf{A}^T \, \boldsymbol{f}' \, , \tag{11.15}$$

which is a linear, regular system of dimension $d = 2\,m + 1$.

With the notations

$$\begin{aligned}
\mathbf{N} &= \mathbf{A}^T \mathbf{A} \\
\boldsymbol{b} &= \mathbf{A}^T \boldsymbol{f}' \\
\mathbf{Q} &= \mathbf{N}^{-1}
\end{aligned} \tag{11.16}$$

its solution may be written in the form

$$\boldsymbol{p} = \mathbf{Q}\,\boldsymbol{b} \, . \tag{11.17}$$

The a posteriori value for the mean error m_i of the parameters p_i is given by

$$m_i = m_0 \, \sqrt{Q_{ii}} \, , \quad i = 1, 2, \ldots, 2\,m + 1 \, , \tag{11.18}$$

where the mean error m_0 of the individual observation is computed as

$$m_0 = \sqrt{\frac{\boldsymbol{v}^T \boldsymbol{v}}{N - (2\,m + 1)}} = \sqrt{\frac{\boldsymbol{f}'^T \boldsymbol{f}' - \boldsymbol{b}^T \boldsymbol{p}}{N - (2\,m + 1)}} \, . \tag{11.19}$$

The mean error a posteriori can only be computed if the number of observations N exceeds the number of parametes, i.e., if $N > 2\,m + 1$. An error assessment may also be made for $N = 2\,m + 1$, provided we know an a priori value σ_0 for m_0. In this case formula (11.18) has to be replaced by

$$m_i = \sigma_0 \, \sqrt{Q_{ii}} \, . \tag{11.20}$$

If the minimum principle (11.13) is replaced by the principle (11.14), the resulting equations are only slightly more complicated. By defining the diagonal weight matrix as

$$\mathbf{P} = \sigma_0^2 \begin{pmatrix}
\sigma_1^{-2} & 0 & 0 & \ldots & 0 & 0 \\
0 & \sigma_2^{-2} & 0 & \ldots & 0 & 0 \\
\ldots & \ldots & \ldots & \ldots & \ldots & \ldots \\
\ldots & \ldots & \ldots & \ldots & \ldots & \ldots \\
0 & 0 & 0 & \ldots & \sigma_{2m}^{-2} & 0 \\
0 & 0 & 0 & \ldots & 0 & \sigma_N^{-2}
\end{pmatrix} , \tag{11.21}$$

the normal equation system assumes the form

$$\mathbf{A}^T \mathbf{P} \, \mathbf{A} \, \boldsymbol{p} = \mathbf{A}^T \, \mathbf{P} \, \boldsymbol{f}', \tag{11.22}$$

The notations

$$\mathbf{N} = \mathbf{A}^T \mathbf{P} \mathbf{A}$$
$$\boldsymbol{b} = \mathbf{A}^T \mathbf{P} \boldsymbol{f}'$$
$$\mathbf{Q} = \mathbf{N}^{-1}$$

(11.23)

allow it to write the solution in the form (11.17) and the mean errors m_i a posteriori of the parameters in the form (11.18). The mean error m_0 of the observation of unit weight has to be computed according to the modified formula

$$m_0 = \sqrt{\frac{\boldsymbol{v}^T \mathbf{P} \boldsymbol{v}}{N - (2\,m+1)}} = \sqrt{\frac{\boldsymbol{f}'^T \mathbf{P} \boldsymbol{f}' - \boldsymbol{b}^T \boldsymbol{p}}{N - (2\,m+1)}} \; .$$

(11.24)

If a time series with an arbitrary distribution of epochs t_k and possibly with epoch-dependent weights σ_k has to be analyzed, the LSQ method for the determination of the Fourier-coefficients a_i and b_i leads to eqns. of type (11.17) with a regular and symmetric, but otherwise arbitrary matrix $\mathbf{Q} = \mathbf{N}^{-1}$. As the procedure includes the inversion of a symmetric matrix of dimension $d = 2\,m+1$, and as the processing time required associated with matrix inversion grows proportional to d^3, the LSQ method becomes almost prohibitive for $m > 10000$.

A more efficient procedure results if the data points t_k are equally spaced, and if, moreover, all observations are of the same accuracy.

11.3 Classical Discrete Fourier Analysis

Assuming a unit weight matrix $\mathbf{P} = \mathbf{E}$ (see eqn. (11.1)), the elements of the normal equation matrix $\mathbf{N}$ are defined by (see eqns. (11.11), (11.12), and (11.16)):

$$N_{i+1,l+1} = \sum_{k=1}^{N} \cos(i\omega t_k)\cos(l\omega t_k) \; , \;\; i = 0,1,\ldots,m \; , \;\; l = 0,1,\ldots,m$$

$$N_{i+m+1,l+m+1} = \sum_{k=1}^{N} \sin(i\omega t_k)\sin(l\omega t_k) \; , \;\; i = 1,2,\ldots,m \; , \;\; l = 1,2,\ldots,m$$

$$N_{i+1,l+m+1} = \sum_{k=1}^{N} \cos(i\omega t_k)\sin(l\omega t_k) \; , \;\; i = 0,1,\ldots,m \; , \;\; l = 1,2,\ldots,m \; .$$

(11.25)

Let the epochs t_k be defined by

$$h = \frac{P}{2m+1}$$
$$t_k = \left(k - \tfrac{1}{2}\right)h , \quad k = 1, 2, \ldots, 2m+1 \tag{11.26}$$
$$f'_k = f(t_k) .$$

Using the trigonometric relations

$$
\begin{aligned}
\cos(i\omega t_k)\cos(l\omega t_k) &= \tfrac{1}{2}\left\{+\cos\left((i+l)\omega t_k\right) + \cos\left((i-l)\omega t_k\right)\right\} \\
\sin(i\omega t_k)\sin(l\omega t_k) &= \tfrac{1}{2}\left\{-\cos\left((i+l)\omega t_k\right) + \cos\left((i-l)\omega t_k\right)\right\} \\
\cos(i\omega t_k)\sin(l\omega t_k) &= \tfrac{1}{2}\left\{+\sin\left((i+l)\omega t_k\right) - \sin\left((i-l)\omega t_k\right)\right\}
\end{aligned}
\tag{11.27}
$$

we obtain:

$$
\begin{aligned}
N_{i+1,\,l+1} &= \tfrac{1}{2}\sum_{k=1}^{N}\left\{+\cos\left((i+l)\omega t_k\right) + \cos\left((i-l)\omega t_k\right)\right\} , \\
&\quad i = 0, 1, \ldots, m , \; l = 0, 1, \ldots, m
\end{aligned}
$$

$$
\begin{aligned}
N_{i+m+1,\,l+m+1} &= \tfrac{1}{2}\sum_{k=1}^{N}\left\{-\cos\left((i+l)\omega t_k\right) + \cos\left((i-l)\omega t_k\right)\right\} , \\
&\quad i = 1, 2, \ldots, m , \; l = 1, 2, \ldots, m
\end{aligned}
\tag{11.28}
$$

$$
\begin{aligned}
N_{i+1,\,l+m+1} &= \tfrac{1}{2}\sum_{k=1}^{N}\left\{+\sin\left((i+l)\omega t_k\right) - \sin\left((i-l)\omega t_k\right)\right\} , \\
&\quad i = 0, 1, \ldots, m , \; l = 1, 2, \ldots, m .
\end{aligned}
$$

One easily verifies that in eqns. (11.28) only the terms with $i = l$ are different from zero. The matrix N is therefore diagonal

$$
N = \frac{N}{2}\begin{pmatrix}
2 & 0 & 0 & \ldots & 0 & 0 \\
0 & 1 & 0 & \ldots & 0 & 0 \\
& & \cdots & & & \\
& & \cdots & & & \\
0 & 0 & 0 & \ldots & 1 & 0 \\
0 & 0 & 0 & \ldots & 0 & 1
\end{pmatrix} .
\tag{11.29}
$$

This makes the inversion of matrix N a trivial, processing time and space-saving process:

$$Q = N^{-1} = \frac{1}{N} \begin{pmatrix} 1 & 0 & 0 & \ldots & 0 & 0 \\ 0 & 2 & 0 & \ldots & 0 & 0 \\ & & \cdots & \cdots & & \\ & & \cdots & \cdots & & \\ 0 & 0 & 0 & \ldots & 2 & 0 \\ 0 & 0 & 0 & \ldots & 0 & 2 \end{pmatrix} . \tag{11.30}$$

In view of eqns. (11.16) the Fourier coefficients may be given explicitly

$$a_0 = \frac{1}{N} \sum_{k=1}^{N} f'_k$$

$$a_i = \frac{2}{N} \sum_{k=1}^{N} \cos(i\omega t_k) f'_k , \quad i = 1, 2, \ldots, m \tag{11.31}$$

$$b_i = \frac{2}{N} \sum_{k=1}^{N} \sin(i\omega t_k) f'_k , \quad i = 1, 2, \ldots, m .$$

These equations, contained in any textbook of spectral analysis, are attributed to Fourier.

With Personal Computers available nowadays, problems with $m \approx 5000$ may be easily solved within a few minutes. If time series with tens of thousands of epochs must be analyzed, the direct use of eqns. (11.31) again becomes prohibitive. For such cases, the methods of FFT, to be discussed in the next section, are the appropriate tool. One has to be aware of the fact, however, that not only time, but also the memory to store the coefficients resp. the data points becomes an important issue.

11.3.1 Amplitude Spectra and Power Spectra

The representation (11.4) for the approximating function is ideally suited for the determination of the amplitudes a_i and b_i, because the resulting observation equations (11.9) are linear in the unknowns. For interpreting the data it is usually preferable to combine the sin- and cos-terms into one cos-term. This is possible by introducing a new amplitude $\tilde{a}_i$ and a phase angle ϕ_i for each term

$$f(t) = a_0 + \sum_{i=1}^{\infty} \tilde{a}_i \cos(i\omega t - \phi_i) , \tag{11.32}$$

where

$$\tilde{a}_i \quad = \sqrt{a_i^2 + b_i^2}$$
$$\cos\phi_i = \frac{a_i}{\sqrt{a_i^2 + b_i^2}}$$
$$\sin\phi_i = \frac{b_i}{\sqrt{a_i^2 + b_i^2}} \tag{11.33}$$
$$i = 0, 1, \ldots, m \; .$$

The amplitudes $\tilde{a}_i$ are very well suited to characterize the signal with the angular frequencies $i\omega$, whereas the phase angle ϕ_i is usually only of interest, if the approximating function (11.32) has to be used subsequently.

If the amplitudes $\tilde{a}_i$, $i = 0, 1, \ldots m$, are represented as a function of the angular frequencies $i\omega$ or of the corresponding periods $2\pi/i\omega$, the resulting graph is called an *amplitude spectrum*.

Often, one is not interested in the amplitudes of a harmonic series, but in the relative distribution of the "energy" in a given spectrum. It is a well known fact in physics (optics) that the energy associated with a sin (cos) wave is proportional to the square of the amplitude of the signal. This explains that

$$e_i = \frac{\tilde{a}_i^2}{e_t}, \quad \text{where} \quad e_t = \sum_{i=0}^{m} \tilde{a}_i^2 \; . \tag{11.34}$$

are defined as the so-called *power spectral densities* at frequency $i\omega$. e_t is the total power contained in a spectrum.

If the power spectral densities are displayed as a function of the frequencies or the corresponding periods, the resulting graph is called the *power spectrum*.

11.4 Fast Fourier Analysis

The only goal of this section consists of finding an efficient algorithm for the computation of the Fourier coefficients defined by eqns. (11.31). With the understanding that $b_0 \stackrel{\text{def}}{=} 0$ we may slightly re-arrange these equations:

$$a_i = \frac{\kappa_i}{N} \sum_{k=1}^{N} \cos(i\omega t_k)\, f'_k\,, \quad i = 0, 1, 2, \ldots, n$$

$$b_i = \frac{2}{N} \sum_{k=1}^{N} \sin(i\omega t_k)\, f'_k\,, \quad i = 0, 1, 2, \ldots, n \tag{11.35}$$

$$\kappa_0 = 1\,,$$

$$\kappa_i = 2\,, \quad \text{for} \quad i = 1, 2, \ldots, \frac{N-1}{2} \stackrel{\text{def}}{=} n\,.$$

Compared to a general LSQ algorithm, the computation of the Fourier coefficients with eqn. (11.35) is already orders of magnitude more efficient by avoiding a matrix inversion and laborious matrix multiplications. Note, that the order $n = (N-1)/2$ has to be interpreted as the maximum integer contained in $(N-1)/2$. n is the maximum order of the harmonic series achievable with a time series of N epochs.

In order to further optimize the computation of the Fourier coefficients explicit use has to be made of the definition (11.26) for the epochs t_k. Furthermore we make use of the relation

$$\omega = \frac{2\pi}{P} = \frac{2\pi}{Nh}\,, \quad \text{where} \quad h = t_{i+1} - t_i \tag{11.36}$$

is the spacing between subsequent observation epochs.

This allows us to re-write eqns. (11.35) as

$$a_i = \frac{\kappa_i}{N} \sum_{k=1}^{N} \cos\left(\frac{i\,(k-1+\frac{1}{2})}{N}\, 2\pi\right) f'_k$$

$$b_i = \frac{2}{N} \sum_{k=1}^{N} \sin\left(\frac{i\,(k-1+\frac{1}{2})}{N}\, 2\pi\right) f'_k \tag{11.37}$$

$$i = 1, 2, \ldots, n\,.$$

Using the well-known addition theorems of trigonometry the computation of the Fourier coefficients may be further modified

$$a_i = +\frac{\kappa_i}{N} \cos\left(\frac{i}{N}\pi\right) \sum_{k=1}^{N} \cos\left(\frac{i(k-1)}{N} 2\pi\right) f'_k$$

$$-\frac{\kappa_i}{N} \sin\left(\frac{i}{N}\pi\right) \sum_{k=1}^{N} \sin\left(\frac{i(k-1)}{N} 2\pi\right) f'_k$$

$$b_i = +\frac{2}{N} \cos\left(\frac{i}{N}\pi\right) \sum_{k=1}^{N} \sin\left(\frac{i(k-1)}{N} 2\pi\right) f'_k$$

$$+\frac{2}{N} \sin\left(\frac{i}{N}\pi\right) \sum_{k=1}^{N} \cos\left(\frac{i(k-1)}{N} 2\pi\right) f'_k .$$

$$(11.38)$$

Introducing the auxiliary coefficients

$$A_{iN} = \sum_{k=1}^{N} \cos\left(\frac{i(k-1)}{N} 2\pi\right) f'_k$$

$$B_{iN} = \sum_{k=1}^{N} \sin\left(\frac{i(k-1)}{N} 2\pi\right) f'_k ,$$

$$(11.39)$$

we may write the original unknowns as

$$a_i = \frac{\kappa_i}{N} \cos\left(\frac{i}{N}\pi\right) A_{iN} - \frac{\kappa_i}{N} \sin\left(\frac{i}{N}\pi\right) B_{iN}$$

$$b_i = \frac{2}{N} \cos\left(\frac{i}{N}\pi\right) B_{iN} + \frac{2}{N} \sin\left(\frac{i}{N}\pi\right) B_{iN} .$$

$$(11.40)$$

From now on we only have to deal with the computation of the auxiliary coefficients A_{iN}, B_{iN}, where the index N indicates that N values of the function $f(t)$ are required for the computation.

It is possible to further optimize the computation if

$$N = 2M \tag{11.41}$$

is an even number. This allows us to split up the right-hand sides of eqns. (11.39) into two partial sums, each with the same number M of terms:

$$A_{iN} = \sum_{k=1}^{M} \cos\left(\frac{2i(k-1)}{N} 2\pi\right) f'_{2k-1} + \sum_{k=1}^{M} \cos\left(\frac{i(2k-1)}{N} 2\pi\right) f'_{2k}$$

$$B_{iN} = \sum_{k=1}^{M} \sin\left(\frac{2i(k-1)}{N} 2\pi\right) f'_{2k-1} + \sum_{k=1}^{M} \sin\left(\frac{i(2k-1)}{N} 2\pi\right) f'_{2k} .$$

$$(11.42)$$

In order to have the same arguments in all sums of the sin- and cos-functions in all partial sums, we again make use of the addition theorems of trigonometry and obtain

$$\sum_{k=1}^{M} \cos\left(\frac{i\,(2\,k-1)}{N}\,2\pi\right) f'_{2k} = +\cos\left(\frac{i}{N}\,2\pi\right) \sum_{k=1}^{M} \cos\left(\frac{2\,i\,(k-1)}{N}\,2\pi\right) f'_{2k}$$

$$- \sin\left(\frac{i}{N}\,2\pi\right) \sum_{k=1}^{M} \sin\left(\frac{2\,i\,(k-1)}{N}\,2\pi\right) f'_{2k}$$

$$\sum_{k=1}^{M} \sin\left(\frac{i\,(2\,k-1)}{N}\,2\pi\right) f'_{2k} = +\cos\left(\frac{i}{N}\,2\pi\right) \sum_{k=1}^{M} \sin\left(\frac{2\,i\,(k-1)}{N}\,2\pi\right) f'_{2k}$$

$$+ \sin\left(\frac{i}{N}\,2\pi\right) \sum_{k=1}^{M} \cos\left(\frac{2\,i\,(k-1)}{N}\,2\pi\right) f'_{2k} \ .$$

$$(11.43)$$

Considering eqn (11.41) and substituting eqn. (11.43) into the formula (11.42) leads to the basic recursion formula for the computation of the Fourier coefficients:

$$A_{iN} = A_{iM_o} + \cos\left(\frac{i}{M}\,\pi\right) A_{iM_e} - \sin\left(\frac{i}{M}\,\pi\right) B_{iM_e}$$

$$B_{iN} = B_{iM_o} + \cos\left(\frac{i}{M}\,\pi\right) B_{iM_e} + \sin\left(\frac{i}{M}\,\pi\right) A_{iM_e} \qquad (11.44)$$

$$i = 0, 1, \ldots, n = \frac{N}{2} - 1 \ .$$

The indices "e" and "o" were used on the right-hand sides of the recursion formula (11.44) to indicate that the corresponding coefficients were computed with the $M = N/2$ even and odd function values f'_k, respectively. Note, that the right-hand sides of eqns. (11.44) are linear combinations of the four auxiliary coefficients A_{iM_o}, B_{iM_o}, A_{iM_e}, B_{iM_e}, which in turn have to be computed as sums with M terms each.

Equations (11.44) are the equivalent to the *Lemma by Danielson and Lanczos* of the complex Fourier analysis (see, e.g., [88]).

What is won by computing the terms A_{iN} and B_{iN} using the recursion formula (11.44) instead of the "direct" formula (11.42)? For each index i we have replaced the computation of two sums of type (11.39) consisting of $N = 2\,M$ terms by the computation of four sums of the same type but only with half the number $M = N/2$ of terms. As the processing time for each of the sums is proportional to M^2, we gain in essence a factor of 2 in processing time by using the recursion formula (11.44) instead of the direct formula!

If M is an even number as well, the same recursion formula may be used again on the next lower level. The best performance is achieved, if $N = 2^n$. For $N = 2^{15} = 32768$ we may therefore expect a reduction of processing time by about four orders of magnitude! This just may be the difference between a solvable and unsolvable problem for the practitioner.

The drawback of FFT lies in the assumption that the number of data points must be a power of two. There are several more or less quick, but "dirty" ways to cope with this problem. If the actual number of epochs in a time series may be written as

$$N + x = 2^n , \tag{11.45}$$

where n is the minimum exponent for which $2^n > N$, it is possible to fill in the missing (last) x epochs with zeroes. This method is referred to as "zero padding". Alternatively, one may skip x data points and reduce series to $\tilde{N} \stackrel{\text{def}}{=} 2^{n-1}$ points – but the result might be an inappropriate loss of data (in the worst case close to a factor of 2).

This is why in the program FOURIER (see section 11.6) we let the program user select the number of times the recursion formula (11.44) has to be applied. In program FOURIER, the length of the series is written as

$$N \stackrel{\text{def}}{=} m \cdot 2^n + x , \tag{11.46}$$

and the user is allowed to choose n. For $n = 10$ one looses at maximum $x = 2^n - 1 \approx 1000$. For $n = 8$ this number is reduced by a factor of four. For $n = 10$ the procedure is still about a factor of 1000 more rapid than a classical Fourier analysis of the same time series.

Let us conclude this section by the remark that the coefficients on the right-hand sides of eqn. (11.44) for the indices $i = M/2 + i'$, $i' = 0, 1, \ldots, M/2$, may be computed very efficiently together with the coefficients with indices $i' = 0, 1, \ldots, M/2$. For

$$i = \frac{M}{2} + i' , \quad i' = 0, 1, \ldots, \frac{M}{2} - 1 , \tag{11.47}$$

eqn. (11.39) may be modified in the following way:

$$\begin{aligned}
A_{iM_o} &= \sum_{k=1}^{M} \cos\left(\frac{i\,(k-1)}{M}\,2\pi\right) f'_{2k-1} \\
&= \sum_{k=1}^{M} \cos\left((k-1)\,\pi + \frac{i'\,(k-1)}{M}\,2\pi\right) f'_{2k-1} \\
&= \sum_{k=1}^{M} (-1)^{k-1} \cos\left(\frac{i'\,(k-1)}{M}\,2\pi\right) f'_{2k-1}
\end{aligned} \tag{11.48}$$

$$B_{iM_o} = \sum_{k=1}^{M} \sin\left(\frac{i\,(k-1)}{M}\,2\pi\right) f'_{2k-1}$$

$$= \sum_{k=1}^{M} \sin\left((k-1)\,\pi + \frac{i'\,(k-1)}{M}\,2\pi\right) f'_{2k-1} \qquad (11.49)$$

$$= \sum_{k=1}^{M} (-1)^{k-1} \sin\left(\frac{i'\,(k-1)}{M}\,2\pi\right) f'_{2k-1}\;.$$

By an analogue transformation we obtain

$$A_{iM_e} = \sum_{k=1}^{M} (-1)^{k-1} \cos\left(\frac{i'\,(k-1)}{M}\,2\pi\right) f'_{2k}$$

$$\qquad (11.50)$$

$$B_{iM_e} = \sum_{k=1}^{M} (-1)^{k-1} \sin\left(\frac{i'\,(k-1)}{M}\,2\pi\right) f'_{2k}\;.$$

If the terms with even and odd indices k are summed up separately in these equations, the number of operations to compute A_{iN} is practically reduced by a factor of two, compared to the direct computation of the expressions A_{iN}.

11.5 Prograde and Retrograde Motions of Vectors

So far, only scalar functions were spectrally analyzed. On the other hand we know, that many (if not most) objects of interest in Celestial Mechanics are vectors. How are vectors treated in spectral analysis?

In practice the answer to this question is frighteningly simple: The Cartesian components of a vector are separately and independently subject to spectral analysis. Only the formal aspect, whether the representations (11.4) for the coordinates is the best possible, remains to be considered.

Before addressing this formal point, an aspect of mathematical purity (and of bad conscience) must be considered: The simple answer given above lacks of mathematical correctness, if there were mathematical correlations between the vector components pertaining to the individual epochs. Let us, e.g., assume that $x(t)$ and $y(t)$ are two components of a vector and that the time series

$$t_i\,,\; x'_i\,,\; y'_i\,,\quad i = 1, 2, \ldots, N = 2n+1\;, \qquad (11.51)$$

was the result of N independent parameter estimation processes. The polar motion series, as established by modern space geodetic methods (see Chapter 3), may serve as an example.

Usually, the estimates x_i and y_i pertaining to one and the same epoch t_i are not independent, but correlated by the variance-covariance matrix associated with the estimated quantities x_i and y_i:

$$\mathbf{cov}(x_i', y_i') = m_0^2 \, \mathbf{Q}_i \, , \qquad (11.52)$$

where the symmetric 2×2 matrix $\mathbf{Q}_i$, as the inverse of a normal equation matrix in a least squares estimation, is fully populated.

It is possible to handle this problem correctly from the mathematical point of view: The time series (11.51) are considered as observations in *one* least squares parameter estimation process, where the $2\,(2\,n{+}1)$ Fourier coefficients of the x- *and* y-components are estimated together. The weight matrix $\mathbf{P}$ associated with this process will, however, not be diagonal, but block-diagonal with the matrices (11.52) as non-zero blocks.

The resulting parameter estimation process is a generalization of the least-squares approach to spectral analysis, outlined in section 11.2, to more than one dimension. The procedure is straight forward, but in most cases close to unpracticable. This is the "justification" that in spectral analysis the correlations between the components of a vectorial time series are generally (and graciously) ignored.

After having addressed this aspect of mathematical correctness and of bad conscience, we return to the simple (and solvable) aspects of life: Is the representation (11.4) the best possible for vector components?

Let $x(t)$ and $y(t)$ be two Cartesian components of a vector which were spectrally analyzed independently, using one of the techniques discussed in the previous sections. The results are assumed to be available in the following form (compare eqns. (11.4)):

$$
\begin{aligned}
x(t) &= a_{x0} + \sum_{i=1}^{n} \big(a_{xi} \cos(i\omega t) + b_{xi} \sin(i\omega t)\big) \\
y(t) &= a_{y0} + \sum_{i=1}^{n} \big(a_{yi} \cos(i\omega t) + b_{yi} \sin(i\omega t)\big) \, .
\end{aligned}
\qquad (11.53)
$$

Equations (11.53) in principle fully describe the motion of the vector

$$\boldsymbol{\rho}(t) \stackrel{\text{def}}{=} \begin{pmatrix} x(t) \\ y(t) \end{pmatrix} \qquad (11.54)$$

in the (x, y)-plane as a function of time t.

The motion of the vector defined by eqns. (11.53) may, however, also be represented as a superposition of pure circular motions

$$\boldsymbol{\rho}(t) = \begin{pmatrix} a_{x0} \\ a_{y0} \end{pmatrix} + \sum_{i=1}^{n} \left\{ \rho_{pi} \begin{pmatrix} \cos\left(i\omega t - \phi_{pi}\right) \\ \sin\left(i\omega t - \phi_{pi}\right) \end{pmatrix} + \rho_{ri} \begin{pmatrix} \cos\left(-i\omega t - \phi_{ri}\right) \\ \sin\left(-i\omega t - \phi_{ri}\right) \end{pmatrix} \right\} \, .$$

$$(11.55)$$

Obviously, for each frequency there may be a counterclockwise motion (positive sense of rotation) with radius ρ_{pi} with the (constant) angular velocity $i\omega$ and a clock-wise motion (negative sense of rotation) with radius ρ_{ri} and angular velocity $-i\omega$. The counterclockwise motion is called *prograde motion*, the clockwise motion *retrograde motion*.

The result (11.55) is a general one and it is interesting: An arbitrary motion of a vector in $\mathbb{E}^2$ may be written as a superposition of circular motions. *Cum grano salis* this may be viewed as a revival (and generalization) of the old epicycle theory.

The representations (11.53) and (11.53) are equivalent from the mathematic point of view (the formal proof is sketched below). The representation (11.55) gives, however, much better insight into the actual motion of the vector, if this motion is of a periodic or quasi-periodic nature. Usually it is possible to recognize easily the key characteristics of the motion, in particular if there is a hierarchy in the terms.

The radius ρ_{pi} of the prograde motion, the radius ρ_{ri} of the retrograde motion, and the corresponding phase angles ϕ_{pi} and ϕ_{ri} are defined by the equations

$$
\begin{aligned}
\rho_{pi} &= \tfrac{1}{2}\sqrt{(a_{xi}+b_{yi})^2 + (a_{yi}-b_{xi})^2} \\
\rho_{ri} &= \tfrac{1}{2}\sqrt{(a_{yi}+b_{xi})^2 + (a_{xi}-b_{yi})^2} \\
\phi_{pi} &= \arctan\left(\frac{-(a_{yi}-b_{xi})}{a_{xi}+b_{yi}}\right) \\
\phi_{ri} &= \arctan\left(\frac{-(a_{yi}+b_{xi})}{a_{xi}-b_{yi}}\right) .
\end{aligned}
\tag{11.56}
$$

The relations (11.56) are easily verified by using the addition theorems of trigonometry in eqns. (11.55). The resulting equations are of the same structure as eqns. (11.53). The results (11.56) are then obtained by comparing the coefficients of the terms $\cos i\omega$ and $\sin i\omega$ in both equations.

11.6 The Computer Program FOURIER

11.6.1 General Characterization

Program FOURIER produces spectra of time series contained in one input file. One or more spectra may be produced in one program run. If two time series are declared as the x- and y-components of one vector, the spectrum of the vector is also provided in the form (11.55). The program produces either power spectra or amplitude spectra.

Panel FOURIER 1 (Figure 11.1) shows the primary menu of program FOURIER. The file with the constants is the same for all program package. The input file containing the time series to be analyzed has to be defined in the second input field of the primary menu. Observe that the complete path has to be provided in order to avoid copying output files from one sub-directory (e.g., from /NUMINT/ORB in the example of Panel FOURIER 1 to /FOURIER). One file only may be analyzed in one program run. The program produces

- one general output file (two examples, to be discussed below are given in Figures 11.7 and 11.8),
- a file with error messages (if the field is left blank, the messages are written into the general output file), and
- a file with the spectra.

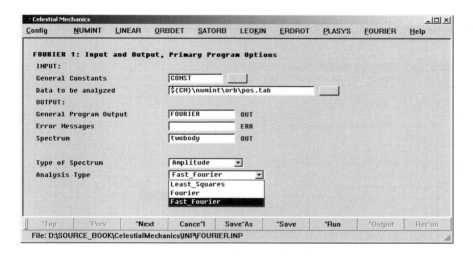

Fig. 11.1. Primary menu of program FOURIER

The third output file contains the main result, the spectra for the time series analyzed. It is a table with the frequency in column 1, the corresponding period in column 2, and either the amplitudes $\tilde{a}_i$ or the power spectral densities e_i in the columns 3 to $2 + n_s$ (n_s being the number of spectra analyzed). One column per spectrum is produced. Usually, this output file is long and should be inspected using the graphical capabilities of the menu system.

The last two input items in Panel FOURIER 1 (Figure 11.1) concern the type of the spectrum (power or amplitude spectra) and the analysis method (LSQ, FT, FFT). FT needs no further input characterizing the method, LSQ and

FFT requires additional information (in Panels FOURIER 4a, 4b, Figures 11.5, 11.6, respectively). The corresponding panels (commented below) are only presented, if one of the two techniques was selected.

In Panel FOURIER 2 (Figure 11.2) the structure of the input data and the pre-processing of data are defined. Depending on the origin, the input file may contain header records, which must be skipped. Also, one may wish to sample the input data series. If a sampling rate of "1" is specified, every record is used in the analysis, if one of "10" is selected, only every tenth record is used in the analysis – which drastically reduces the length of the data series, but also the resolution of the spectrum. If the data set to be analyzed exceeds the maximum length, an error message is issued and processing is terminated. The maximum number of data records, which can used by the program, is defined in the file /INP/FOURIER.INP, by the variable MAXVAL (in the fist non-blank line of this file). It is set to MAXVAL=700000 on the CD. Should a PC not have enough memory to run program FOURIER, this value should be reduced. If (even) longer data series should be analyzed (and if the PC has enough memory), MAXVAL may be increased. One furthermore has to specify the column number with the independent argument.

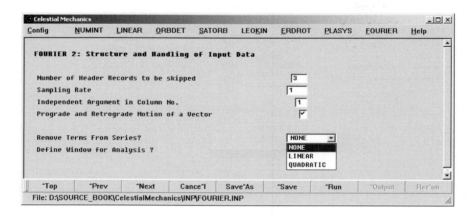

Fig. 11.2. Structure and data handling in program FOURIER

The next decision concerns the interpretation of the individual spectra: Shall the spectra be analyzed separately or shall two spectra be interpreted as the x- and y-components of a two-dimensional vector (in the sense of section 11.5)?

The two last input items in Panel FOURIER 2 (Figure 11.2) concern the pre-processing of data: It may be wise to remove an offset, a drift, or even a polynomial of degree 2 from the input data set before subjecting them to spectral analysis. This procedure is, e.g., recommended if time series like UT1−UTC (difference of UT defined by the rotation of the Earth and by atomic clocks, respectively) or time series of perihelia are analyzed. An unremoved drift would generate many spectral lines, possibly hiding the lines of physical interest. It is are also possible to define a time window within which the data are accepted. If this option is selected, one more panel is presented, where the left and right boundary of the window are defined (not documented here).

Panel FOURIER 3a (Figure 11.3) is only activated, if single spectra (no prograde/retrograde motion) shall be produced. The number of spectra and the column number with the first spectrum have to be defined here. At maximum four spectra can be analyzed in the same program run.

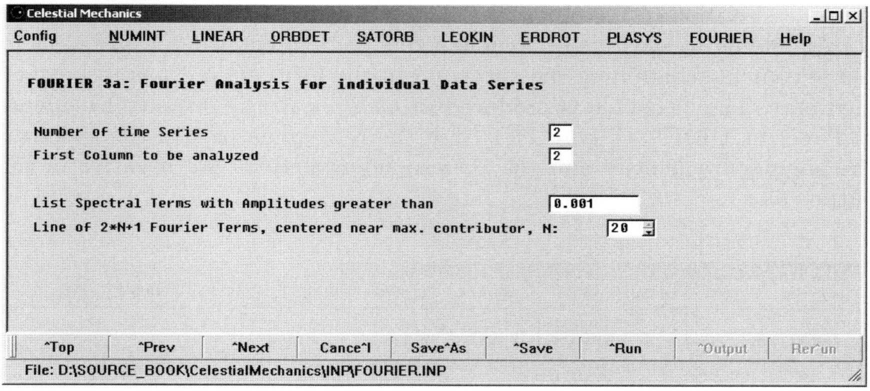

Fig. 11.3. Structure and data handling for single spectra for program FOURIER

When analyzing the spectra independently, one may ask for a list of the prominent spectral lines *assuming that the lines are due to one frequency (of infinitesimal width)*. The program looks for relative maxima of the calculated amplitudes. The program then creates an artificial spectral line by superposing $2n + 1$ lines centered around these maxima.

The procedure is the following: Let us assume that i_m is the index corresponding to such a maximum. The Fourier term with index $i \stackrel{\text{def}}{=} i_m + \Delta i$ contributes as follows to the combined line:

$$a_i \cos(i\omega t) + b_i \sin(i\omega t) = + \{+ a_i \cos(\Delta i\omega t) + b_i \sin(\Delta i\omega t)\} \cos(i_m\omega t)$$
$$+ \{- a_i \sin(\Delta i\omega t) + b_i \cos(\Delta i\omega t)\} \sin(i_m\omega t) .$$
$$(11.57)$$

It is thus possible to write the $2n + 1$ terms around the frequency $i_m\omega$ as a linear combination of $\sin(i_m\omega t)$ and $\cos(i_m\omega t)$. For a particular time τ

the resulting line amplitude simply is the square root of the coefficients of the sine- and the cosine-terms. The above formula shows, however, that the coefficients are time-dependent (with only low frequencies). Therefore, the amplitude of the resulting line is calculated in program FOURIER as the mean value of the resulting amplitudes over the time interval of the entire series.

The number n of contributing frequencies on both sides of the maximum frequency $i_m\omega$ must be defined in Panel FOURIER 3. Normally, one is only interested in lines above a certain threshold amplitude. Therefore, this quantity must be provided in the same panel.

Panel FOURIER 3b (Figure 11.4) is activated if the "Prograde and Retrograde ..." option in Panel FOURIER 2 was selected. The column numbers corresponding to the x- and y-components have to defined and the name of the output file containing the spectrum in the form (11.55) has to be specified here. The latter file is produced *in addition to* the output files defined in Panel FOURIER (Figure 11.1). For the retrograde part of the spectrum the angular frequencies and the corresponding periods are negative in this output file.

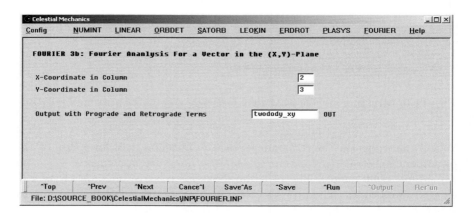

Fig. 11.4. Structure and data handling (cont) for vectors in program FOURIER

If the "Fast Fourier" option is chosen in Panel FOURIER 1 (Figure 11.1), the depth of the strategy has to be selected in the panel shown in Panel FOURIER 4b (Figure 11.5). This strategy depth is given by the number of times n the recursion formula (11.44) is used. For $n = 14$, as in the example of Panel FOURIER 4b (Figure 11.5), the algorithm should be by about a factor of $2^{14} \approx 16000$ faster than a classical Fourier analysis, and fewer than about 16000 data points are lost. The actual number of points lost is given in the general output file. In the example (discussed below) $x = 6160$ data

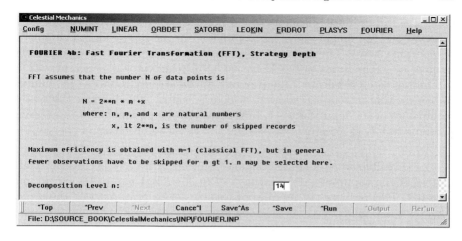

Fig. 11.5. Option for FFT in program FOURIER

points out of about 212000 had to be skipped (see Figure 11.7). If this number of lost data records is unacceptable, the strategy depth (represented by the number n) has to be lowered, if a more rapid processing is required, n has to be increased. If one wishes to perform the "classical" FFT with the maximum number of applications of the recursion formula (11.44), a very high value for n, e.g., $n = 50$ (which would correspond to about 10^{15} data records), can be specified. The program will then automatically reduce n to the maximum number possible for the length of the input data.

If LSQ is selected in Panel FOURIER 1 (Figure 11.1), Panel FOURIER 4a (Figure 11.6) is presented. The epochs may be distributed arbitrarily when

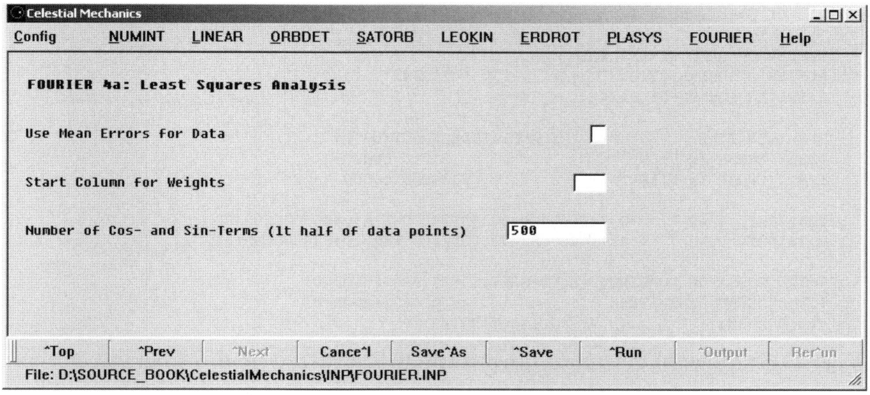

Fig. 11.6. Options for LSQ in program FOURIER

using LSQ and epoch-specific weights may be taken into account for each observation (the weights have to be provided in the input file in columns to be specified here). The number n_s of the sine-terms (which is equal to the number of cosine-terms minus 1) has to be provided in this panel, as well. If too high a value is selected, the program automatically reduces this number to the maximum number allowed, which is "only" 600. If more than 1200 records are in the input data set, LSQ is therefore *not* capable of performing a complete FT. It does only determine the 600 leading terms of the Fourier series.

It is not possible to get a high spectral resolution with LSQ, because a square matrix of the (approximate) dimension $(2n_s) \times (2n_s)$ has to be set up and inverted. This reduces the number of feasible terms to about one thousand, whereas the number of terms may be as high as about 250000 for the other two options. The processing times also provides a strong argument against using this option (see next section).

Let us now briefly browse through an example of general output file in Figure 11.7 corresponding to the application of the FFT. The technicalities of the processing (like the number of data points skipped, the processing time, the number of data points, the number of spectra produced) may be found in the general output. The names of the input and output files are included for reference. If the offsets and trends were removed from the data sets, the corresponding values may be found here.

```
SPECTRAL ANALYSIS USING FAST FOURIER TRANSFORM
DATE: 15-OCT-03    TIME: 09:34
****************************************************************
FOR FFT: NUMBER OF EPOCHS L = 2**( 14) * M + X
         M : NUMBER OF POINTS FOR LOW-LEVEL FTs:      13
         X : NUMBER OF DATA POINTS SKIPPED:         6160

INPUT FILE                  : ${CM}\numint\orb\pos.tab
OUTPUT FILE (SPECTRA)       : ${CM}\FOURIER\twobody.OUT
OUTPUT FILE (PRO/RETRO)     : ${CM}\FOURIER\twodody_xy.OUT

NUMBER OF SPECTRA ANALYZED :            2
NUMBER OF OBSERVATIONS      :       212992
SAMPLING FOR INPUT DATA     :            1

SPECTRUM TYPE               : AMPLITUDE SPECTRUM
FIRST EPOCH IN FILE         :        51910.00000
LAST  EPOCH IN FILE         :      5376685.00000

DATA SET  1 AO   =-0.74985740D+00 TOTAL POWER= 0.66868529D+01
DATA SET  2 AO   = 0.11448009D-03 TOTAL POWER= 0.59995669D+01

TOTAL POWER OF COMBINED (X,Y)-SERIES = 0.63432099D+01
PSD OF ZERO ORDER TERM               = 0.44321892D-01
CPU FOR RUN:            0.140 MINUTES
```

Fig. 11.7. General output of program FOURIER (pro- and retrograde motion)

It is important that the zero-order terms a_0 are included for all series analyzed: The other amplitudes $\tilde{a}_i$, $i = 1, 2, \ldots, 2n+1$, (and ρ_{pi}, ρ_{ri}) are given in the output file with the spectra. The zero-order term is not contained in those files, because the corresponding period would be infinite $P_0 = 2\pi/0 \cdot \omega$.

If the series are analyzed separately, the output file contains in addition a list of spectral terms (if there were any above the limit specified). The relevant part of the output may be inspected in Figure 11.8.

```
SPECTRAL ANALYSIS USING FAST FOURIER TRANSFORM
DATE: 15-OCT-03    TIME: 13:14
*******************************************************************************
FOR FFT: NUMBER OF EPOCHS L = 2**( 14) * M + X
         M : NUMBER OF POINTS FOR LOW-LEVEL FTs:       13
         X : NUMBER OF DATA POINTS SKIPPED:          6160
         .............................................

Spectra Lines for Spectrum 1 with Amplitudes ge  0.1000D-01
-----------------------------------------------------------
Line    OMEGA             PERIOD          AMPLITUDE
   1 0.43519687D-02     1443.7570          2.3321
   2 0.87043415D-02      721.8450          0.2118
   3 0.13056481D-01      481.2311          0.0347

Spectra Lines for Spectrum 2 with Amplitudes ge  0.1000D-01
-----------------------------------------------------------
Line    OMEGA             PERIOD          AMPLITUDE
   1 0.43519687D-02     1443.7570          2.3079
   2 0.87043267D-02      721.8462          0.2099
   3 0.13056344D-01      481.2362          0.0341
```

Fig. 11.8. General output of program FOURIER (analysis of two separate spectra)

11.6.2 Examples

The following example shall illustrate some of the technicalities of generating amplitude (or power) spectra. Let us analyze a series of (x,y)-coordinates of a two-body orbit in the ecliptic. Using program NUMINT (see Chapter 6), the orbit of a "minor planet" with inclination $i = 0°$, a revolution period of $P \stackrel{\text{def}}{=} P_4/3$ years (corresponding to a semi-major axis $a \approx 2.502$ AU), and an eccentricity $e = 0.2$ was integrated over a time period of 15000 years (corresponding to about 3800 revolutions). One set of (x, y)−coordinates was stored every 25 days. The entire time series consists of 219152 pairs of (x, y)-coordinates.

A small fraction of the entire amplitude spectrum is reproduced in Figure 11.9. The retrograde part of the spectrum is not reproduced because it is virtually empty. The dominating spectral line is, as expected, centered at the body's revolution period $P \approx 1443.7$ days (corresponding to 3.95 years). If a circular orbit would have been analyzed, one would expect exactly one

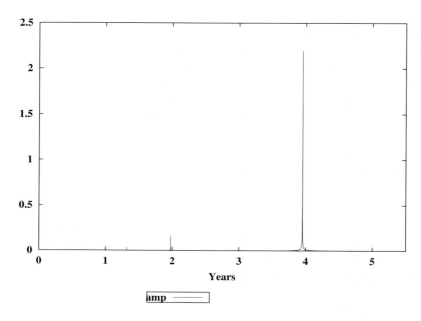

Fig. 11.9. Amplitude spectrum of two-body motion for periods $0 < P < 5$ Years ($e = 0.2$)

spectral line with an amplitude equal to the semi-major axis of $a \approx 2.502$ AU centered at the revolution period. The same amplitude is expected for eccentric orbits, as well. The reason for the significant deviation from the expected amplitude value of $a \approx 2.502$ will be addressed below.

Apart from the dominating line we see lines centered at periods corresponding to one half, one third, etc. of the body's revolution period. This result has to be expected. Using Kepler's equation $E = \sigma + e \sin E$, which may be approximated up to terms of order one in e by

$$E \approx \sigma + e \sin \sigma , \qquad (11.58)$$

one easily verifies that (to the same approximation)

$$
\begin{aligned}
x = a\,(\cos E - e) &\approx a \left(-\frac{3\,e}{2} + \cos \sigma + \frac{e}{2} \cos 2\sigma \right) \\
y = a\,\sqrt{1 - e^2}\,\sin E &\approx a \left(\sin \sigma + \frac{e}{2} \sin 2\sigma \right) .
\end{aligned}
\qquad (11.59)
$$

These equations show that for an elliptic orbit with its major-axis on the x-axis the elliptical motion up to order 1 in the eccentricity e may be approximated by two circular, prograde motions. The main circle has the radius $\rho = a$ of the semi-major axis, and the center lies on the negative x-axis with coordinates $(x, y) = a\left(-\frac{3e}{2}, 0\right)$. In the coordinate system centered at the focus of the ellipse, the center of the primary circle is halfway between the center of the ellipse and the "empty focus". The secondary circle has a radius of $\rho_2 = ae$. The center of the secondary circle is identical with the current location on the primary circle. The angular velocity in the secondary circle is twice the velocity in the primary circle. The attributes primary and secondary are justified by the radii of the circles. The output reproduced in Figure 11.7 shows that the program at least estimates the correct orders of magnitudes for the radii of these circles. Qualitatively, the finding (11.59) are confirmed in Figure 11.9 – but we should try to explain, why the amplitudes are somewhat too small.

Figure 11.10 helps to explain the effect. The solid line is a detailed view of the spectral line at $P = 3.95\ldots$ years. As one can see, the peak is not as sharp as Figure 11.9 suggests. The spectral line has a finite width, and one can see that (at least) three frequencies (epochs) significantly contribute to that line. The result would have been different, if we would have selected the interval $[t_1, t_N]$ as an entire number of revolution periods: The revolution period P would have been one of the grid points of the spectrum, and the entire power (amplitude) would have been attributed to period P. This would have created a wrong impression, however: In a long time series one generally does not know the location of all spectral lines and it is just not achievable in practice to have the centers of all lines coincide with grid points.

The problem encountered here is a general one. Therefore the program FOURIER offers the possibility to calculate the theoretical amplitude of a spectral line, assuming that this line is actually due to one frequency of "infinitesimal" width. The method was developed previously.

The dashed curve in Figure 11.10 illustrates that the line width is greatly reduced by increasing the number of grid point per time unit: The dashed curve corresponds to a data set covering a time interval of 150000 years (instead of 15000 years) and to a data spacing of 100 instead of 25 days. The spectral line is (as expected) much narrower and the peak amplitude is approaching the value of $a = 2.5$ AU as expected by formula 11.59. The exact location of the line and the exact amplitude could be reconstructed from the amplitudes *and* the phases associated with the grid point near the spectral line.

Figures 11.11 and 11.12 are those parts of the power spectrum corresponding to the parts of the amplitude spectra shown in Figures 11.9 and 11.10. Figure 11.11 gives the impression that almost the entire power is contained in one sharp line. Figure 11.12 shows (naturally) again that the finite width of the line must be taken into account. It is allowed to add up the power spectral densities of the individual contributors (the phase angles do not matter in this case). If this is done, one ends up with a power density of about 0.95 which can be attributed to the line at $P \approx 3.9528$ years. One easily sees

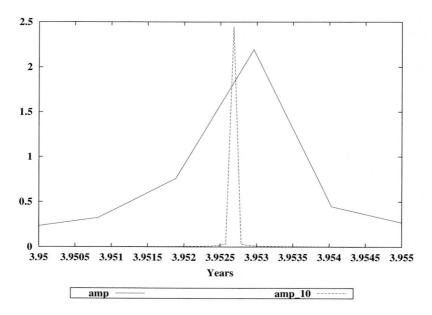

Fig. 11.10. Amplitude spectrum of two-body motion for periods $3.95 < P < 3.96$ years ($e = 0.2$)

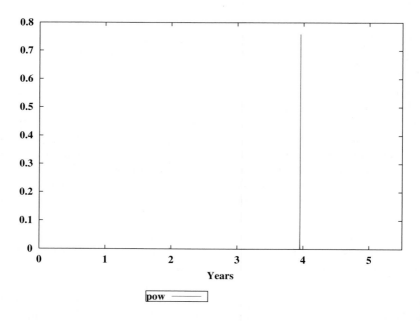

Fig. 11.11. Power spectrum of two-body motion for periods $0 < P < 5$ years ($e = 0.2$)

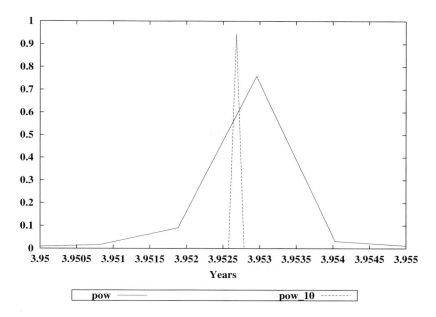

Fig. 11.12. Power spectrum of two-body motion for periods $3.95 < P < 3.96$ years $(e = 0.2)$

that the other lines at $P/2$, $P/3$ only give marginal contributions. Where is the missing power (the sum over all power spectral densities must be 1)? The general output file in Figure 11.7 gives the answer: It is contained in the zero-order term! (The PSD of the zero order term was determined to be 0.04).

Table 11.1 gives an overview of the performance of the three methods for spectral analysis in program FOURIER. The same input data set is used for all experiments. Two independent spectra were derived for the x- and y-components of the heliocentric position vector of the test particle, and the associated spectrum in the (x,y)-plane was generated in each program run. The independent parameters are

N, the number of epochs analyzed (varied with the data sampling option),

n_p, the number of parameters (sum of the number of sin- and cos-terms), and, in the case of the FFT,

n, the number of times the recursion formula (11.44) is used.

The performance may be judged on the basis of the processing time (CPU) and x, the number of data points skipped.

For the FT and the FFT, the first two parameters are identical, i.e., $N = n_p$. The latter two parameters only matter for the FFT. From Table 11.1 the

Table 11.1. Performance of three methods for spectral analysis

Method	N	n_p	n	x	CPU [s]
LSQ	2739	1199	–	–	617.658
LSQ	5478	1199	–	–	1124.197
LSQ	10957	1199	–	–	2138.375
LSQ	10957	301	–	–	132.611
LSQ	10957	601	–	–	528.019
LSQ	10957	1199	–	–	2138.375
FT	10957	10957	–	–	59.025
FT	21915	21915	–	–	246.775
FT	43830	43830	–	–	983.024
FFT	16384	16384	14	11009	2.303
FFT	49152	49152	14	5635	3.054
FFT	98304	98304	14	11241	4.537
FFT	212992	212992	14	6159	9.764
FFT	219136	219136	7	15	221.589
FFT	219136	219136	8	15	112.492
FFT	219136	219136	9	15	57.833
FFT	219136	219136	10	15	42.451
FFT	219136	219136	11	15	34.289
FFT	217988	217988	12	2063	20.820
FFT	212992	212992	13	6159	13.499
FFT	212992	212992	14	6159	9.764
FFT	196608	196608	15	22543	7.541
FFT	196608	196608	16	22543	6.710
FFT	131072	131072	17	88079	4.987

following conclusions may be drawn:

- The performance differs by orders of magnitude for the three methods. Clearly, FFT will be given the preference wherever possible. Exceptions may make sense

  – if the length of the time series does not exceed a few tens of thousands of data points, and if (for one reason or another) one wants to be sure that all data points are used,

  – if a time series with arbitrarily spaced epochs has to be analyzed.

- The processing time for the LSQ method grows roughly linearly with the length N of the series and quadratically with the number n_p of parameters. If a conventional spectral analysis is performed using LSQ, the number of parameters is equal to the length of the series, $n_p = N$, which is why the processing time grows with N^3. The price for being able to use epoch-specific weights is exorbitant.

- For the classical Fourier analysis, the processing time grows roughly quadratically with the length of the series, i.e., $\sim N^2$. If only few spectra have to be analyzed, and if $N < 20000$, the method is a valuable choice. It makes sure that all data in the input data set were used.

- The performance of the FFT method is truly remarkable. When varying the length of the series with a constant n, the growth of processing time is "close to linear". Actually, a law of type $N \ln N$ results (see, e.g., [88]).

- In section 11.4 we argued that a factor of 2 in processing time should be gained with each application of the recursion formula (11.44). Table 11.1 tells that this law is a good approximation for $n < 8$. Afterwards the gain is not as pronounced. This is actually good news: By lowering n from an originally high order, one looses not too much efficiency.

- Table 11.1 supports the argument to produce time series with a length of $N = 2^n$. Selecting the n in Table 11.1 accordingly results in the best performance. Unfortunately this advice is usually of no value because the length of a time series usually is dictated by the experiment and not by the mathematician's preference.

Let us point out, that this chapter is intended to serve as a first introduction into harmonic analysis. Many details and subtleties were not touched here. For further reading (e.g., for analyzing unevenly spaced data efficiently) we recommend [88].

References

1. M. Abramowitz, I. A. Stegun: *Handbook of Mathematical Functions* (Dover Publ., New York 1965)
2. T. J. Ahrens (ed.): *Global Earth Physics – A Handbook of Physical Constants* (American Geophysical Union, Washington D.C. 1995)
3. R. R. Allen, G. E. Cook: 'The long-period motion of the plane of a distant circular orbit', Proc. R. Soc. Lond., Ser. A, **280**, 97–109, (1964)
4. E. F. Arias, P. Charlot, M. Feissel, J.-F. Lestrade: 'The extragalactic reference system of the International Earth Rotation Service, ICRS', Astron. Astrophys., **303**, 604–608 (1995)
5. R. T. H. Barnes, R. Hide, A. A. White, C. A. Wilson: 'Atmospheric angular momentum fluctuations, length-of-day changes and polar motion', Proc. R. Soc. Lond., Ser. A, **387**, 31–73 (1983)
6. A. L. Berger: 'Obliquity and Precession for the Last 5 000 000 Years', Astron. Astrophys., **51**, 127–135 (1976)
7. A. L. Berger: 'Long-Term Variations of the Earth's Orbital Elements', Cel. Mech., **15**, 53–74 (1977)
8. A. L. Berger: 'The Milankovitch Astronomical Theory of Paleoclimates: A Modern Review', Vistas in Astron., **24**, 103–122 (1980)
9. G. Beutler: *Integrale Auswertung von Satellitenbeobachtungen*, (Schweizerische Geodätische Kommission, Zürich 1977), Astronomisch-geodätische Arbeiten in der Schweiz, **33**
10. G. Beutler: *Lösung von Parameterbestimmungsproblemen in Himmelsmechanik und Satellitengeodäsie mit modernen Hilfsmitteln*, (Schweizerische Geodätische Kommission, Zürich 1982), Astronomisch-geodätische Arbeiten in der Schweiz, **34**
11. G. Beutler: *Himmelsmechanik II: Der erdnahe Raum. Mit einem Anhang von Andreas Verdun*, (Astronomisches Institut, Universität Bern, Bern 1992), Mitteilungen der Satelliten-Beobachtungsstation Zimmerwald, **28**
12. G. Beutler, E. Brockmann, W. Gurtner, U. Hugentobler, L. Mervart, M. Rothacher, A. Verdun: 'Extended orbit modeling techniques at the CODE processing center of the international GPS service for geodynamics (IGS): theory and initial results', Manuscr. Geod., **19**, 367–386 (1994)
13. G. Beutler, J. Kouba, T. Springer: 'Combining the orbits of the IGS Analysis Centers', Bull. Géod., **69**, 200–222 (1995)
14. G. Beutler: Rotation der Erde: Theorie, Methoden, Resultate aus Satellitengeodäsie und Astrometrie. Lecture Notes, Astronomical Institute, University of Bern, Bern (1997)
15. G. Beutler: Numerische Integration gewöhnlicher Differentialgleichungssysteme. Lecture Notes, Astronomical Institute, University of Bern, Bern (1998)
16. G. Beutler: Himmelsmechanik des Planetensystems. Lecture Notes, Astronomical Institute, University of Bern, Bern (1999)

17. G. Beutler, M. Rothacher, S. Schaer, T. A. Springer, J. Kouba, R. E. Neilan: 'The International GPS Service (IGS): An interdisciplinary service in support of Earth sciences', Adv. Space Res., **23**, 631–653 (1999)
18. G. Beutler: Himmelsmechanik des erdnahen Raumes. Lecture Notes, Astronomical Institute, University of Bern, Bern (2000)
19. G. Beutler, M. Rothacher, J. Kouba, R. Weber: 'Polar Motion with Daily and Sub-daily Time Resolution'. In: *Polar Motion: Historical and Scientific Problems, IAU Colloquium 178, Cagliari, Sardinia, Italy, 27–30 September 1999*, ed. by S. Dick, D. McCarthy, B. Luzum (Astronomical Society of the Pacific, San Francisco 2000), ASP Conference Series, **208**, pp. 513–525
20. J. Binney, S. Tremaine: *Galactic Dynamics* (Princeton University Press, Princeton 1987)
21. H. Bock, G. Beutler, S. Schaer, T. A. Springer, M. Rothacher: 'Processing aspects related to permanent GPS arrays', Earth Planets Space, **52**, 657–662 (2000)
22. H. Bock, U. Hugentobler, T. S. Springer, G. Beutler: 'Efficient Precise Orbit Determination of LEO Satellites Using GPS', Adv. Space Res., **30**, 295–300 (2002)
23. H. Bock, G. Beutler, U. Hugentobler: 'Kinematic Orbit Determination for Low Earth Orbiter (LEOs)'. In: *Vistas for Geodesy in the New Millennium – IAG 2001 Scientific Assembly, Budapest, Hungary, September 2–7, 2001*, ed. by J. Ádám, K.-P. Schwarz (Springer, Berlin, Heidelberg 2002), International Association of Geodesy Symposia, **125**, pp. 303–308
24. H. Bock: Efficient Methods for Determining Precise Orbits of Low Earth Orbiters Using the Global Positioning System, Ph.D. Thesis, Astronomical Institute, University of Bern, Bern (2003)
25. I. N. Bronstein, K. A. Semendjajew, G. Musiol, H. Mühlig: *Taschenbuch der Mathematik*, 5th edn. (Harri Deutsch, Thun, Frankfurt/Main 2000)
26. D. Brouwer: 'On the accumulation of errors in numerical integration', Astron. J., **46**, 149–153 (1937)
27. D. Brouwer, G. M. Clemence: *Methods of Celestial Mechanics*, 2nd impr. (Academic Press, Orlando, San Diego, New York 1985)
28. H. E. Coffey (ed.): 'Data for November and December 1999', Solar-Geophys. Data, **665**, Part I, 127–135 (2000)
29. C. J. Cohen, E. C. Hubbard, C. Oesterwinter: *Elements Of The Outer Planets For One Million Years*, (The Nautical Almanac Office, U.S. Naval Observatory, Washington D.C. 1973) Astronomical Papers Prepared for the Use of the American Ephemeris and Nautical Almanac, **22**, Part 1
30. C. J. Cohen, E. C. Hubbart, C. Oesterwinter: 'Planetary elements for 10 000 000 years', Cel. Mech., **7**, 438–448 (1973)
31. J. M. A. Danby: *Fundamentals of Celestial Mechanics*, 2nd edn. (Willmann-Bell, Richmond, Virginia 1989)
32. A. T. Doodson: 'The harmonic development of the tide-generating potential', Proc. R. Soc. Lond., Ser. A, **100**, 305–329 (1922)
33. L. Euler: 'Recherches sur le mouvement des corps célestes en général', Hist. l'Acad. Roy. Sci. et belles lettres (Berlin), **3**, 93–143 (1749); E. 112, O. II, **25**, 1–44 (1960)
34. L. Euler: 'Découverte d'un nouveau principe de mécanique', Hist. l'Acad. Roy. Sci. et belles lettres (Berlin), **6**, 185–217 (1752); E. 177, O. II, **5**, 81–108 (1957)
35. L. Euler: 'Principes généraux du mouvement des fluides', Hist. l'Acad. Roy. Sci. et belles lettres (Berlin), **11**, 274–315 (1757); E. 226, O. II, **12**, 54–91 (1954)

36. L. Euler: 'Du mouvement de rotation des corps solides autour d'un axe variable'. Hist. l'Acad. Roy. Sci. et belles lettres (Berlin), **14**, 154–193 (1765); E. 292, O. II, **8**, 200–235 (1965)

37. L. Euler: *Institutionum calculi integralis volumen primum in quo methodus integrandi a primis principiis usque ad integrationem aequationum differentialium primi gradus pertractatur* (Acad. imp. sc., Petropoli 1768), pp. 493–508; E. 342, O. I, **11**, 424–434 (1913)

38. E. Fehlberg: Classical fifth-,sixth-, seventh-, and eight-order Runge-Kutta formulas with stepsize control. NASA Technical Report R-287 (NASA, Huntsville 1968)

39. S. Ferras-Mello, T. A. Michtchenko, D. Nesvorný, F. Roig, A. Simula: 'The depletion of the Hecuba gap vs the long-lasting Hilda group', Planet. Space Sci., **46**, 1425–1432 (1998)

40. H. F. Fliegel, T. E. Galini, E. R. Swift: 'Global Positioning System Radiation Force Model for Geodetic Applications', J. Geophys. Res., **97**, 559–568 (1992)

41. W. Flury: Raumfahrtmechanik. Lecture Notes, Technical University of Darmstadt, Darmstadt (1994)

42. C. Froeschlé, Ch. Froeschlé: 'Order and chaos in the solar system'. In: *Proceedings of the Third International Workshop on Positional Astronomy and Celestial Mechanics, University of Valencia, Cuenca, Spain, October 17–21, 1994*, ed. by A. López Garcia, E. I. Yagudina, M. J. Martinez Usó, A. Cordero Barbero (Universitat de Valencia, Observatorio Astronomico, Valencia 1996) pp. 155–171

43. L.-L. Fu, E. J. Christensen, C. A. Yamarone, M. Lefebvre, Y. Ménard, M. Dorrer, P. Escudier: 'TOPEX/POSEIDON mission overview', J. Geophys. Res. **99**, 24369–24381 (1994)

44. C. F. Gauss: 'Summarische Übersicht der zur Bestimmung der Bahnen der beyden neuen Hauptplaneten angewandten Methoden', Monatl. Corresp., **20**, 197–224 (1809)

45. C. W. Gear: *Numerical Initial Value Problems in Ordinary Differential Equations* (Prentice-Hall, Englewood Cliffs, New Jersey 1971)

46. J. M. Gipson: 'Very long baseline interferometry determination of neglected tidal terms in high-frequency Earth orientation variation', J. Geophys. Res., **101**, 28051–28064 (1996)

47. W. B. Gragg: 'On extrapolation algorithms for ordinary initial value problems', J. SIAM, Ser. B, **2**, 384–403 (1965)

48. W. Gurtner: 'RINEX: The Receiver-Independent Exchange Format', GPS World, **5**, No. 7, 48–52 (1994)

49. A. Guthmann: *Einführung in die Himmelsmechanik und Ephemeridenrechnung – Theorie, Algorithmen, Numerik*, 2. Aufl. (Spektrum Akademischer Verlag, Heidelberg, Berlin, 2000)

50. A. E. Hedin: 'MSIS-86 Model', J. Geophys. Res., **92**, 4649–4662 (1987)

51. A. E. Hedin: 'Extension of the MSIS Thermosphere Model into the Middle and Lower Atmosphere', J. Geophys. Res., **96**, 1159–1172 (1991)

52. W. A. Heiskanen, H. Moritz: *Physical Geodesy* (W. H. Freeman Comp., San Francisco, London 1967)

53. P. Henrici: *Discrete Variable Methods in Ordinary Differential Equations* (John Wiley & Sons, New York, London Sidney 1968)

54. P. Herget: 'Computation of Preliminary Orbits', Astron. J., **70**, 1–3 (1965)

55. T. A. Herring: 'An a priori model for the reduction of nutation observations: KSV$_{1994.3}$ nutation series', Highlights of Astronomy, **10**, 222–227 (1995)

56. K. Hirayama: 'Groups of Asteroids Probably of Common Origin', Astron. J., **31**, 185–188 (1918)

57. U. Hugentobler: *Astrometry and Satellite Orbits: Theoretical Considerations and Typical Applications*, (Schweizerische Geodätische Kommission, Zürich 1998), Geodätisch-geophysikalische Arbeiten in der Schweiz, **57**

58. U. Hugentobler, S. Schaer, P. Fridez: *Bernese GPS Software Version 4.2* (Astronomical Institute, University of Bern, Bern 2001)

59. U. Hugentobler, D. Ineichen, G. Beutler: 'GPS satellites: Radiation pressure, attitude and resonance', Adv. Space Res. **31**, 1917–1926 (2003)

60. J. Imbrie: 'Astronomical Theory of the Pleistocene Ice Ages: A Brief Historical Review', Icarus, **50**, 408–422 (1982)

61. A. Jäggi: 'Efficient Stochastic Orbit Modeling Techniques using Least Squares Estimators', International Association of Geodesy Symposia, **127** (in press)

62. W. M. Kaula: *Theory of Satellite Geodesy – Applications of Satellites to Geodesy* (Blaisdell Publ. Comp., Waltham, Toronto, London 1966)

63. Z. Kopal: *Numerical Analysis* (Chapman & Hall, London 1955)

64. J. Kouba: 'A review of geodetic and geodynamic satellite Doppler positioning', Rev. Geophys. Space Phys., **21**, 27–40 (1983)

65. J. Kouba, G. Beutler, M. Rothacher: 'IGS Combined and Contributed Earth Rotation Parameter Solutions'. In: *Polar Motion: Historical and Scientific Problems, IAU Colloquium 178, Cagliari, Sardinia, Italy, 27–30 September 1999*, ed. by S. Dick, D. McCarthy, B. Luzum (Astronomical Society of the Pacific, San Francisco 2000), ASP Conference Series, **208**, pp. 277–302

66. K. Lambeck: *Geophysical Geodesy – The Slow Deformations of the Earth* (Clarendon Press, Oxford 1988)

67. L D. Landau, J. M. Lifschitz: *Lehrbuch der theoretischen Physik*, Bd. 6 (Hydrodynamik), 5. Aufl., (Verlag Harri Deutsch, Frankfurt/Main 1991)

68. K. R. Lang: *Astrophysical Data: Planets and Stars* (Springer, New York, Berlin, Heidelberg 1992)

69. W. Lowrie: *Fundamentals of Geophysics* (Cambridge University Press, Cambridge 1997)

70. D. D. McCarthy: *IERS Conventions (1996)*, (Central Bureau of IERS, Observatoire de Paris, Paris 1996) IERS Technical Note, **21**

71. D. D. McCarthy: *IERS Conventions (2000)*, (Central Bureau of IERS, Observatoire de Paris, Paris 2004) IERS Technical Note, **32** (in press)

72. J. Meeus: *Astronomical Algorithms*, 2nd edn. (Willmann-Bell, Richmond, Virginia 1999)

73. W. G. Melbourne, E. S. Davis, C. B. Duncan, G. A. Hajj, K. R. Hardy, E. R. Kursinski, T. K. Meehan, L. E. Young, T. P. Yunck: *The Application of Spaceborne GPS to Atmospheric Limb Sounding and Global Change Monitoring* (NASA, JPL, Pasadena, 1994), JPL Publication 94-18

74. A. Milani, A. M. Nobili, P. Farinella: *Non-Gravitational Perturbations and Satellite Geodesy* (Adam Hilger, Bristol 1987)

75. O. Montenbruck, E. Gill: *Satellite Orbits – Models, Methods, and Appications*, corr. 2nd. print. (Springer, Berlin, Heidelberg 2001)

76. H. Moritz, I. I. Mueller: *Earth Rotation – Theory and Observation* (Ungar Publ. Comp., New York 1988)

77. F. R. Moulton: *An Introduction to Celestial Mechanics*, 2nd rev. edn. (Dover Publ., New York 1970)

78. W. H. Munk, G. J. F. Macdonald: *The Rotation of the Earth – A Geophysical Discussion*, 2nd edn. (Cambridge University Press, Cambridge 1975)

79. P. Murdin (ed.): *Encyclopedia of Astronomy and Astrophysics* (Institute of Physics Publ., Bristol, Philadelphia 2001)

80. C. D. Murray, S. F. Dermott: *Solar System Dynamics* (Cambridge University Press, Cambridge 1999)

81. NAg-Library: *The NAG Fortran Library Introductory Guide, Mark 17* (NAG-Ltd, Wilkinson House, Oxford 1995)

82. X. X. Newhall, E. M. Standish, J. G. Williams: 'DE 102: a numerically integrated ephemeris of the Moon and planets spanning forty-four centuries', Astron. Astrophys., **125**, 150–167 (1983)

83. I. Newton: *Philosophiae naturalis principia mathematica* (Jussu Societatis Regiae ac Typis Josephi Streater, Londini 1687)

84. I. Newton: *The Principia – Mathematical Principles of Natural Philosophy*, Transl. by I. B. Cohen and A. Whitman (University of California Press, Berkeley, Los Angeles, London 1999)

85. I. Peterson: *Newton's Clock – Chaos in the Solar System* (W. H. Freeman Comp., New York 1993)

86. H. Poincaré: *New Methods of Celestial Mechanics*, ed. by D. L. Goroff (American Institute of Physics, USA, 1993)

87. H. Poincaré: 'Sur la précession des corps déformables', Bull. Astron. (Paris), **27**, 321–356 (1910)

88. W. H. Press, S. A. Teukolsky, W. T. Vetterling, B. P. Flannery: *Numerical Recipes in Fortran 77 – The Art of Scientific Computing*, 2nd edn. (Cambridge University Press, Cambridge 1996)

89. C. Reigber: 'Gravity Field Recovery from Satellite Tracking Data'. In: *Theory of Satellite Geodesy and Gravity Field Determination*, ed. by F. Sansò, R. Rummel (Springer, Berlin, Heidelberg 1989), Lecture Notes in Earth Sciences, **25**, pp. 197–234

90. F. P. J. Rimrott: *Introductory Orbit Dynamics* (Vieweg & Sohn, Braunschweig, Wiesbaden 1989)

91. N. T. Roseveare: *Mercury's perihelion from Le Verrier to Einstein* (Clarendon Press, Oxford 1982)

92. M. Rothacher, G. Beutler, T. A. Herring, R. Weber: 'Estimation of nutation using the Global Positioning System' J. Geophys. Res., **104**, 4835–4859 (1999)

93. M. Rothacher, G. Beutler, R. Weber, J. Hefty: 'High-frequency variations in Earth rotation from Global Positioning System data', J. Geophys. Res., **106**, 13711–13738 (2001)

94. A. E. Roy: *Orbital Motion*, 3rd edn. (Adam Hilger, Bristol, Philadelphia 1988)

95. A. E. Roy, I. W. Walker, A. J. Macdonald, I. P. Williams, K. Fox, C. D. Murray, A. Milani, A. M. Nobili, P. J. Message, A. T. Sinclair, M. Carpino: 'Project LONGSTOP', Vistas in Astron., **32**, 95–116 (1988)

96. D. A. Salstein, D. M. Kann, A. J. Miller, R. D. Rosen: 'The Sub-bureau for Atmospheric Angular Momentum of the International Earth Rotation Service: A Meteorological Data Center with Geodetic Applications', Bull. Amer. Meteor. Soc., **74**, 67–80 (1993)

97. S. Schaer, G. Beutler, M. Rothacher: 'Mapping and Predicting the Ionosphere'. In: *Proceedings of the IGS 1998 Analysis Center Workshop, ESA/ESOC, Darmstadt, Germany, February 9–11, 1998*, ed. by J. M. Dow, J. Kouba, T. Springer (ESA/ESOC, Darmstadt 1998) pp. 307–318

98. S. Schaer: *Mapping and Predicting the Earth's Ionosphere Using the Global Positioning System*, (Schweizerische Geodätische Kommission, Zürich 1999), Geodätisch-geophysikalische Arbeiten in der Schweiz, **59**

99. T. Schildknecht, U. Hugentobler, M. Ploner: 'Optical surveys of space debris in GEO', Adv. Space Res., **23**, 45–54 (1999)

100. T. Schildknecht, M. Ploner, U. Hugentobler: 'The search for debris in GEO', Adv. Space Res., **28**, 1291–1299 (2001)

101. T. Schildknecht, R. Musci, M. Ploner, S. Preisig, J. de Leon Cruz, H. Krag: 'Optical Observation of Space Debris in the Geostationary Ring'. In: *Proceed-*

ings of the Third European Conference on Space Debris, March 19-21, 2001, ESOC, Darmstadt, Germany, (ESA Publ. Div., ESTEC, Noordwijk 2001), SP-473, Vol. 1, pp. 89–93

102. L. D. Schmadel: *Dictionary of Minor Planet Names*, 4th rev. and enl. edn. (Springer, Berlin, Heidelberg 1999)

103. M. Schneider: *Satellitengeodäsie* (B.I. Wissenschaftsverlag, Mannheim, Wien, Zürich 1988)

104. J. Schubart: 'Three Characteristic Parameters of Orbits of Hilda-type Asteroids', Astron. Astrophys., **114**, 200–204 (1982)

105. J. Schubart: 'Additional results on orbits of Hilda-type asteroids', Astron. Astrophys., **241**, 2997–302 (1991)

106. G. Seeber: *Satellite Geodesy – Foundations, Methods, and Applications*, 2nd edn., (Walter de Gruyter, Berlin, New York, 2003)

107. P. K. Seidelmann (ed.): *Explanatory Supplement to the Astronomical Almanac* (University Science Books, Mill Valley, California, 1992)

108. L. F. Shampine, M. K. Gordon: *Computer Solution of Ordinary Differential Equations – The Initial Value Problem*, (W. H. Freeman Comp., San Francisco 1975)

109. M. H. Soffel: *Relativity in Astrometry, Celestial Mechanics and Geodesy* (Springer, Berlin, Heidelberg 1989)

110. T. A. Springer: *Modeling and Validating Orbits and Clocks Using the Global Positioning System*, (Schweizerische Geodätische Kommission, Zürich 2000), Geodätisch-geophysikalische Arbeiten in der Schweiz, **60**

111. E. M. Standish: 'The observational basis for JPL's DE 200, the planetary ephemerides of the Astronomical Almanac', Astron. Astrophys., **233**, 252–271 (1990)

112. J. Stoer, R. Bulirsch: *Einführung in die Numerische Mathematik*, 2. Aufl. (Springer, Berlin, Heidelberg 1976-1978), Heidelberger Taschenbücher, **105, 114**

113. J. Stoer, R. Bulirsch: *Introduction to numerical analysis*, 3rd edn., (Springer, New York 2002), Texts in applied mathematics, **12**

114. K. Stumpff: *Himmelsmechanik* (VEB Deutscher Verlag der Wissenschaften, Berlin 1959–1974)

115. D. Švehla, M. Rothacher: 'Kinematic Orbit Determination of LEOs Based on Zero or Double-difference Algorithms Using Simulated and Real SST GPS Data'. In: *Vistas for Geodesy in the New Millennium – IAG 2001 Scientific Assembly, Budapest, Hungary, September 2–7, 2001*, ed. by J. Ádám, K.-P. Schwarz (Springer, Berlin, Heidelberg 2002), International Association of Geodesy Symposia, **125**, pp. 322–328

116. D. Švehla, M. Rothacher: 'CHAMP Double-Difference Kinematic POD with Ambiguity Resolution'. In: *First CHAMP Mission Results for Gravity, Magnetic and Atmospheric Studies*, ed. by C. Reigber, H. Lühr, P. Schwintzer (Springer, Berlin, Heidelberg 2003) pp. 70–77

117. V. Szebehely: *Theory of Orbits – The Restricted Problem of Three Bodies* (Academic Press, New York, London 1967)

118. L. G. Taff: *Celestial Mechanics – A Computational Guide for the Practitioner* (John Wiley & Sons, New York, Chichester 1985)

119. B. D. Tapley: 'Fundamentals of Orbit Determination'. In: *Theory of Satellite Geodesy and Gravity Field Determination*, ed. by F. Sansò, R. Rummel (Springer, Berlin, Heidelberg 1989), Lecture Notes in Earth Sciences, **25**, pp. 235–260

120. B. D. Tapley, M. M. Watkins, J. C. Ries, G. W. Davies, R. J. Eanes, S. R. Poole, H. J. Rim, B. E. Schutz, C. K. Shum, R. S. Nerem, F. J. Lerch,

J. A. Marshall, S. M. Klosko, N. K. Pavlis, R. G. Williamson: 'The Joint Gravity Model 3', J. Geophys. Res., **101**, 28029–28049 (1996)

121. F. Tissérand: *Traité de Mécanique Céleste*, Nouv. tirage (Gauthier-Villars, Paris 1960)

122. P. J. G. Teunissen, A. Kleusberg (eds.): *GPS for Geodesy*, 2nd edn., (Springer, Berlin, Heidelberg 1998)

123. 'TOPEX/POSEIDON: Geophysical Evaluation', J. Geophys. Res., **99**, 24369–25062 (1994)

124. W. Torge: *Geodesy*, 3rd edn., (Walter de Gruyter, Berlin, New York, 2001)

125. W. Torge: *Geodäsie*, 2. Aufl., (Walter de Gruyter, Berlin, New York 2003)

126. A. Verdun, G. Beutler: 'Early Observational Evidence of Polar Motion'. In: *Polar Motion: Historical and Scientific Problems, IAU Colloquium 178, Cagliari, Sardinia, Italy, 27–30 September 1999*, ed. by S. Dick, D. McCarthy, B. Luzum (Astronomical Society of the Pacific, San Francisco 2000), ASP Conference Series, **208**, pp. 67–81

127. H. G. Walter, O. J. Sovers: *Astrometry of Fundamental Catalogues – The Evolution from Optical to Radio Reference Frames* (Springer, Berlin, Heidelberg, New York 2000)

128. R. Weber, M. Rothacher: 'The Quality of Sub-daily Polar Motion Estimates based on GPS Observations'. In: *Polar Motion: Historical and Scientific Problems, IAU Colloquium 178, Cagliari, Sardinia, Italy, 27–30 September 1999*, ed. by S. Dick, D. McCarthy, B. Luzum (Astronomical Society of the Pacific, San Francisco 2000), ASP Conference Series, **208**, pp. 527–532

129. A. Wegener: *Die Entstehung der Kontinente und Ozenane* (Friedrich Vieweg & Sohn, Braunschweig 1915), Sammlung Vieweg, **23**

130. J. Wisdom: 'Chaotic Behavior and the Origin of the 3/1 Kirkwood Gap', Icarus, **56**, 51–74 (1983)

131. J. Wisdom: 'Chaotic behavior in the solar system'. Nucl. Phys. B (Proc. Suppl.), **2**, 391–414 (1987)

132. E. W. Woolard: *Theory of the Rotation of the Earth around its Center of Mass* (The Nautical Almanac Office, U.S. Naval Observatory, Washington D.C. 1953) Astronomical Papers Prepared for the Use of the American Ephemeris and Nautical Almanac, **15**, Part 1

133. V. N. Zharkov, S. M. Molodensky, A. Brzeziński, E. Groten, P. Varga: *The Earth and its Rotation – Low Frequency Geodynamics* (H. Wichmann Verlag, Heidelberg 1996)

134. J. F. Zumberge, M. B. Heflin, D. C. Jefferson, M. M. Watkins, F. H. Webb: 'Precise point positioning for the efficient and robust analysis of GPS data from large networks', J. Geophys. Res., **102**, 5005–5017 (1997)

Abbreviations and Acronyms

AAM	Atmospheric Angular Momentum
AIUB	Astronomical Institute, University of Bern
AMF	Angular Momentum Functions
AU	Astronomical Unit
BIH	Bureau International de l'Heure
BKG	Bundesamt für Kartographie und Geodäsie
CCD	Charge Coupled Device
CHAMP	CHAllenging Minisatellite Payload
CIRA	COSPAR International Reference Atmosphere
CODE	Center for Orbit Determinattion in Europe
COSPAR	Committee on Space Research
CPU	Central Processing Unit
CSTG	Commission on Coordination of Space Techniques
ΔLOD	Excess LOD
DE200	Development Ephemeris 200
DORIS	Doppler Orbitography by Radiopositioning Integrated on Satellite
DoY	Day of Year
ECMWF	European Center for Medium-Range Weather Forecasts
EOP	Earth Orientation Parameter
ERP	Earth Rotation Parameters
ERS-2	Earth Remote Sensing 2
ESA	European Space Agency
ET	Ephemeris Time
FCN	Free Core Nutation
FFT	Fast Fourier Transformation
FT	Fourier Transformation
GARP	Global Atmospheric Research Program
GFZ	GeoForschungsZentrum
GOCE	Gravity field and steady-state Ocean Circulation Experiment
GPS	Global Positioning System
GPS/MET	GPS Meteorology using limb sounding

GRACE	Gravity Recover and Climate Experiment
HIPPARCOS	HIgh Precision PARallax COllecting Satellite
IAG	International Association of Geodesy
IAU	International Astronomical Union
ICRF	International Celestial Reference Frame
ICRS	International Celestial Reference System
IERS	International Earth Rotation and Reference Systems Service
IGN	Institut Géographique National
IGS	International GPS Service
ILRS	International Laser Ranging Service
ILS	International Latitude Service
IPMS	International Polar Motion Service
ITRF	International Terrestrial Reference Frame
ITRS	International Terrestrial Reference System
IVS	International VLBI Service for Astrometry and Geodesy
JD	Julian Date
JGM3	Joint Gravity Model 3
JPL	Jet Propulsion Laboratory
LAGEOS	LAser GEOdetic Satellite
LEO	Low Earth Orbiter
LLR	Lunar Laser ranging
LOD	Length of Day
LSQ	Least Squares
Laser	Light Amplification through Stimulated Emission of Radiation
mas	milliarcseconds
μs/day	microseconds per day
MJD	Modified Julian Date
MPC	Minor Planet Center
MSIS	Mass Spectrometer and Incoherent Scatter
NCEP	U.S. National Centers for Environmental Prediction
NDFW	Nearly-Diurnal Free Wobble
NEA	Near Earth Asteroids
NEO	Near Earth Objects
NNSS	U.S. Navy Navigation Satellite System
NOAA	National Oceanic and Atmospheric Administration
ns	nanoseconds
PAGEOS	PAssive GEOdetic Satellite
PAI	Principal Axes of Inertia
PC	Personal Computer
PM	Polar Motion
POD	Precise Orbit Determination

PPN	Parametrized Post-Newtonian
PPP	Precise Point Positioning
PRARE	Precise Range And Range-rate Equipment
Quasars	Quasistellar Radio Sources
RINEX	Receiver Independent Exchange Format
SI	International System of units
SLR	Satellite Laser Ranging
ST	Sidereal Time
swisstopo	Swiss Federal Office of Topography
TAI	International Atomic Time
TDB	Barycentric Dynamical Time
TNO	Trans-Neptunian Objects
TOPEX	TOPography EXperiment for Ocean Circulation
TT	Terrestrial Time
TUM	Technical University of Munich
UT	Universal Time
UT1	UT corrected for polar motion effects
UTC	Universal Time Coordinated
VLBI	Very Long Baseline Interferometry
WGS-84	World Geodetic System 1984

Name Index

Subject Index

Printing: Saladruck, Berlin
Binding: Stein+Lehmann, Berlin